outlines ™

Probability

Second Edition

Seymour Lipschutz, Ph.D.
Professor of Mathematics
Temple University

Marc Lipson, Ph.D.
University of Virginia

Schaum's Outline Series

New York   Chicago   San Francisco   Lisbon   London   Madrid
Mexico City   Milan   New Delhi   San Juan   Seoul
Singapore   Sydney   Toronto

The *McGraw-Hill* Companies

SEYMOUR LIPSCHUTZ, who is presently on the mathematics faculty at Temple University, formerly taught at the Polytechnic Institute of Brooklyn and was visiting professor in the Computer Science Department of Brooklyn College. He received his Ph.D. in 1960 at the Courant Institute of Mathematical Sciences of New York University. Some of his other books in the Schaum's Outline Series are *Beginning Linear Algebra, Discrete Mathematics*, and *Linear Algebra*.

MARC LARS LIPSON is on the faculty at the University of Virginia and formerly taught at Northeastern University, Boston University, and the University of Georgia. He received his Ph.D. in finance in 1994 from the University of Michigan. He is also coauthor of *Schaum's Outline of Discrete Mathematics* with Seymour Lipschutz.

1 2 3 4 5 6 7 8 9 10   ROV/ROV   1 9 8 7 6 5 4 3 2 1

ISBN   978-0-07-175561-0
MHID   0-07-175561-6

PREFACE

Probability theory had its beginnings in the early seventeenth century as a result of investigations of various games of chance. Since then many leading mathematicians and scientists made contributions to this theory. However, despite its long and active history, probability theory was not axiomatized until the twentieth century. This axiomatic development, called modern probability theory, was then able to make the concepts of probability precise and place them on a firm mathematical foundation.

This book is designed for an introductory course in probability with high school algebra as the main prerequisite. It can serve as a text for such a course, or as a supplement to all current comparable texts. The book should also prove to be useful as a supplement to texts and courses in statistics. Furthermore, as the book is complete and self-contained it can easily be used for self-study.

This new edition includes and expands the content of the first edition. It begins with a chapter on sets and their operations, and then with a chapter on techniques of counting. Next comes a chapter on probability spaces, and then a chapter on conditional probability and independence. The fifth and main chapter is on random variables where we define expectation, variance, and standard deviation, and prove Chebyshev's inequality and the law of large numbers. Although calculus is not a prerequisite, both discrete and continuous random variables are considered. We follow with a separate chapter on specific distributions, mainly the binomial, normal, and Poisson distributions. Here the central limit theorem is given in the context of the normal approximation to the binomial distribution. The seventh and last chapter offers a thorough elementary treatment of Markov chains with applications.

This new edition also has two new appendixes. The first is on descriptive statistics where expectation, variance, and standard deviation are again defined, but now in the context of statistics. This appendix also treats bivariate data, including scatterplots, the correlation coefficient, and methods of least squares. The second appendix discusses the chi-square distribution and various applications in the context of testing hypotheses. These two new appendixes motivate many of the concepts which appear in the chapters on probability, and also make the book even more useful as a supplement to texts and courses in statistics.

The positive qualities that distinguished the first edition have been retained. Each chapter begins with clear statements of pertinent definitions, principles, and theorems together with illustrative and other descriptive material. This is followed by graded sets of solved and supplementary problems. The solved problems serve to illustrate and amplify the theory, and provide the repetition of basic principles so vital to effective learning. Proof of most of the theorems is included among the solved problems. The supplementary problems serve as a complete review of the material of each chapter.

Finally, we wish to thank the staff of McGraw-Hill, especially Barbara Gilson and Maureen Walker, for their excellent cooperation.

SEYMOUR LIPSCHUTZ
Temple University

MARC LARS LIPSON
University of Georgia

CONTENTS

CHAPTER 1

Set Theory

1.1 INTRODUCTION

This chapter treats some of the elementary ideas and concepts of set theory which are necessary for a modern introduction to probability theory.

1.2 SETS AND ELEMENTS, SUBSETS

A *set* may be viewed as any well-defined collection of objects, and they are called the *elements* or *members* of the set. We usually use capital letters, A, B, X, Y, ... to denote sets, and lowercase letters, a, b, x, y, ... to denote elements of sets. Synonyms for set are *class*, *collection*, and *family*.

The statement that an element a belongs to a set S is written

$$a \in S$$

(Here $\in$ is the symbol meaning "is an element of".) We also write

$$a, b \in S$$

when both a and b belong to S.

Suppose every element of a set A also belongs to a set B, that is, suppose $a \in A$ implies $a \in B$. Then A is called a *subset* of B, or A is said to be *contained* in B, which is written as

$$A \subseteq B \quad \text{or} \quad B \supseteq A$$

Two sets are equal if they both have the same elements or, equivalently, if each is contained in the other. That is,

$$\boxed{A = B \text{ if and only if } A \subseteq B \text{ and } B \subseteq A}$$

The negations of $a \in A$, $A \subseteq B$, and $A = B$ are written $a \notin A$, $A \nsubseteq B$, and $A \neq B$, respectively.

Remark 1: It is common practice in mathematics to put a vertical line "|" or slanted line "/" through a symbol to indicate the opposite or negative meaning of the symbol.

1

Remark 2:   The statement $A \subseteq B$ does not exclude the possibility that $A = B$.   In fact, for any set A, we have $A \subseteq A$ since, trivially, every element in A belongs to A.   However, if $A \subseteq B$ and $A \neq B$, then we say that A is a *proper subset* of A (sometimes written $A \subset B$).

Remark 3:   Suppose every element of a set A belongs to a set B, and every element of B belongs to a set C.   Then clearly every element of A belongs to C.   In other words, if $A \subseteq B$ and $B \subseteq C$, then $A \subseteq C$.

The above remarks yield the following theorem.

Theorem 1.1:   Let A, B, C be any sets.   Then:

     (i) $A \subseteq A$.

     (ii)   If $A \subseteq B$ and $B \subseteq A$, then $A = B$.

     (iii)   If $A \subseteq B$ and $B \subseteq C$, then $A \subseteq C$.

Specifying Sets

There are essentially two ways to specify a particular set.   One way, if possible, is to list its elements.   For example,

$$A = \{1, 3, 5, 7, 9\}$$

means A is the set consisting of the numbers 1, 3, 5, 7, and 9.   Note that the elements of the set are separated by commas and enclosed in braces { }.   This is called the *tabular form* or *roster method* of a set.

The second way, called the *set-builder form* or *property method*, is to state those properties which characterize the elements in the set, that is, properties held by the members of the set but not by nonmembers.   Consider, for example, the expression

$$B = \{x : x \text{ is an even integer, } x > 0\}$$

which is read:

"B is the set of x such that x is an even integer and $x > 0$"

It denotes the set B whose elements are positive even integers.   A letter, usually x, is used to denote a typical member of the set; the colon is read as "such that" and the comma as "and."

EXAMPLE 1.1

(*a*)   The above set A can also be written as

$$A = \{x : x \text{ is an odd positive integer, } x < 10\}$$

We cannot list all the elements of the above set B, but we frequently specify the set by writing

$$B = \{2, 4, 6, \ldots\}$$

where we assume everyone knows what we mean.   Observe that $9 \in A$ but $9 \notin B$.   Also $6 \in B$, but $6 \notin A$.

(*b*)   Consider the sets

$$A = \{1, 3, 5, 7, 9\}, \qquad B = \{1, 2, 3, 4, 5\}, \qquad C = \{3, 5\}$$

Then $C \subseteq A$ and $C \subseteq B$ since 3 and 5, the elements C, are also members of A and B.   On the other hand, $A \nsubseteq B$ since $7 \in A$ but $7 \notin B$, and $B \nsubseteq A$ since $2 \in B$ but $2 \notin A$.

(*c*)   Suppose a die is tossed.   The possible "number" or "points" which appears on the uppermost face of the die belongs to the set $\{1, 2, 3, 4, 5, 6\}$.   Now suppose a die is tossed and an even number appears.   Then the outcome is a member of the set $\{2, 4, 6\}$ which is a (proper) subset of the set $\{1, 2, 3, 4, 5, 6\}$ of all possible outcomes.

Special Symbols, Real Line R, Intervals

Some sets occur very often in mathematics, and so we use special symbols for them. Some such symbols follow:

N = the *natural numbers* or positive integers:

$$\{1, 2, 3, \ldots\}$$

Z = all integers, positive, negative, and zero:

$$\{\ldots, -2, -1, 0, 1, 2, \ldots\}$$

R = the real numbers

Thus we have $\mathbf{N} \subseteq \mathbf{Z} \subseteq \mathbf{R}$.

The set **R** of real numbers plays an important role in probability theory since such numbers are used for numerical data. We assume the reader is familiar with the graphical representation of **R** as points on a straight line, as pictured in Fig. 1-1. We refer to such a line as the *real line* or the *real line* **R**.

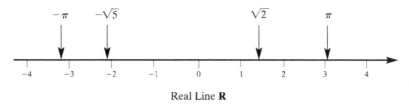

Real Line **R**

Fig. 1-1

Important subsets of **R** are the *intervals* which are denoted and defined as follows (where a and b are real numbers with $a < b$):

Open interval from a to $b = (a, b) = \{x : a < x < b\}$

Closed interval from a to $b = [a, b] = \{x : a \le x \le b\}$

Open-closed interval from a to $b = (a, b] = \{x : a < x \le b\}$

Closed-open interval from a to $b = [a, b) = \{x : a \le x < b\}$

The numbers a and b are called the *endpoints* of the interval. The word "open" and a parenthesis "(" or ")" are used to indicate that an endpoint does not belong to the interval, whereas the word "closed" and a bracket "[" or "]" are used to indicate that an endpoint belongs to the interval.

Universal Set and Empty Set

All sets under investigation in any application of set theory are assumed to be contained in some large fixed set called the *universal set* or *universe of discourse*. For example, in plane geometry, the universal set consists of all the points in the plane; in human population studies, the universal set consists of all the people in the world. We will let

U

denote the universal set unless otherwise stated or implied.

Given a universal set **U** and a property P, there may be no elements in **U** which have the property P. For example, the set

$$S = \{x : x \text{ is a positive integer}, x^2 = 3\}$$

has no elements since no positive integer has the required property. Such a set with no elements is called the *empty set* or *null set*, and is denoted by

$$\varnothing$$

There is only one empty set: If S and T are both empty, then $S = T$ since they have exactly the same elements, namely, none.

The empty set $\varnothing$ is also regarded as a subset of every other set. Accordingly, we have the following simple result which we state formally:

Theorem 1.2: For any set A, we have $\varnothing \subseteq A \subseteq \mathbf{U}$.

Disjoint Sets

Two sets A and B are said to be *disjoint* if they have no elements in common. Consider, for example, the sets

$$A = \{1, 2\}, \qquad B = \{2, 4, 6\}, \qquad C = \{4, 5, 6, 7\}$$

Observe that A and B are not disjoint since each contains the element 2, and B and C are not disjoint since each contains the element 4, among others. On the other hand, A and C are disjoint since they have no element in common. We note that if A and B are disjoint, then neither is a subset of the other (unless one is the empty set).

1.3   VENN DIAGRAMS

A *Venn diagram* is a pictorial representation of sets where sets are represented by enclosed areas in the plane. The universal set $\mathbf{U}$ is represented by the points in a rectangle, and the other sets are represented by disks lying within the rectangle. If $A \subseteq B$, then the disk representing A will be entirely within the disk representing B, as in Fig. 1-2(*a*). If A and B are disjoint, that is, have no elements in common, then the disk representing A will be separated from the disk representing B, as in Fig. 1-2(*b*).

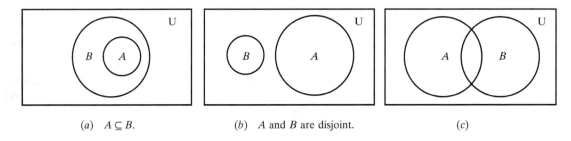

(*a*) $A \subseteq B$.   (*b*) A and B are disjoint.   (*c*)

Fig. 1-2

On the other hand, if A and B are two arbitrary sets, it is possible that some elements are in A but not in B, some elements are in B but not in A, some are in both A and B, and some are in neither A nor B; hence, in general, we represent A and B as in Fig. 1-2(*c*).

1.4   SET OPERATIONS

This section defines a number of set operations, including the basic operations of union, intersection, and complement.

Union and Intersection

The *union* of two sets A and B, denoted by $A \cup B$, is the set of all elements which belong to A or to B, that is,

$$A \cup B = \{x : x \in A \quad \text{or} \quad x \in B\}$$

Here, "or" is used in the sense of and/or. Figure 1-3(*a*) is a Venn diagram in which $A \cup B$ is shaded.

The *intersection* of two sets A and B, denoted by $A \cap B$, is the set of all elements which belong to both A and B, that is,

$$A \cap B = \{x : x \in A \quad \text{and} \quad x \in B\}$$

Figure 1-3(*b*) is a Venn diagram in which $A \cap B$ is shaded.

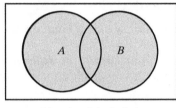
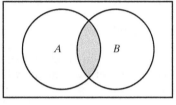

(*a*) $A \cup B$ is shaded.   (*b*) $A \cap B$ is shaded.

Fig. 1-3

Recall that sets A and B are said to be disjoint if they have no elements in common or, using the definition of intersection, if $A \cap B = \varnothing$, the empty set.   If

$$S = A \cup B \quad \text{and} \quad A \cap B = \varnothing$$

then S is called the *disjoint union* of A and B.

EXAMPLE 1.2

(*a*)   Let $A = \{1, 2, 3, 4\}$, $B = \{3, 4, 5, 6, 7\}$, $C = \{2, 3, 8, 9\}$.   Then

$$A \cup B = \{1, 2, 3, 4, 5, 6, 7\}, \qquad A \cup C = \{1, 2, 3, 4, 8, 9\}, \qquad B \cup C = \{3, 4, 5, 6, 7, 8, 9\},$$
$$A \cap B = \{3, 4\}, \qquad A \cap C = \{2, 3\}, \qquad B \cap C = \{3\}$$

(*b*)   Let **U** be the set of students at a university, and let M and F denote, respectively, the sets of male and female students.   Then **U** is the disjoint union of M and F, that is,

$$\mathbf{U} = M \cup F \quad \text{and} \quad M \cap F = \varnothing$$

This comes from the fact that every student in **U** is either in M or in F, and clearly no students belong to both M and F, that is, M and F are disjoint.

The following properties of the union and intersection should be noted:

(i)   Every element x in $A \cap B$ belongs to both A and B; hence, x belongs to A and x belongs to B.   Thus, $A \cap B$ is a subset of A and of B, that is,

$$A \cap B \subseteq A \quad \text{and} \quad A \cap B \subseteq B$$

(ii)   An element x belongs to the union $A \cup B$ if x belongs to A or x belongs to B; hence, every element in A belongs to $A \cup B$, and every element in B belongs to $A \cup B$.   That is,

$$A \subseteq A \cup B \quad \text{and} \quad B \subseteq A \cup B$$

We state the above results formally.

Theorem 1.3: For any sets A and B, we have

$$A \cap B \subseteq A \subseteq A \cup B \quad \text{and} \quad A \cap B \subseteq B \subseteq A \cup B$$

The operations of set inclusion is closely related to the operations of union and intersection, as shown by the following theorem (proved in Problem 1.16).

Theorem 1.4: The following are equivalent: $A \subseteq B$, $\quad A \cap B = A$, $\quad A \cup B = B$.

Other conditions equivalent to $A \subseteq B$ are given in Problem 1.55.

Complements, Difference, Symmetric Difference

Recall that all sets under consideration at a particular time are subsets of a fixed universal set **U**. The *absolute complement* or, simply, *complement* of a set A, denoted by A^c, is the set of elements which belong to **U** but which do not belong to A, that is,

$$A^c = \{x : x \in \mathbf{U}, x \notin A\}$$

Some texts denote the complement of A by A' or $\bar{A}$. Figure 1-4(a) is a Venn diagram in which A^c is shaded.

The *relative complement* of a set B with respect to a set A or, simply, the *difference* between A and B, denoted by $A \setminus B$, is the set of elements which belong to A but which do not belong to B, that is,

$$A \setminus B = \{x : x \in A, x \notin B\}$$

The set $A \setminus B$ is read "A minus B". Some texts denote $A \setminus B$ by $A - B$ or $A \sim B$. Figure 1-4(b) is a Venn diagram in which $A \setminus B$ is shaded.

The *symmetric difference* of the sets A and B, denoted by $A \oplus B$, consists of those elements which belong to A or B, but not both. That is,

$$A \oplus B = (A \cup B) \setminus (A \setminus B) \quad \text{or} \quad A \oplus B = (A \setminus B) \cup (B \setminus A)$$

Figure 1-4(c) is a Venn diagram in which $A \oplus B$ is shaded.

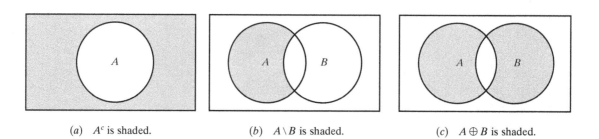

(a) A^c is shaded.   (b) $A \setminus B$ is shaded.   (c) $A \oplus B$ is shaded.

Fig. 1-4

EXAMPLE 1.3 Let $\mathbf{U} = \mathbf{N} = \{1, 2, 3, \ldots\}$ be the universal set, and let

$$A = \{1, 2, 3, 4\}, \qquad B = \{3, 4, 5, 6, 7\}, \qquad C = \{2, 3, 8, 9\}, \qquad E = \{2, 4, 6, \ldots\}$$

[Here E is the set of even positive integers.] Then

$$A^c = \{5, 6, 7, \ldots\}, \qquad B^c = \{1, 2, 8, 9, 10, \ldots\}, \qquad E^c = \{1, 3, 5, \ldots\}$$

That is, E^c is the set of odd integers. Also

$$A \setminus B = \{1, 2\}, \qquad A \setminus C = \{1, 4\}, \qquad B \setminus C = \{4, 5, 6, 7\}, \qquad A \setminus E = \{1, 3\},$$
$$B \setminus A = \{5, 6, 7\}, \qquad C \setminus A = \{8, 9\}, \qquad C \setminus B = \{2, 8, 9\}, \qquad E \setminus A = \{6, 8, 10, \ldots\}$$

Furthermore

$$A \oplus B = (A \setminus B) \cup (B \setminus A) = \{1, 2, 5, 6, 7\}, \qquad B \oplus C = \{2, 4, 5, 6, 7, 8, 9\},$$
$$A \oplus C = (A \setminus C) \cup (C \setminus A) = \{1, 4, 8, 9\}, \qquad A \oplus E = \{1, 3, 6, 8, 10, \ldots\}$$

Algebra of Sets

Sets under the operations of union, intersection, and complement satisfy various laws (identities) which are listed in Table 1-1. In fact, we formally state:

Theorem 1.5: Sets satisfy the laws in Table 1-1.

Table 1-1 Laws of the Algebra of Sets

Idempotent Laws	
1a. $A \cup A = A$	1b. $A \cap A = A$

Associative Laws	
2a. $(A \cup B) \cup C = A \cup (B \cup C)$	2b. $(A \cap B) \cap C = A \cap (B \cap C)$

Commutative Laws	
3a. $A \cup B = B \cup A$	3b. $A \cap B = B \cap A$

Distributive Laws	
4a. $A \cup (B \cap C) = (A \cup B) \cap (A \cup C)$	4b. $A \cap (B \cup C) = (A \cap B) \cup (A \cap C)$

Identity Laws	
5a. $A \cup \varnothing = A$	5b. $A \cap \mathbf{U} = A$
6a. $A \cup \mathbf{U} = \mathbf{U}$	6b. $A \cap \varnothing = \varnothing$

Involution Law	
7. $(A^c)^c = A$	

Complement Laws	
8a. $A \cup A^c = \mathbf{U}$	8b. $A \cap A^c = \varnothing$
9a. $\mathbf{U}^c = \varnothing$	9b. $\varnothing^c = \mathbf{U}$

DeMorgan's Laws	
10a. $(A \cup B)^c = A^c \cap B^c$	10b. $(A \cap B)^c = A^c \cup B^c$

Remark: Each law in Table 1-1 follows from an equivalent logical law. Consider, for example, the proof of DeMorgan's law:

$$(A \cup B)^c = \{x : x \notin (A \text{ or } B)\} = \{x : x \notin A \text{ and } x \notin B\} = A^c \cap B^c$$

Here we use the equivalent (DeMorgan's) logical law:

$$\neg (p \vee q) = \neg p \wedge \neg q$$

where $\neg$ means "not", $\vee$ means "or", and $\wedge$ means "and". (Sometimes Venn diagrams are used to illustrate the laws in Table 1-1 as in Problem 1.17.)

Duality

The identities in Table 1-1 are arranged in pairs, as, for example, 2a and 2b. We now consider the principle behind this arrangement. Let E be an equation of set algebra. The *dual E^** of E is the equation obtained by replacing each occurrence of $\cup$, $\cap$, $\mathbf{U}$, $\varnothing$ in E by $\cap$, $\cup$, $\varnothing$, $\mathbf{U}$, respectively. For example, the dual of

$$(\mathbf{U} \cap A) \cup (B \cap A) = A \quad \text{is} \quad (\varnothing \cup A) \cap (B \cup A) = A$$

Observe that the pairs of laws in Table 1-1 are duals of each other. It is a fact of set algebra, called the *principle of duality*, that, if any equation E is an identity, then its dual E^* is also an identity.

1.5 FINITE AND COUNTABLE SETS

Sets can be finite or infinite. A set S is *finite* if S is empty or if S consists of exactly m elements where m is a positive integer; otherwise S is infinite.

EXAMPLE 1.4

(a) Let A denote the letters in the English alphabet, and let D denote the days of the week, that is, let

$$A = \{a, b, c, \ldots, y, z\} \quad \text{and} \quad D = \{\text{Monday, Tuesday, ..., Sunday}\}$$

Then A and D are finite sets. Specifically, A has 26 elements and D has 7 elements.

(b) Let $R = \{x : x \text{ is a river on the earth}\}$. Although it may be difficult to count the number of rivers on the earth, R is still a finite set.

(c) Let E be the set of even positive integers, and let $\mathbf{I}$ be the *unit interval*; that is, let

$$E = \{2, 4, 6, \ldots\} \quad \text{and} \quad \mathbf{I} = [0, 1] = \{x : 0 \le x \le 1\}$$

Then both E and $\mathbf{I}$ are infinite sets.

Countable Sets

A set S is *countable* if S is finite or if the elements of S can be arranged in the form of a sequence, in which case S is said to be *countably infinite*. A set is *uncountable* if it is not countable. The above set E of even integers is countably infinite, whereas it can be proven that the unit interval $\mathbf{I} = [0, 1]$ is uncountable.

1.6 COUNTING ELEMENTS IN FINITE SETS, INCLUSION-EXCLUSION PRINCIPLE

The notation $n(S)$ or $|S|$ will denote the number of elements in a set S. Thus $n(A) = 26$ where A consists of the letters in the English alphabet, and $n(D) = 7$ where D consists of the days of the week. Also $n(\varnothing) = 0$, since the empty set has no elements.

The following lemma applies.

Lemma 1.6: Suppose A and B are finite disjoint sets. Then $A \cup B$ is finite and

$$n(A \cup B) = n(A) + n(B)$$

This lemma may be restated as follows:

Lemma 1.6: Suppose S is the disjoint union of finite sets A and B. Then S is finite and

$$n(S) = n(A) + n(B)$$

Proof: In counting the elements of $A \cup B$, first count the elements of A. There are $n(A)$ of these. The only other elements in $A \cup B$ are those that are in B but not in A. Since A and B are disjoint, no element of B is in A. Thus, there are $n(B)$ elements which are in B but not in A. Accordingly, $n(A \cup B) = n(A) + n(B)$.

For any sets A and B, the set A is the disjoint union of $A \setminus B$ and $A \cap B$ (Problem 1.45). Thus, Lemma 1.6 gives us the following useful result.

Corollary 1.7: Let A and B be finite sets. Then

$$n(A \setminus B) = n(A) - n(A \cap B)$$

That is, the number of elements in A but not in B is the number of elements in A minus the number of elements in both A and B. For example, suppose an art class A has 20 students and 8 of the students are also taking a biology class B. Then there are

$$20 - 8 = 12$$

students in the class A which are not in the class B.

Given any set A, we note that the universal set $\mathbf{U}$ is the disjoint union of A and A^c. Accordingly, Lemma 1.6 also gives us the following result.

Corollary 1.8: Suppose A is a subset of a finite universal set $\mathbf{U}$. Then

$$n(A^c) = n(\mathbf{U}) - n(A)$$

For example, suppose a class $\mathbf{U}$ of 30 students has 18 full-time students. Then there are

$$30 - 18 = 12$$

part-time students in the class.

Inclusion-Exclusion Principle

There is also a formula for $n(A \cup B)$, even when they are not disjoint, called the *inclusion-exclusion principle*. Namely,

Theorem (Inclusion-Exclusion Principle) 1.9: Suppose A and B are finite sets. Then $A \cap B$ and $A \cup B$ are finite and

$$n(A \cup B) = n(A) + n(B) - n(A \cap B)$$

That is, we find the number of elements in A or B (or both) by first adding $n(A)$ and $n(B)$ (inclusion) and then subtracting $n(A \cap B)$ (exclusion) since its elements were counted twice.

We can apply this result to get a similar result for three sets.

Corollary 1.10: Suppose A, B, C are finite sets. Then $A \cup B \cup C$ is finite and

$$n(A \cup B \cup C) = n(A) + n(B) + n(C) - n(A \cap B) - n(A \cap C) - n(B \cap C) + n(A \cap B \cap C)$$

Mathematical induction (Section 1.9) may be used to further generalize this result to any finite number of finite sets.

EXAMPLE 1.5 Suppose list A contains the 30 students in a mathematics class and list B contains the 35 students in an English class, and suppose there are 20 names on both lists. Find the number of students:

(a) Only on list A (c) On list A or B (or both)

(b) Only on list B (d) On exactly one of the two lists

(a) List A contains 30 names and 20 of them are on list B; hence $30 - 20 = 10$ names are only on list A. That is, by Corollary 1.7,

$$n(A \setminus B) = n(A) - n(A \cap B) = 30 - 20 = 10$$

(b) Similarly, there are $35 - 20 = 15$ names only on list B. That is,

$$n(B \setminus A) = n(B) - n(A \cap B) = 35 - 20 = 15$$

(c) We seek $n(A \cup B)$. Note we are given that $n(A \cap B) = 20$.

One way is to use the fact that $A \cup B$ is the disjoint union of $A \setminus B$, $A \cap B$, and $B \setminus A$ (Problem 1.54), which is pictured in Fig. 1-5 where we have also inserted the number of elements in each of the three sets $A \setminus B, A \cap B, B \setminus A$. Thus

$$n(A \cup B) = 10 + 20 + 15 = 45$$

Alternately, by Theorem 1.8,

$$n(A \cup B) = n(A) + n(B) - n(A \cap B) = 30 + 35 - 20 = 45$$

In other words, we combine the two lists and then cross out the 20 names which appear twice.

(d) By (a) and (b), there are $10 + 15 = 25$ names on exactly one of the two lists; so $n(A \oplus B) = 25$. Alternately, by the Venn diagram in Fig. 1-5, there are 10 elements in $A \setminus B$, and 15 elements in $B \setminus A$; hence

$$n(A \oplus B) = 10 + 15 = 25$$

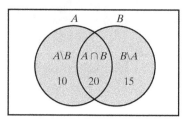

$A \cup B$ is shaded.

Fig. 1-5

1.7 PRODUCT SETS

Consider two arbitrary sets A and B. The set of all ordered pairs (a, b) where $a \in A$ and $b \in B$ is called the *product*, or *Cartesian product*, of A and B. A short designation of this product is $A \times B$, which is read "A cross B". By definition,

$$A \times B = \{(a, b) : a \in A, b \in B\}$$

One frequently writes A^2 instead of $A \times A$.

We note that ordered pairs (a, b) and (c, d) are equal if and only if their *first* elements, a and c, are equal and their *second* elements, b and d, are equal. That is,

$$\boxed{(a, b) = (c, d) \qquad \text{if and only if} \qquad a = c \text{ and } b = d}$$

EXAMPLE 1.6 $\mathbf{R}$ denotes the set of real numbers, and so $\mathbf{R}^2 = \mathbf{R} \times \mathbf{R}$ is the set of ordered pairs of real numbers. The reader is familiar with the geometrical representation of $\mathbf{R}^2$ as points in the plane, as in Fig. 1-6. Here each point P represents an ordered pair (a, b) of real numbers, and vice versa; the vertical line through P meets the x axis at a, and the horizontal line through P meets the y axis at b. $\mathbf{R}^2$ is frequently called the *Cartesian plane*.

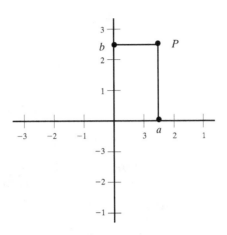

Cartesian Plane R^2

Fig. 1-6

EXAMPLE 1.7 Let $A = \{1, 2\}$ and $B = \{a, b, c\}$. Then

$$A \times B = \{(1, a), (1, b), (1, c), (2, a), (2, b), (2, c)\}$$
$$B \times A = \{(a, 1), (a, 2), (b, 1), (b, 2), (c, 1), (c, 2)\}$$

Also,
$$A \times A = \{(1, 1), (1, 2), (2, 1), (2, 2)\}$$

There are two things worth noting in the above Example 1.7. First of all, $A \times B \neq B \times A$. The Cartesian product deals with ordered pairs, so naturally the order in which the sets are considered is important.

Secondly, using $n(S)$ for the number of elements in a set S, we have:

$$n(A \times B) = 6 = 2 \cdot 3 = n(A) \cdot n(B)$$

In fact, $n(A \times B) = n(A) \cdot n(B)$ for any finite sets A and B. This follows from the observation that, for each $a \in A$, there will be $n(B)$ ordered pairs in $A \times B$ beginning with a. Hence, altogether there will be $n(A)$ times $n(B)$ ordered pairs in $A \times B$.

We state the above result formally.

Theorem 1.11: Suppose A and B are finite. Then $A \times B$ is finite and

$$n(A \times B) = n(A) \cdot n(B)$$

The concept of a product of sets can be extended to any finite number of sets in a natural way. That is, for any sets $A_1, A_2, \ldots, A_m$, the set of all ordered m-tuples $(a_1, a_2, \ldots, a_m)$, where $a_1 \in A_1$, $a_2 \in A_2, \ldots, a_m \in A_m$, is called the *product* of the sets $A_1, A_2, \ldots, A_m$ and is denoted by

$$A_1 \times A_2 \times \cdots \times A_m \qquad \text{or} \qquad \prod_{i=1}^{m} A_i$$

Just as we write A^2 instead of $A \times A$, so we write A^m for $A \times A \times \cdots \times A$, where there are m factors.

Furthermore, for finite sets $A_1, A_2, \ldots, A_m$, we have

$$n(A_1 \times A_2 \times \cdots \times A_m) = n(A_1)n(A_2)\cdots n(A_m)$$

That is, Theorem 1.11 may be easily extended, by induction, to the product of m sets.

1.8 CLASSES OF SETS, POWER SETS, PARTITIONS

Given a set S, we may wish to talk about some of its subsets. Thus, we would be considering a "set of sets". Whenever such a situation arises, to avoid confusion, we will speak of a *class* of sets or a *collection* of sets. The words "subclass" and "subcollection" have meanings analogous to subset.

EXAMPLE 1.8 Suppose $S = \{1, 2, 3, 4\}$. Let $\mathscr{A}$ be the class of subsets of S which contains exactly three elements of S. Then

$$\mathscr{A} = [\{1, 2, 3\}, \{1, 2, 4\}, \{1, 3, 4\}, \{2, 3, 4\}]$$

The elements of $\mathscr{A}$ are the sets $\{1, 2, 3\}$, $\{1, 2, 4\}$, $\{1, 3, 4\}$, and $\{2, 3, 4\}$.

Let $\mathscr{B}$ be the class of subsets of S which contains the numeral 2 and two other elements of S. Then

$$\mathscr{B} = [\{1, 2, 3\}, \{1, 2, 4\}, \{2, 3, 4\}]$$

The elements of $\mathscr{B}$ are $\{1, 2, 3\}$, $\{1, 2, 4\}$, and $\{2, 3, 4\}$. Thus $\mathscr{B}$ is a subclass of $\mathscr{A}$. (To avoid confusion, we will usually enclose the sets of a class in brackets instead of braces.)

Power Sets

For a given set S, we may consider the class of all subsets of S. This class is called the *power set* of S, and it will be denoted by $\mathscr{P}(S)$. If S is finite, then so is $\mathscr{P}(S)$. In fact, the number of elements in $\mathscr{P}(S)$ is 2 raised to the power of S; that is,

$$n(\mathscr{P}(S)) = 2^{n(S)}$$

(For this reason, the power set of S is sometimes denoted by 2^S.) We emphasize that S and the empty set $\varnothing$ belong to $\mathscr{P}(S)$ since they are subsets of S.

EXAMPLE 1.9 Suppose $S = \{1, 2, 3\}$. Then

$$\mathscr{P}(S) = [\varnothing, \{1\}, \{2\}, \{3\}, \{1, 2\}, \{1, 3\}, \{2, 3\}, S]$$

As expected from the above remark, $\mathscr{P}(S)$ has $2^3 = 8$ elements.

Partitions

Let S be a nonempty set. A *partition* of S is a *subdivision* of S into nonoverlapping, nonempty subsets. Precisely, a *partition* of S is a collection $\{A_i\}$ of nonempty subsets of S such that

   (i)   Each a in S belongs to one of the A_i.

   (ii) The sets of $\{A_i\}$ are mutually disjoint; that is, if

$$A_i \neq A_j, \text{ the } A_i \cap A_j = \varnothing.$$

The subsets in a partition are called *cells*. Figure 1-7 is a Venn diagram of a partition of the rectangular set S of points into five cells, A_1, A_2, A_3, A_4, A_5.

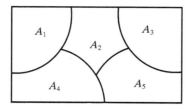

Fig. 1-7

EXAMPLE 1.10   Consider the following collections of subsets of $S = \{1, 2, 3, \ldots, 8, 9\}$:

  (i) $[\{1, 3, 5\}, \{2, 6\}, \{4, 8, 9\}]$

  (ii) $[\{1, 3, 5\}, \{2, 4, 6, 8\}, \{5, 7, 9\}]$

  (iii) $[\{1, 3, 5\}, \{2, 4, 6, 8\}, \{7, 9\}]$

Then (i) is not a partition of S since 7 in S does not belong to any of the subsets.   Furthermore, (ii) is not a partition of S since $\{1, 3, 5\}$ and $\{5, 7, 9\}$ are not disjoint.   On the other hand, (iii) is a partition of S.

Indexed Classes of Sets

An *indexed class of sets*, usually presented in the form

$$\{A_i : i \in I\} \qquad \text{or simply} \qquad \{A_i\}$$

means that there is a set A_i assigned to each element $i \in I$.   The set I is called the *indexing set* and the sets A_i are said to be indexed by I.   The *union* of the sets A_i, written $\cup_{i \in I} A_i$ or simply $\cup_i A_i$, consists of those elements which belong to at least one of the A_i; and the *intersection* of the sets A_i, written $\cap_{i \in I} A_i$ or simply $\cap_i A_i$, consists of those elements which belong to every A_i.

When the indexing set is the set **N** of positive integers, the indexed class $\{A_1, A_2, \ldots\}$ is called a *sequence* of sets.   In such a case, we also write

$$\cup_{i=1}^{\infty} A_i = A_1 \cup A_2 \cup \cdots \qquad \text{and} \qquad \cup_{i=1}^{\infty} A_i = A_1 \cap A_2 \cap \cdots$$

for the union and intersection, respectively, of a sequence of sets.

Definition: A nonempty class $\mathscr{A}$ of subsets of **U** is called an *algebra* (σ-*algebra*) of sets if it has the following two properties:

  (i)   The complement of any set in $\mathscr{A}$ belongs to $\mathscr{A}$.

  (ii)   The union of any finite (countable) number of sets in $\mathscr{A}$ belongs to $\mathscr{A}$.

  That is, $\mathscr{A}$ is closed under complements and finite (countable) unions.

It is simple to show (Problem 1.40) that any algebra (σ-algebra) of sets contains **U** and $\emptyset$ and is closed under finite (countable) intersections.

1.9 MATHEMATICAL INDUCTION

An essential property of the set $\mathbf{N} = \{1, 2, 3, \ldots\}$ of positive integers which is used in many proofs follows:

Principle of Mathematical Induction I: Let $A(n)$ be an assertion about the set **N** of positive integers, that is, $A(n)$ is true or false for each integer $n \geq 1$.   Suppose $A(n)$ has the following two properties:

  (i) $A(1)$ is true.

  (ii) $A(n + 1)$ is true whenever $A(n)$ is true.

Then $A(n)$ is true for every positive integer.

We shall not prove this principle.   In fact, this principle is usually given as one of the axioms when **N** is developed axiomatically.

EXAMPLE 1.11   Let $A(n)$ be the assertion that the sum of the first n odd numbers is n^2; that is,

$$A(n): 1 + 3 + 5 + \cdots + (2n - 1) = n^2$$

(The nth odd number is $2n - 1$ and the next odd number is $2n + 1$.)
    Observe that $A(n)$ is true for $n = 1$ since

$$A(1): 1 = 1^2$$

Assuming $A(n)$ is true, we add $2n + 1$ to both sides of $A(n)$, obtaining

$$1 + 3 + 5 + \cdots + (2n - 1) + (2n + 1) = n^2 + (2n + 1) = (n + 1)^2$$

However, this is $A(n + 1)$.   That is, $A(n + 1)$ is true assuming $A(n)$ is true.   By the principle of mathematical induction, $A(n)$ is true for all $n \geq 1$.

    There is another form of the principle of mathematical induction which is sometimes more convenient to use.   Although it appears different, it is really equivalent to the above principle of induction.

Principle of Mathematical Induction II: Let $A(n)$ be an assertion about the set $\mathbf{N}$ of positive integers with the following two properties:

   (i) $A(1)$ is true.

   (ii) $A(n)$ is true whenever $A(k)$ is true for $1 \leq k \leq n$.

    Then $A(n)$ is true for every positive integer.

    Remark: Sometimes one wants to prove that an assertion A is true for a set of integers of the form

$$\{a, a + 1, a + 2, \ldots\}$$

where a is any integer, possibly 0.   This can be done by simply replacing 1 by a in either of the above Principles of Mathematical Induction.

Solved Problems

SETS, ELEMENTS, SUBSETS

1.1.   List the elements of the following sets; here $\mathbf{N} = \{1, 2, 3, \ldots\}$:

   (a) $A = \{x : x \in \mathbf{N}, 2 < x < 9\}$,   (c) $C = \{x : x \in \mathbf{N}, x + 5 = 2\}$,

   (b) $B = \{x : x \in \mathbf{N}, x \text{ is even}, x \leq 15\}$,   (d) $D = \{x : x \in \mathbf{N}, x \text{ is a multiple of 5}\}$

   (a) A consists of the positive integers between 2 and 9; hence $A = \{3, 4, 5, 6, 7, 8, 9\}$.

   (b) B consists of the even positive integers less than or equal to 15; hence $B = \{2, 4, 6, 8, 10, 12, 14\}$.

   (c)   There are no positive integers which satisfy the condition $x + 5 = 2$; hence C contains no elements.   In other words, $C = \varnothing$, the empty set.

   (d) D is infinite, so we cannot list all its elements.   However, sometimes we write $D = \{5, 10, 15, 20, \ldots\}$ assuming everyone understands that we mean the multiples of 5.

1.2.   Which of these sets are equal: $\{r, s, t\}, \{t, s, r\}, \{s, r, t\}, \{t, r, s\}$?

    They are all equal.   Order does not change a set.

1.3.   Describe in words how you would prove each of the following:

    (*a*) A is equal to B.   (*c*) A is a proper subset of B.

    (*b*) A is a subset of B.   (*d*) A is not a subset of B.

    (*a*) Show that each element of A also belongs to B, and then show that each element of B also belongs to A.

    (*b*) Show that each element of A also belongs to B.

    (*c*) Show that each element of A also belongs to B, and then show that at least one element of B is not in A. (Note that it is not necessary to show that more than one element of B is not in A.)

    (*d*) Show that one element of A is not in B.

1.4.   Show that $A = \{2, 3, 4, 5\}$ is not a subset of $B = \{x : x \in \mathbf{N},\ x \text{ is even}\}$.

    It is necessary to show that at least one element in A does not belong to B. Now $3 \in A$, but $3 \notin B$ since B only consists of even integers. Hence A is not a subset of B.

1.5.   Show that $A = \{3, 4, 5, 6\}$ is a proper subset of $C = \{1, 2, 3, \ldots, 8, 9\}$.

    Each element of A belongs to C; hence $A \subseteq C$. On the other hand, $1 \in C$ but $1 \notin A$; hence $A \neq C$. Therefore, A is a proper subset of C.

1.6.   Consider the following sets where $\mathbf{U} = \{1, 2, 3, \ldots, 8, 9\}$:

$$\varnothing,\ A = \{1\}, \quad B = \{1, 3\}, \quad C = \{1, 5, 9\}, \quad D = \{1, 2, 3, 4, 5\}, \quad E = \{1, 3, 5, 7, 9\}$$

Insert the correct symbol $\subseteq$ or $\nsubseteq$ between each pair of sets:

    (*a*) $\varnothing, A$ (*c*) B, C (*e*) C, D (*g*) D, E

    (*b*) A, B (*d*) B, E (*f*) C, E (*h*) $D, \mathbf{U}$

    (*a*) $\varnothing \subseteq A$ since $\varnothing$ is a subset of every set.

    (*b*) $A \subseteq B$ since 1 is the only element of A and it belongs to B.

    (*c*) $B \nsubseteq C$ since $3 \in B$ but $3 \notin C$.

    (*d*) $B \subseteq E$ since the elements of B also belong to E.

    (*e*) $C \nsubseteq D$ since $9 \in C$ but $9 \notin D$.

    (*f*) $C \subseteq E$ since the elements of C also belong to E.

    (*g*) $D \nsubseteq E$ since $2 \in D$ but $2 \notin E$.

    (*h*) $D \subseteq \mathbf{U}$ since the elements of D also belong to $\mathbf{U}$.

1.7.   Determine which of the following sets are equal: $\varnothing, \{0\}, \{\varnothing\}$.

    Each is different from the other. The set $\{0\}$ contains one element, the number zero. The set $\varnothing$ contains no element; it is the empty set. The set $\{\varnothing\}$ also contains one element, the null set.

1.8.   A pair of dice are tossed and the sum of the faces are recorded. Find the smallest set S which includes all possible outcomes.

    The faces of the die are the numbers 1 to 6. Thus, no sum can be less than 2 nor greater than 12. Also, every number between 2 and 12 could occur. Thus

$$S = \{2, 3, 4, 5, 6, 7, 8, 9, 10, 11, 12\}$$

SET OPERATIONS

1.9. Let $\mathbf{U} = \{1, 2, \ldots, 9\}$ be the universal set, and let

$A = \{1, 2, 3, 4, 5\}$, $C = (4, 5, 6, 7, 8, 9\}$, $E = \{2, 4, 6, 8\}$,

$B = \{4, 5, 6, 7\}$, $D = \{1, 3, 5, 7, 9\}$, $F = \{1, 5, 9\}$

Find:

(a) $A \cup B$ and $A \cap B$ (c) $A \cup C$ and $A \cap C$ (e) $E \cup E$ and $E \cap E$
(b) $B \cup D$ and $B \cap D$ (d) $D \cup E$ and $D \cap E$ (f) $D \cup F$ and $D \cap F$

Recall that the union $X \cup Y$ consists of those elements in either X or in Y (or both), and the intersection $X \cap Y$ consists of those elements in both X and Y.

(a) $A \cup B = \{1, 2, 3, 4, 5, 6, 7\}$, $A \cap B = \{4, 5\}$
(b) $B \cup D = \{1, 3, 4, 5, 6, 7, 9\}$, $B \cap D = \{5, 7\}$
(c) $A \cup C = (1, 2, 3, 4, 5, 6, 7, 8, 9\} = \mathbf{U}$, $A \cap C = \{4, 5\}$
(d) $D \cup E = \{1, 2, 3, 4, 5, 6, 7, 8, 9\} = \mathbf{U}$, $D \cap E = \varnothing$
(e) $E \cup E = \{2, 4, 6, 8\} = E$, $E \cap E = \{2, 4, 6, 8\} = E$
(f) $D \cup F = \{1, 3, 5, 7, 9\} = D$, $D \cap F = \{1, 5, 9\} = F$

(Observe that $F \subseteq D$; hence, by Theorem 1.4, we must have $D \cup F = D$ and $D \cap F = F$.)

1.10. Consider the sets in the preceding Problem 1.9.   Find:

(a) A^c, B^c, D^c, E^c;   (b) $A \backslash B$, $B \backslash A$, $D \backslash E$, $F \backslash D$;   (c) $A \oplus B$, $C \oplus D$, $E \oplus F$.

(a) The complement X^c consists of those elements in the universal set $\mathbf{U}$ which do not belong to X.   Hence:

$A^c = \{6, 7, 8, 9\}$, $B^c = \{1, 2, 3, 8, 9\}$, $D^c = \{2, 4, 6, 8\} = E$, $E^c = \{1, 3, 5, 7, 9\} = D$

(Note D and E are complements; that is, $D \cup E = \mathbf{U}$ and $D \cap E = \varnothing$.)

(b) The difference $X \backslash Y$ consists of the elements in X which do not belong to Y.   Therefore

$A \backslash B = \{1, 2, 3\}$, $B \backslash A = \{6, 7\}$, $D \backslash E = \{1, 3, 5, 7, 9\} = D$, $F \backslash D = \varnothing$

(Since D and E are disjoint, we must have $D \backslash E = D$; and since $F \subseteq D$, we must have $F \backslash D = \varnothing$.)

(c) The symmetric difference $X \oplus Y$ consists of the elements in X or in Y but not in both X and Y.   In other words, $X \oplus Y = (X \backslash Y) \cup (Y \backslash X)$.   Hence:

$A \oplus B = \{1, 2, 3, 6, 7\}$, $C \oplus D = \{1, 3, 8, 9\}$, $E \oplus F = \{2, 4, 6, 8, 1, 5, 9\} = E \cup F$

(Since E and F are disjoint, we must have $E \oplus F = E \cup F$.)

1.11. Show that we can have $A \cap B = A \cap C$ without $B = C$.

Let $A = \{1, 2\}$, $B = \{2, 3\}$, $C = \{2, 4\}$.   Then $A \cap B = \{2\}$ and $A \cap C = \{2\}$; hence $A \cap B = A \cap C$.
However, $B \neq C$.

1.12. Show that we can have $A \cup B = A \cup C$ without $B = C$.

Let $A = \{1, 2\}$, $B = \{1, 3\}$, $C = \{2, 3\}$.   Then $A \cup B = \{1, 2, 3\}$ and $A \cup C = \{1, 2, 3\}$; hence $A \cup B = A \cup C$.
However, $B \neq C$.

1.13. Prove: $B \setminus A = B \cap A^c$. Thus, the set operation of difference can be written in terms of the operations of intersection and complement.

$$B \setminus A = \{x : x \in B,\ x \notin A\} = \{x : X \in B,\ x \in A^c\} = B \cap A^c$$

1.14. Consider the following intervals:

$$A = [-3, 5), \qquad B = (3, 8), \qquad C = [0, 4], \qquad D = (-7, -3]$$

(a) Rewrite each interval in set-builder form.

(b) Find: $A \cap B,\quad A \cap C,\quad A \cap D,\quad B \cap C,\quad B \cap D,\quad C \cap D$.

(a) Recall that a parenthesis means that the endpoint does not belong to the interval, and that a bracket means that the endpoint does belong to the interval. Thus:

$$A = \{x : -3 \le x < 5\}, \qquad C = \{x : 0 \le x \le 4\},$$
$$B = (x : 3 < x < 8\}, \qquad D = \{x : -7 < x \le -3\}$$

(b) Using the short notation for intervals, we have:

$$A \cap B = [-3, 8), \qquad A \cap C = [-3, 5), \qquad A \cap D = \{-3\},$$
$$B \cap C = [0, 8), \qquad B \cap D = \emptyset, \qquad C \cap D = \emptyset$$

1.15. Under what condition will the intersection of two intervals be an interval?

The intersection of two intervals will always be an interval, or a singleton set $\{a\}$, or the empty set $\emptyset$. Thus, if we view

$$[a, a] = \{x : a \le x \le a\} = \{a\} \qquad \text{and} \qquad (a, a) = \{x : a < x < a\} = \emptyset$$

as intervals, then the intersection of any two intervals is always an interval.

1.16. Prove Theorem 1.4: The following are equivalent:

$$A \subseteq B, \qquad A \cap B = A, \qquad A \cup B = B$$

The theorem can be reduced to the following two cases:

(a) $A \subseteq B$ is equivalent to $A \cap B = A$.

(b) $A \subseteq B$ is equivalent to $A \cup B = B$.

(a) Suppose $A \subseteq B$ and let $x \in A$. Then $x \in B$, and so $x \in A \cap B$. Thus, $A \subseteq A \cap B$. Moreover, by Theorem 1.3, $(A \cap B) \subseteq A$. Accordingly, $A \cap B = A$.

   On the other hand, suppose $A \cap B = A$ and let $x \in A$. Then $x \in A \cap B$; hence $x \in A$ and $x \in B$. Therefore, $A \subseteq B$.

   Both results show that $A \subseteq B$ is equivalent to $A \cap B = A$.

(b) Suppose again that $A \subseteq B$. Let $x \in (A \cup B)$. Then $x \in A$ or $x \in B$. If $x \in A$, then $x \in B$ because $A \subseteq B$. In either case $x \in B$. Thus, $A \cup B \subseteq B$. By Theorem 1.3, $B \subseteq A \cup B$. Accordingly, $A \cup B = B$.

   On the other hand, suppose $A \cup B = B$ and let $x \in A$. Then $x \in A \cup B$, by definition of union of sets. However, $A \cup B = B$; hence $x \in B$. Thus, $A \subseteq B$.

   Both results show that $A \subseteq B$ is equivalent to $A \cup B = B$.

Thus, all three statements, $A \subseteq B,\quad A \cap B = A,\quad A \cup B = B$, are equivalent.

VENN DIAGRAMS, ALGEBRA OF SETS, DUALITY

1.17. Illustrate DeMorgan's Law $(A \cup B)^c = A^c \cap B^c$ (proved in Section 1.4) using Venn diagrams.

Shade the area outside $A \cup B$ in a Venn diagram of sets A and B. This is shown in Fig. 1-8(a); hence the shaded area represents $(A \cup B)^c$. Now shade the area outside A in a Venn diagram of A and B with strokes in one direction ($/ / /$), and then shade the area outside B with strokes in another direction ($\backslash \backslash \backslash$). This is shown in Fig. 1-8(b); hence the cross-hatched area (area where both lines are present) represents the intersection of A^c and B^c, that is, $A^c \cap B^c$. Both $(A \cup B)^c$ and $A^c \cap B^c$ are represented by the same area; thus the Venn diagrams indicate $(A \cup B)^c = A^c \cap B^c$. (We emphasize that a Venn diagram is not a formal proof but it can indicate relationships between sets.)

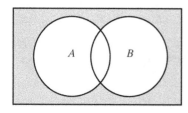   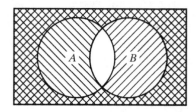

(a)   Shaded area: $(A \cup B)^c$ (b)   Cross-hatched area: $A^c \cap B^c$

Fig. 1-8

1.18. Prove the Distributive Law: $A \cap (B \cup C) = (A \cap B) \cup (A \cap C)$ [Theorem 1.5 (4b)].

By definition of union and intersection,

$$A \cap (B \cup C) = \{x : x \in A, x \in B \cup C\}$$
$$= \{x : x \in A, x \in B \text{ or } x \in A, x \in C\} = \{A \cap B) \cup (A \cap C)$$

Here we use the analogous logical law $p \wedge (q \vee r) \equiv (p \wedge q) \vee (p \wedge r)$ where $\wedge$ denotes "and" and $\vee$ denotes "or".

1.19. Describe in words: (a) $(A \cup B) \setminus (A \cap B)$ and (b) $(A \setminus B) \cup (B \setminus A)$. Then prove they are the same set. (Thus, either one may be used to define the symmetric difference $A \oplus B$.)

(a) $(A \cup B) \setminus (A \cap B)$ consists of the elements in A or B but not in both A and B.

(b) $(A \setminus B) \cup (B \setminus A)$ consists of the elements in A which are not in B, or the elements in B which are not in A.

Using $X \setminus Y = X \cap Y^c$ and the laws in Table 1-1, including DeMorgan's law, we obtain:

$$(A \cup B) \setminus (A \cap B) = (A \cup B) \cap (A \cap B)^c = (A \cup B) \cap (A^c \cap B^c)$$
$$= (A \cap A^c) \cup (A \cap B^c) \cup (B \cap A^c) \cup (B \cap B^c)$$
$$= \varnothing \cup (A \cap B^c) \cup (B \cap A^c) \cup \varnothing$$
$$= (A \cap B^c) \cap (B \cap A^c) = (A \setminus B) \cup (B \setminus A)$$

1.20. Write the dual of each set equation:

(a) $(\mathbf{U} \cap A) \cup (B \cap A) = A$ (c) $(A \cap \mathbf{U}) \cap)\varnothing \cup A^c) = \varnothing$

(b) $(A \cup B \cup C)^c = (A \cup C)^c \cap (A \cup B)^c$ (d) $(A \cap \mathbf{U})^c \cap A = \varnothing$

Interchange $\cap$ and $\cup$ and also $\mathbf{U}$ and $\varnothing$ in each set equation:

(a) $(\varnothing \cup A) \cap (B \cup A) = A$ (c) $(A \cup \varnothing) \cup (\mathbf{U} \cup A^c) = \mathbf{U}$

(b) $(A \cap B \cap C)^c = (A \cap C)^c \cup (A \cap B)^c$ (d) $(A \cup \mathbf{U})^c \cup A = \mathbf{U}$

FINITE SETS AND COUNTING PRINCIPLE, COUNTABLE SETS

1.21. Determine which of the following sets are finite:

(a) $A = \{$seasons in the year$\}$ (d) $D = \{$odd integers$\}$

(b) $B = \{$states in the United States$\}$ (e) $E = \{$positive integral divisors of 12$\}$

(c) $C = \{$positive integers less than 1$\}$ (f) $F = \{$cats living in the United States$\}$

(a) A is finite since there are four seasons in the year, that is, $n(A) = 4$.

(b) B is finite because there are 50 states in the United States, that is, $n(B) = 50$.

(c) There are no positive integers less than 1; hence C is empty.   Thus, C is finite and $n(C) = \varnothing$.

(d) D is infinite.

(e) The positive integer divisors of 12 are 1, 2, 3, 4, 6, 12.   Hence E is finite and $n(E) = 6$.

(f) Although it may be difficult to find the number of cats living in the United States, there is still a finite number of them at any point in time.   Hence F is finite.

1.22. Suppose 50 science students are polled to see whether or not they have studied French (F) or German (G), yielding the following data:

25 studied French, 20 studied German, 5 studied both

Find the number of students who: (a) studied only French, (b) did not study German, (c) studied French or German, (d) studied neither language.

(a) Here 25 studied French, and 5 of them also studied German; hence $25 - 5 = 20$ students only studied French.   That is, by Corollary 1.7,

$$n(F \setminus G) = n(F) - N(F \cap G) = 25 - 5 = 20$$

(b) There are 50 students of whom 20 studied German; hence $50 - 20 = 30$ did not study German.   That is, by Corollary 1.8,

$$n(G^c) = n(\mathbf{U}) - n(G) = 50 - 20 = 30$$

(c) By the inclusion-exclusion principle in Theorem 1.9,

$$n(F \cup G) = n(F) + n(G) - n(F \cap G) = 25 + 20 - 5 = 40$$

That is, 40 students studied French or German.

(d) The set $F^c \cap G^c$ consists of the students who studied neither language. By DeMorgan's law, $F^c \cap G^c = (F \cup G)^c$. By (c), 40 studied at least one of the languages; hence

$$n(F^c \cap G^c) = n(\mathbf{U}) - n(F \cup G) = 50 - 40 = 10$$

That is, 10 students studied neither language.

1.23. Each student at some college has a mathematics requirement M (to take at least one mathematics course) and a science requirement S (to take at least one science course).   A poll of 140 sophomore students shows that:

60 completed M, 45 completed S, 20 completed both M and S

Use a Venn diagram to find the number of students who had completed:

(a) At least one of the two requirements

(b) Exactly one of the two requirements

(c) Neither requirement

Translating the above data into set notation yields:

$$n(M) = 60, \qquad n(S) = 45, \qquad n(M \cap S) = 20, \qquad n(\mathbf{U}) = 140$$

Draw a Venn diagram of sets M and S with four regions, as in Fig. 1-9(a).   Then, as in Fig. 1-9(b), assign numbers to the four regions as follows:

    20 completed both M and S, so $n(M \cap S) = 20$

    $60 - 20 = 40$ completed M but not S, so $n(M \setminus S) = 40$

    $45 - 20 = 25$ completed S but not M, so $n(S \setminus M) = 25$

    $140 - 20 - 40 - 25 = 55$ completed neither M nor S

By the Venn diagram:

(a) $20 + 40 + 25 = 85$ completed M or S.   Alternately, we can find $n(M \cup S)$ without the Venn diagram by using the inclusion-exclusion principle:

$$n(M \cup S) = n(M) + n(S) - n(M \cap S) = 60 + 45 - 20 = 85$$

(b) $40 + 25 = 65$ completed exactly one of the requirements.   That is, $n(M \oplus S) = 65$.

(c)   55 completed neither requirement.   That is, $n(M^c \cap S^c) = 55$.

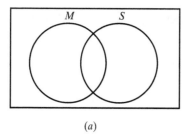
(a)
                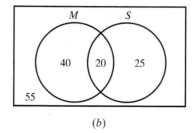
(b)

Fig. 1-9

1.24.   Prove Theorem 1.9 (Inclusion-exclusion principle): Suppose A and B are finite sets.   Then $A \cup B$ and $A \cap B$ are finite and

$$n(A \cup B) = n(A) + n(B) - n(A \cap B)$$

Suppose A and B are finite. Then clearly $A \cap B$ and $A \cup B$ are finite.

Suppose we count the elements of A and then count the elements of B.   Then, every element in $A \cap B$ would be counted twice, once in A and once in B.   Hence, as required,

$$n(A \cup B) = n(A) + n(B) - n(A \cap B)$$

Alternately (Problem 1.54):

   (i) A is the disjoint union of $A \setminus B$ and $A \cap B$.

  (ii) B is the disjoint union of $B \setminus A$ and $A \cap B$.

 (iii) $A \cup B$ is the disjoint union of $A \setminus B$, $A \cap B$, and $B \setminus A$.

Therefore, by Lemma 1.6 and Corollary 1.7,

$$
\begin{aligned}
n(A \cup B) &= n(A \setminus B) + n(A \cap B) + n(B \setminus A) \\
&= [n(A) - n(A \cap B)] + n(A \cap B) + [n(B) - n(A \cap B)] \\
&= n(A) + n(B) - n(A \cap B)
\end{aligned}
$$

1.25. Show that each set is countable: (a) **Z**, the set of integers, (b) **N** × **N**.

A set S is countable if: (a) S is finite or (b) the elements of S can be listed in the form of a sequence or, in other words, there is a one-to-one correspondence between the positive integers (counting numbers) **N** = {1, 2, 3, . . .} and S.

Neither set is finite.

(a) The following shows a one-to-one correspondence between **N** and **Z**:

Counting numbers **N**:   1   2   3   4   5   6   7   8   . . .

↓   ↓   ↓   ↓   ↓   ↓   ↓   ↓

Integers **Z**:   0   1   −1   2   −2   3   −3   4   . . .

That is, $n \in$ **N** corresponds to either $n/2$, when n is even, or $(1 − n)/2$, when n is odd:

$$f(n) = \begin{cases} n/2 & \text{for } n \text{ even,} \\ (1 - n)/2 & \text{for } n \text{ odd.} \end{cases}$$

Thus **Z** is countable.

(b) Figure 1-10 shows that **N** × **N** can be written as an infinite sequence as follows:

$$(1, 1), (2, 1), (1, 2), (1, 3), (2, 2), . . .$$

Specifically, the sequence is determined by "following the arrows" in Fig. 1-10.

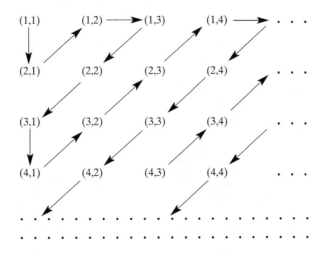

Fig. 1-10

ORDERED PAIRS AND PRODUCT SETS

1.26. Find x and y given that $(2x, x − 3y) = (6, −9)$.

Two ordered pairs are equal if and only if the corresponding entries are equal. This leads to the equations

$$2x = 6 \quad \text{and} \quad x - 3y = -9$$

Solving the equations yields $x = 3$, $y = 4$.

1.27. Given: $A = \{1, 2, 3\}$ and $B = \{a, b\}$. Find: (a) $A \times B$, (b) $B \times A$, (c) $B \times B$.

(a) $A \times B$ consists of all ordered pairs (x, y) where $x \in A$ and $y \in B$. Thus

$$A \times B = \{(1, a), (1, b), (2, a), (2, b), (3, a), (3, b)\}$$

(b) $B \times A$ consists of all ordered pairs (x, y) where $x \in B$ and $y \in A$.   Thus

$$B \times A = \{(a, 1), (a, 2), (a, 3), (b, 1), (b, 2), (b, 3)\}$$

(c) $B \times B$ consists of all ordered pairs (x, y) where $x, y \in B$.   Thus

$$B \times B = \{(a, a), (a, b), (b, a), (b, b)\}$$

Note that, as expected from Theorem 1.11, $n(A \times B) = 6$, $n(B \times A) = 6$, $n(B \times B) = 4$; that is, the number of elements in a product set is equal to the product of the numbers of elements in the factor sets.

1.28.   Given $A = \{1, 2\}$, $B = \{x, y, z\}$, $C = \{3, 4\}$.   Find $A \times B \times C$.

$A \times B \times C$ consists of all ordered triples (a, b, c) where $a \in A$, $b \in B$, $c \in C$.   These elements of $A \times B \times C$ can be systematically obtained by a so-called "tree diagram" as in Fig. 1-11.   The elements of $A \times B \times C$ are precisely the 12 ordered triplets to the right of the diagram.

Observe that $n(A) = 2$, $n(B) = 3$, $n(C) = 2$ and, as expected,

$$n(A \times B \times C) = 12 = n(A) \cdot n(B \cdot n(C)$$

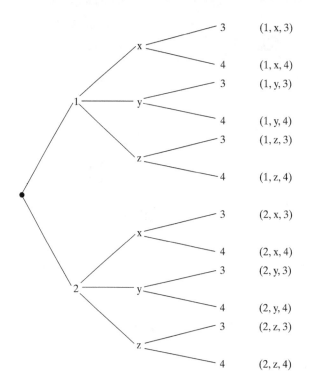

Fig. 1-11

1.29.   Each toss of a coin will yield either a head or a tail.   Let $C = \{H, T\}$ denote the set of outcomes.   Find C^3, $n(C^3)$, and explain what C^3 represents.

Since $n(C) = 2$, we have $n(C^3) = 2^3 = 8$.   Omitting certain commas and parenthesis for notational convenience,

$$C^3 = \{HHH, HHT, HTH, HTT, THH, THT, TTH, TTT\}$$

C^3 represents all possible sequences of outcomes of three tosses of the coin.

1.30. Prove: $A \times (B \cap C) = (A \times B) \cap (A \times C)$.

$$
\begin{aligned}
A \times (B \cap C) &= \{(x, y) : x \in A, y \in B \cap C\} \\
&= \{(x, y) : x \in A, y \in B, y \in C\} \\
&= \{(x, y) : x \in A, y \in B, x \in A, y \in C\} \\
&= \{(x, y) : (x, y) \in A \times B, (x, y) \in A \times C\} \\
&= (A \times B) \cap (A \times C)
\end{aligned}
$$

CLASSES OF SETS AND PARTITIONS

1.31. Consider the set $A = [\{1, 2, 3\}, \{4, 5\}, \{6, 7, 8\}]$. (a) Find the elements of A. (b) Find $n(A)$.

　(a)　A is a collection of sets; its elements are the sets $\{1, 2, 3\}$, $\{4, 5\}$, and $\{6, 7, 8\}$.

　(b)　A has only three elements; hence $n(A) = 3$.

1.32. Consider the class A of sets in Problem 1.31. Determine whether or not each of the following is true or false:

　(a)　$1 \in A$　　　　　(c)　$\{6, 7, 8\} \in A$　　　(e)　$\varnothing \in A$

　(b)　$\{1, 2, 3\} \subseteq A$　　(d)　$\{\{4, 5\}\} \subseteq A$　　(f)　$\varnothing \subseteq A$

　(a)　False.　1 is not one of the three elements of A.

　(b)　False.　$\{1, 2, 3\}$ is not a subset of A; it is one of the elements of A.

　(c)　True.　$\{6, 7, 8\}$ is one of the elements of A.

　(d)　True.　$\{\{4, 5\}\}$, the set consisting of the element $\{4, 5\}$, is a subset of A.

　(e)　False.　The empty set $\varnothing$ is not an element of A, that is, it is not one of the three sets listed as elements of A.

　(f)　True.　The empty set $\varnothing$ is a subset of every set; even a class of sets.

1.33. List the elements of the power set $\mathscr{P}(A)$ of $A = \{a, b, c, d\}$.

　　The elements of $\mathscr{P}(A)$ are the subsets of A. Hence:

$$
\begin{aligned}
\mathscr{P}(A) = [&A, \quad \{a, b, c\}, \quad \{a, b, d\}, \quad \{a, c, d\}, \quad \{b, c, d\}, \quad \{a, b\}, \quad \{a, c\}, \\
&\{a, d\}, \quad \{b, c\}, \quad \{b, d\}, \quad \{c, d\}, \quad \{a\}, \quad \{b\}, \quad \{c\}, \quad \{d\}, \quad \varnothing]
\end{aligned}
$$

As expected, $\mathscr{P}(A)$ has $2^4 = 16$ elements.

1.34. Let $S = \{a, b, c, d, e, f, g\}$. Determine which of the following are partitions of S:

　(a)　$P_1 = [\{a, c, e\}, \{b\}, \{d, g\}]$　　　(c)　$P_3 = [\{a, b, e, g\}, \{c\}, \{d, f\}]$

　(b)　$P_2 = [\{a, e, g\}, \{c, d\}, \{b, e, f\}]$　　(d)　$P_4 = [\{a, b, c, d, e, f, g\}]$

　(a)　P_1 is not a partition of S since $f \in S$ does not belong to any of the cells.

　(b)　P_2 is not a partition of S since $e \in S$ belongs to two of the cells, $\{a, e, g\}$ and $\{b, e, f\}$.

　(c)　P_3 is a partition of S since each element in S belongs to exactly one cell.

　(d)　P_4 is a partition of S into one cell, S itself.

1.35. Find all partitions of $S = \{a, b, c, d\}$.

Note first that each partition of S contains either one, two, three, or four distinct cells. The partitions are as follows:

(1) $[\{a, b, c, d\}] = [S]$

(2a) $[\{a\}, \{b, c, d\}]$, $[\{b\}, \{a, c, d\}]$, $[\{c\}, \{a, b, d\}]$, $[\{d\}, \{a, b, c\}]$

(2b) $[\{a, b\}, \{c, d\}]$, $[\{a, c\}, \{b, d\}]$, $[\{a, d\}, \{b, c\}]$

(3) $[\{a\}, \{b\}, \{c, d\}]$, $[\{a\}, \{c\}, \{b, d\}]$, $[\{a\}, \{d\}, \{b, c\}]$, $[\{b\}, \{c\}, \{a, d\}]$,
     $[\{b\}, \{d\}, \{a, c\}]$, $[\{c\}, \{d\}, \{a, b\}]$

(4) $[\{a\}, \{b\}, \{c\}, \{d\}]$

[Note (2a) refers to partitions with one-element and three-element cells, whereas (2a) refers to partitions with two two-element cells.] There are $1 + 4 + 3 + 6 + 1 = 15$ different partitions of S.

1.36. Let $\mathbf{N} = \{1, 2, 3, \ldots\}$ and, for each $n \in \mathbf{N}$, let

$$A_n = \{x : x \text{ is a multiple of } n\} = \{n, 2n, 3n, \ldots\}$$

Find: (a) $A_3 \cap A_5$, (b) $A_4 \cap A_6$, (c) $\cup_{i \in Q} A_i$, where $Q = \{2, 3, 5, 7, 11, \ldots\}$ is the set of prime numbers.

(a) Those numbers which are multiples of both 3 and 5 are the multiples of 15; hence

$$A_3 \cap A_5 = A_{15}$$

(b) The multiples of 12 and no other numbers belong to both A_4 and A_6; hence

$$A_4 \cap A_6 = A_{12}$$

(c) Every positive integer except 1 is a multiple of at least one prime number; hence

$$\cup_{i \in Q} A_i = \{2, 3, 4, \ldots\} = \mathbf{N} \setminus \{1\}$$

1.37. For each $n \in \mathbf{N}$, let $B_n = (0, 1/n)$, the open interval from 0 to $1/n$. [For example, $B_1 = (0, 1)$, $B_2 = (0, 1/2)$, $B_5 = (0, 1/5)$.] Find:

(a) $B_3 \cup B_7$ and $B_3 \cap B_7$

(b) $\cup_{n \in A} B_n$ where A is a nonempty subset of $\mathbf{N}$

(c) $\cap_{n \in \mathbf{N}} B_n$

(a) Since B_7 is a subset of B_3, we have $B_3 \cup B_7 = B_3$ and $B_3 \cap B_7 = B_7$.

(b) Let k be the smallest element of A. Then $\cup_{n \in A} B_n = B_k$.

(c) Let x be any real number. Then there is at least one $k \in N$ such that $x \notin (0, 1/k) = B_k$. Thus $\cap_{n \in \mathbf{N}} B_n = \varnothing$.

1.38. Prove: Let $\{A_i : i \in I\}$ be an indexed collection of sets, and let $i_0 \in I$. Then

$$\cap_{i \in I} A_i \subseteq A_{i_0} \subseteq \cup_{i \in I} A_i.$$

Let $x \in \cap_{i \in I} A_i$; then $x \in A_i$ for every $i \in I$. In particular $x \in A_{i_0}$. Hence $\cap_{i \in I} A_i \subseteq A_{i_0}$. Now let $y \in A_{i_0}$. Since $i_0 \in I$, $y \in \cup_{i \in I} A_i$. Hence $A_{i_0} \subseteq \cup_{i \in I} A_i$.

1.39. Prove (DeMorgan's law): For any indexed collection $\{A_i : i \in I\}$ of sets,

$$(\cup_i A_i)^c = \cap_i A_i^c$$

Using the definitions of union and intersection of indexed classes of sets, we get:

$$(\cup_i A_i)^c = \{x : x \notin \cup_i A_i\} = \{x : x \notin A_i, \text{ for every } i\}$$
$$= \{x : x \in A_i^c, \text{ for every } i\} = \cap_i A_i^c$$

1.40. Let $\mathscr{A}$ be an algebra (σ-algebra) of subsets of **U**. Show that:

(a) **U** and $\varnothing$ belong to $\mathscr{A}$. (b) $\mathscr{A}$ is closed under finite (countable) intersections.

Recall that, by definition, $\mathscr{A}$ is nonempty and $\mathscr{A}$ is closed under complements and finite (countable) intersections.

(a) Since $\mathscr{A}$ is nonempty, there is a set $A \in \mathscr{A}$. Hence the complement A^c belongs to $\mathscr{A}$. Therefore, the union and complement,

$$A \cup A^c = \mathbf{U} \qquad \text{and} \qquad \mathbf{U^c} = \varnothing$$

belong to $\mathscr{A}$, as required.

(b) Let $\{A_i\}$ be a finite (countable) collection of sets belonging to $\mathscr{A}$. Therefore, by DeMorgan's law (Problem 1.39),

$$(\cup_i A_i^c)^c = \cap_i A_i^{cc} = \cap_i A_i$$

Hence $\cap_i A_i$ belongs to $\mathscr{A}$, as required.

MATHEMATICAL INDUCTION

1.41. Prove the assertion $A(n)$ that the sum of the first n positive integers is $\frac{1}{2}n(n+1)$; that is,

$$A(n) : 1 + 2 + 3 + \cdots + n = \tfrac{1}{2}n(n+1)$$

The assertion holds for $n = 1$ since

$$A(1) : \qquad 1 = \tfrac{1}{2}(1)(1+1)$$

Assuming $A(n)$ is true, we add $n+1$ to both sides of $A(n)$. This yields

$$1 + 2 + 3 + \cdots + n + (n+1) = \tfrac{1}{2}n(n+1) + (n+1)$$
$$= \tfrac{1}{2}[n(n+1) + 2(n+1)]$$
$$= \tfrac{1}{2}[(n+1)(n+2)]$$

which is $A(n+1)$. That is, $A(n+1)$ is true whenever $A(n)$ is true. By the principle of induction, $A(n)$ is true for all $n \geq 1$.

1.42. Prove the following assertion (for $n \geq 0$):

$$A(n) : 1 + 2 + 2^2 + 2^3 + \cdots + 2^n = 2^{n+1} - 1$$

$A(0)$ is true since $1 = 2^1 - 1$. Assuming $A(n)$ is true, we add 2^{n+1} to both sides of $A(n)$. This yields:

$$1 + 2 + 2^2 + 2^3 + \cdots + 2^n + 2^{n+1} = 2^{n+1} - 1 + 2^{n+1}$$
$$= 2(2^{n+1}) - 1$$
$$= 2^{n+2} - 1$$

which is $A(n+1)$. Thus, $A(n+1)$ is true whenever $A(n)$ is true. By the principle of induction, $A(n)$ is true for all $n \geq 0$.

1.43. Prove: $n^2 \geq 2n + 1$ for $n \geq 3$.

Since $3^2 = 9$ and $2(3) + 1 = 7$, the formula is true for $n = 3$. Assuming $n^2 \geq 2n + 1$, we have

$$(n + 1)^2 = n^2 + 2n + 1 \geq (2n + 1) + 2n + 1 = 2n + 2 + 2n \geq 2n + 2 + 1 = 2(n + 1) + 1$$

Thus, the formula is true for $n + 1$. By induction, the formula is true for all $n \geq 3$.

1.44. Prove: $n! \geq 2^n$ for $n \geq 4$.

Since $4! = 1 \cdot 2 \cdot 3 \cdot 4 = 24$ and $2^4 = 16$, the formula is true for $n = 4$. Assuming $n! \geq 2^n$ and $n + 1 \geq 2$, we have

$$(n + 1)! = n!(n + 1) \geq 2^n(n + 1) \geq 2^n(2) = 2^{n+1}$$

Thus, the formula is true for $n + 1$. By induction, the formula is true for all $n \geq 4$.

MISCELLANEOUS PROBLEMS

1.45. Show that A is the disjoint union of $A \setminus B$ and $A \cap B$; that is, show that:

(a) $A = (A \setminus B) \cup (A \cap B)$,   (b) $(A \setminus B) \cap (A \cap B) = \varnothing$.

(a) By Problem 1.13, $A \setminus B = A \cap B^c$. Using the Distributive Law and the Complement Law, we get

$$(A \setminus B) \cup (A \cap B) = (A \cap B^c) \cup (A \cap B) = A \cap (B^c \cup B) = A \cap U = A$$

(b) Also,

$$(A \setminus B) \cap (A \cap B) = (A \cap B^c) \cap (A \cap B) = A \cap (B^c \cap B) = A \cap \varnothing = \varnothing$$

1.46. Prove Corollary 1.10. Suppose A, B, C are finite sets. Then $A \cup B \cup C$ is finite and

$$n(A \cup B \cup C) = n(A) + n(B) + n(C) - n(A \cap B) - n(A \cap C) - n(B \cap C) + n(A \cap B \cap C)$$

Clearly $A \cup B \cup C$ is finite when A, B, C are finite. Using

$$(A \cup B) \cap C = (A \cap C) \cup (B \cap C) \quad \text{and} \quad (A \cap B) \cap (B \cap C) = A \cap B \cap C$$

and using Theorem 1.9 repeatedly, we have

$$\begin{aligned} n(A \cup B \cup C) &= n(A \cup B) + n(C) - n[(A \cap C) \cup (B \cap C)] \\ &= [n(A) + n(B) - n(A \cap B)] + n(C) - [n(A \cap C) + n(B \cap C) - n(A \cap B \cap C)] \\ &= n(A) + n(B) + n(C) - n(A \cap B) - n(A \cap C) - n(B \cap C) + n(A \cap B \cap C) \end{aligned}$$

as required.

Supplementary Problems

SETS, ELEMENTS, SUBSETS

1.47. Which of the following sets are equal?

$A = \{x : x^2 - 4x + 3 = 0\}$, $C = \{x : x \in \mathbf{N}, x < 3\}$, $E = \{1, 2\}$, $G = \{3, 1\}$,

$B = \{x : x^2 - 3x + 2 = 0\}$, $D = \{x : x \in \mathbf{N}, x \text{ is odd}, x < 5\}$, $F = \{1, 2, 1\}$, $H = \{1, 1, 3\}$

1.48. List the elements of the following sets if the universal set is the English alphabet $\mathbf{U} = \{a, b, c, \ldots, y, z\}$. Furthermore, identify which of the sets are equal.

$A = \{x : x \text{ is a vowel}\}$, $C = \{x : x \text{ precedes } f \text{ in the alphabet}\}$,

$B = \{x : x \text{ is a letter in the word "little"}\}$, $D = \{x : x \text{ is a letter in the word "title"}\}$

1.49. Let $A = \{1, 2, \ldots, 8, 9\}$, $B = \{2, 4, 6, 8\}$, $C = \{1, 3, 5, 7, 9\}$, $D = \{3, 4, 5\}$, $E = \{3, 5\}$.

Which of the above sets can equal a set X under each of the given conditions?

(a) X and B are disjoint   (c) $X \subseteq A$ but $X \nsubseteq C$

(b) $X \subseteq D$ but $X \nsubseteq B$ (d) $X \subseteq C$ but $X \nsubseteq A$

SET OPERATIONS

1.50. Given the universal set $\mathbf{U} = \{1, 2, 3, \ldots, 8, 9\}$ and the sets:

$$A = \{1, 2, 5, 6\}, \qquad B = \{2, 5, 7\}, \qquad C = \{1, 3, 5, 7, 9\}$$

Find: (a) $A \cap B$ and $A \cap C$, (b) $A \cup B$ and $A \cup C$, (c) A^c and C^c.

1.51. For the sets in Problem 1.50, find: (a) $A \setminus B$ and $A \setminus C$,   (b) $A \oplus B$ and $A \oplus C$.

1.52. For the sets in Problem 1.50, find: (a) $(A \cup C) \setminus B$,   (b) $(A \cup B)^c$,   (c) $(B \oplus C) \setminus A$.

1.53. Let $A = \{a, b, c, d, e\}$, $B = \{a, b, d, f, g\}$, $C = \{b, c, e, g, h\}$, $D = \{d, e, f, g, h\}$.   Find:

(a) $A \cap (B \cup D)$ (c) $(A \cap D) \cup B$ (e) $B \cap C \cap D$ (g) $(A \oplus C) \cap B$

(b) $B \setminus (C \cup D)$ (d) $(A \cup D) \setminus C$ (f) $(C \setminus A) \setminus D$ (h) $(A \oplus D) \setminus B$

1.54. Let A and B be any sets.   Prove $A \cup B$ is the disjoint union of $A \setminus B$, $A \cap B$, and $B \setminus A$.

1.55. Prove the following:

(a) $A \subseteq B$ if and only if $A \cap B^c = \varnothing$.   (c) $A \subseteq B$ if and only if $B^c \subseteq A^c$.

(b) $A \subseteq B$ if and only if $A^c \cup B = \mathbf{U}$.   (d) $A \subseteq B$ if and only if $A \setminus B = \varnothing$.

(Compare with Theorem 1.4.)

1.56. The formula $A \setminus B = A \cap B^c$ defines the difference operation in terms of the operations of intersection and complement.   Find a formula which defines the union $A \cup B$ in terms of the operations of intersection and complement.

VENN DIAGRAMS, ALGEBRA OF SETS, DUALITY

1.57. The Venn diagram in Fig. 1-12 shows sets A, B, C.   Shade the following sets:

(a) $A \setminus (B \cup C)$ (b) $A^c \cap (B \cup C)$ (c) $(A \cup C) \cap (B \cup C)$

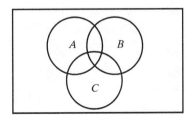

Fig. 1-12

1.58. Write the dual of each equation:

    (a) $A \cup (A \cap B) = A$ (b) $(A \cap B) \cup (A^c \cap B) \cup (A \cap B^c) \cup (A^c \cap B^c) = \mathbf{U}$

1.59. Use the laws in Table 1-1 to prove: $(A \cap B) \cup (A \cap B^c) = A$.

FINITE SETS AND THE COUNTING PRINCIPLE, COUNTABLE SETS

1.60. Determine which of the following sets are finite:

    (a) Lines parallel to the x axis   (c) Animals living on the earth

    (b) Letters in the English alphabet   (d) Circles through the origin $(0,0)$

1.61. Given $n(\mathbf{U}) = 20$, $n(A) = 12$, $n(B) = 9$, $n(A \cap B) = 4$. Find:

    (a) $n(A \cup B)$ (b) $n(A^c)$ (c) $n(B^c)$ (d) $n(A \setminus B)$ (e) $n(\varnothing)$

1.62. Among 120 Freshmen at a college, 40 take mathematics, 50 take English, and 15 take both mathematics and English. Find the number of the Freshmen who:

    (a) Do not take mathematics   (d) Take English but not mathematics

    (b) Take mathematics or English   (e) Take exactly one of the two subjects

    (c) Take mathematics but not English   (f) Take neither mathematics nor English

1.63. In a survey of 60 people, it was found that 25 read *Newsweek* magazine, 26 read *Time*, and 23 read *Fortune*. Also, 9 read both *Newsweek* and *Fortune*, 11 read *Newsweek* and *Time*, 8 read both *Time* and *Fortune*, and 3 read all three magazines.

    (a) Figure 1-13 is a Venn diagram of three sets, N (*Newsweek*), T (*Time*), and F (*Fortune*). Fill in the correct number of people in each of the eight regions of the Venn diagram.

    (b) Find the number of people who read: (i) only *Newsweek*, (ii) only *Time*, (iii) only *Fortune*, (iv) *Newsweek* and *Time*, but not *Fortune*, (v) only one of the magazines, (vi) none of the magazines.

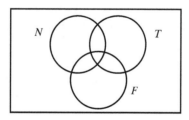

Fig. 1-13

1.64. Let $A_1, A_2, A_3, \ldots$ be a sequence of finite sets. Show that the union $S = \cup_i A_i$ is countable.

1.65. Let $A_1, A_2, A_3, \ldots$ be a sequence of pairwise disjoint countably infinite sets. Show that the union $T = \cup_i A_i$ is countable.

PRODUCT SETS

1.66. Find x and y if: (a) $(x + 3, 3) = (5, 3x + y)$, (b) $(x - 3y, 5) = (7, x - y)$.

1.67. Find x, y, z if $(2x, x + y, x - y - 2z) = (4, -1, 3)$.

1.68. Let $A = \{a, b\}$ and $B = \{1, 2, 3, 4\}$.   Find (a) $A \times B$ and (b) $B \times A$.

1.69. Let $C = \{H, T\}$, the set of possible outcomes if a coin is tossed.   Find:

(a) $C^2 = C \times C$ and (b) $C^4 = C \times C \times C \times C$.

1.70. Suppose $n(A) = 2$ and $n(B) = 6$.   Find the number of elements in:

(a) $A \times B$, $B \times A$,   (b) A^2, B^2, A^3, B^3,   (c) $A \times A \times B \times A$.

CLASSES OF SETS AND PARTITIONS

1.71. Find the power set $\mathscr{P}(A)$ of $A = \{a, b, c, d, e\}$.

1.72. Let $S = \{1, 2, 3, 4, 5, 6\}$.   Determine whether each of the following is a partition of S:

(a) $[\{1, 3, 5\}, \{2, 4\}, \{3, 6\}]$ (d) $[\{1\}, \{3, 6\}, \{2, 4, 5\}, \{3, 6\}]$

(b) $[\{1, 5\}, \{2\}, \{3, 6\}]$ (e) $[\{1, 2, 3, 4, 5, 6\}]$

(c) $[\{1, 5\}, \{2\}, \{4\}, \{3, 6\}]$ (f) $[\{1\}, \{2\}, \{3\}, \{4\}, \{5\}, \{6\}]$

1.73. Find all partitions of $S = \{1, 2, 3\}$.

1.74. For each positive integer $n \in \mathbf{N}$, let $A_n = \{n, 2n, 3n, \ldots\}$, the multiples of n.   Find:

(a) $A_2 \cap A_7$,   (b) $A_6 \cap A_8$,   (c) $A_5 \cup A_{20}$,   (d) $A_5 \cap A_{20}$,   (e) $A_s \cup A_{st}$, where $s, t \in \mathbf{N}$,
(f) $A_s \cap A_{st}$, where $s, t \in \mathbf{N}$.

1.75. Prove: If $J \subseteq \mathbf{N}$ is infinite, then $\cap(A_i : i \in J) = \varnothing$.   (Here the A_i are the sets in Problem 1.74.)

1.76. Let $[A_1, A_2, \ldots, A_m]$ and $[B_1, B_2, \ldots, B_n]$ be partitions of S.   Show that the collection of sets
$$[A_i \cap B_j; \ \ i = 1, \ldots, m, j = 1, \ldots, n] \setminus \varnothing$$
(where the empty set $\varnothing$ is deleted) is also a partition of S.   (It is called the *cross partition*.)

1.77. Prove: For any indexed class of sets $\{A_i : i \in I\}$ and any set B:
$$(a) \ B \cup (\cap_i A_i) = \cap_i (B \cup A_i), \ \ (b) \ B \cap (\cup_i A_i) = \cup_i (B \cap A_i)$$

1.78. Prove (DeMorgan's law): $(\cup_i A_i)^c = \cap_i A_i^c$.

1.79. Show that each of the following is an algebra of subsets of $\mathbf{U}$:
$$(a) \ \mathscr{A} = \{\varnothing, \mathbf{U}\}, \ \ (b) \ \mathscr{B} = \{\varnothing, A, A^c, \mathbf{U}\}, \ \ (c) \ P(\mathbf{U}), \text{ the power set of } \mathbf{U}$$

1.80. Let $\mathscr{A}$ and $\mathscr{B}$ be algebras (σ-algebras) of subsets of $\mathbf{U}$.   Prove that the intersection $\mathscr{A} \cap \mathscr{B}$ is also an algebra (σ-algebra) of subsets of $\mathbf{U}$.

MATHEMATICAL INDUCTION

1.81. Prove: $2 + 4 + 6 + \cdots + 2n = n(n + 1)$.

1.82. Prove: $1 + 4 + 7 + \cdots + (3n - 2) = 2n(3n - 1)$.

1.83. Prove: $1^2 + 2^2 + 3^2 + \cdots + n^2 = \dfrac{n(n + 1)(2n + 1)}{6}$.

1.84. Prove: For $n \geq 3$, we have $2^n \geq n^2$.

1.85. Prove: $\dfrac{1}{1 \cdot 3} + \dfrac{1}{3 \cdot 5} + \dfrac{1}{5 \cdot 7} + \cdots + \dfrac{1}{(2n - 1)(2n + 1)} = \dfrac{1}{2n + 1}$.

Answers to Supplementary Problems

1.47. $B = C = E = F$; $A = D = G = H$.

1.48. $A = \{a, e, i, o, u\}$; $B = D = \{1, i, t, e\}$; $C = \{a, b, c, d, e\}$.

1.49.   (a) C and E; (b) D and E; (c) A, B, D; (d) None.

1.50.   (a) $A \cap B = \{2, 5\}$, $A \cap C = \{1, 5\}$;   (b) $A \cup B = \{1, 2, 5, 6, 7\}$, $B \cup C = \{1, 2, 3, 5, 7, 9\}$;
(c) $A^c = \{3, 4, 7, 8, 9\}$, $C^c = \{2, 4, 6, 8\}$.

1.51.   (a) $A \setminus B = \{1, 6\}$, $A \setminus C = \{2, 6\}$;   (b) $A \oplus B = \{1, 6, 7\}$, $A \oplus C = \{2, 3, 6, 7, 9\}$.

1.52.   (a) $(A \cup C) \setminus B = \{1, 3, 6, 9\}$; (b) $\{A \cup B\}^c = \{3, 4, 8, 9\}$; (c) $\{B \oplus C\} \setminus A = \{3, 9\}$.

1.53.   (a) $A \cap (B \cup D) = \{a, b, d, e\}$; (b) $B \setminus (C \cup D) = \{a\}$; (c) $(A \cap D) \cup B = \{a, b, d, e, f, g\}$;
(d) $(A \cup D) \setminus C = \{a, d, f\}$; (e) $B \cap C \cap D = (g)$; (f) $(C \setminus A) \setminus D = \varnothing$; (g) $(A \oplus C) \cap B = \{a, d, g\}$;
(h) $(A \oplus D) \setminus B = \{c, h\}$.

1.56. $A \cup B = (A^c \cap B^c)^c$.

1.57.   See Fig. 1-14.

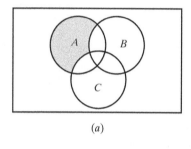   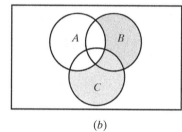   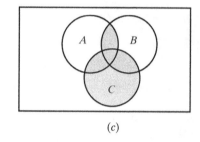

(a)   (b)   (c)

Fig. 1-14

1.58.   (a) $A \cap (A \cup B) = A$; (b) $(A \cup B) \cap (A^c \cup B) \cap (A \cup B^c) \cap (A^c \cup B^c) = \varnothing$.

1.60.   (a) Infinite; (b) finite; (c) finite; (d) infinite.

1.61.   (a) $n(A \cup B) = 17$; (b) $n(A^c) = 8$; (c) $n(B^c) = 11$; (d) $n(A \setminus B) = 8$; (e) $n(\varnothing) = 0$.

1.62.   (a) 80; (b) 75; (c) 25; (d) 35; (e) 60; (f) 45.

1.63.   (a) See Fig. 1-15; (b) (i) 8, (ii) 10, (iii) 9, (iv) 8, (v) 27, (vi) 11.

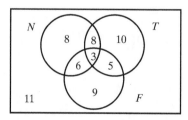

Fig. 1-15

1.64. Let $B_1 = A_1$, $B_2 = A_2 \setminus B_1$, $B_3 = A_3 \setminus B_2$, ..., that is, $B_n = A_k \setminus B_{k-1}$. Then the B_k are finite and pairwise disjoint, and $S = \cup_i A_i = \cup_i B_i$. Say

$$B_k = \{b_{k1}, b_{k2}, \ldots, b_{kn_k}\}$$

Then S can be written as a sequence as follows:

$$S = \{b_{11}, b_{12}, \ldots, b_{1n_1}, b_{21}, b_{22}, \ldots, b_{2n_2}, \ldots\}$$

That is, first write down the elements of B_1, then the elements of B_2, and so on.

1.65. Suppose

$$A_1 = \{a_{11}, a_{12}, a_{13}, \ldots\}, \qquad A_2 = \{a_{21}, a_{22}, a_{23}, \ldots\}, \qquad \ldots$$

For $n > 1$, define $D_n = \{a_{ij} : i + j = n\}$. For example,

$$D_2 = \{a_{11}\}, \qquad D_3 = \{a_{12}, a_{21}\}, \qquad D_4 = \{a_{31}, a_{22}, a_{13}\}, \qquad \ldots$$

Each D_k is finite and $T = \cup_i D_i$. By Problem 1.64, T is countable.

1.66. (a) $x = 2$, $y = -3$; (b) $x = 6$, $y = -1$.

1.67. $x = 2$, $y = -3$, $z = 1$.

1.68. (a) $A \times B = \{(a, 1), (a, 2), (a, 3), (a, 4), (b, 1), (b, 2), (b, 3), (b, 4)\}$;
(b) $B \times A = \{(1, a), (2, a), (3, a), (4, a), (1, a), (2, a), (3, a), (4, a)\}$.

1.69. Note $n(C^2) = 2^2 = 4$ and $n(C^4) = 2^4 = 16$.
(a) $C^2 = C \times C = \{HH, HT, TH, TT\}$;
(b) $C^4 = C \times C \times C \times C = \{HHHH, HHHT, HHTH, HHTT, HTHH, HTHT, HTTH, HTTT, THHH,$
$THHT, THTH, THTT, TTHH, TTHT, TTTH, TTTT\}$.

1.70. (a) 12, 12; (b) 4, 36, 8, 216; (c) 48.

1.71. Note $\mathscr{P}(A)$ has $2^5 = 32$ elements; that is, there are 32 subsets of A. Each subset has at most five elements, and we list them in terms of their numbers of elements:

None (1): $\varnothing$
One (5): $\{a\}, \{b\}, \{c\}, \{d\}, \{e\}$
Two (10): $\{a, b\}, \{a, c\}, \{a, d\}, \{a, e\}, \{b, c\}, \{b, d\}, \{b, e\}, \{c, d\}, \{c, e\}, \{d, e\}$
Three (10): $\{a, b, c\}, \{a, b, d\}, \{a, b, e\}, \{a, c, d\}, \{a, c, e\}, \{a, d, e\}, \{b, c, d\}, \{b, c, e\}, \{b, d, e\}, \{c, d, e\}$
Four (5): $\{a, b, c, d\}, \{a, b, c, e\}, \{a, b, d, e\}, \{a, c, d, e\}, \{b, c, d, e\}$
Five (1): $A = \{a, b, c, d, e\}$

1.72. (a) and (b): No.   Others: Yes.

1.73. There are five: $[S]$, $[\{1, 2\}, \{3\}]$, $[\{1, 3\}, \{2\}]$, $[\{1\}, \{2, 3\}]$, $[\{1\}, \{2\}, \{3\}]$.

1.74. (a) A_{14}; (b) A_{24}; (c) A_{20}; (d) A_5; (e) A_{si}; (f) A_s.

1.75. Let $n \in \mathbf{N}$ and let $B = \cap(A_i : i \in J)$. Since J is infinite, there exists $k \in J$ such that $n < k$. Thus $n \notin A_k$ and so $n \notin B$. That is, for every n, we have shown $n \notin B$. Thus $B = \varnothing$.

Techniques of Counting

2.1 INTRODUCTION

This chapter develops some techniques for determining, without direct enumeration, the number of possible outcomes of a particular experiment or event or the number of elements in a particular set. Such *sophisticated* counting is sometimes called *combinatorial analysis*.

2.2 BASIC COUNTING PRINCIPLES

There are two basic counting principles which are used throughout this chapter. One involves addition and the other involves multiplication.

Sum Rule Principle

The first counting principle follows:

> **Sum Rule Principle**: Suppose some event E can occur in m ways and a second event F can occur in n ways, and suppose both events cannot occur simultaneously. Then E or F can occur in $m + n$ ways.

This principle can be stated in terms of sets, and it is simply a restatement of Lemma 1.4.

> **Sum Rule Principle**: Suppose A and B are disjoint sets. Then:
> $$n(A \cup B) = n(A) + n(B)$$

Clearly, this principle can be extended to three or more events. That is, suppose an event E_1 can occur in n_1 ways, a second event E_2 can occur in n_2 ways, a third event E_3 can occur in n_3 ways, and so on, and suppose no two of the events can occur at the same time.   Then one of the events can occur in $n_1 + n_2 + n_3 + \ldots$ ways.

EXAMPLE 2.1

(a) Suppose there are 8 male professors and 5 female professors teaching a calculus class.   A student can choose a calculus professor in $8 + 5 = 13$ ways.

(b) Suppose there are 3 different mystery novels, 5 different romance novels, and 4 different adventure novels on a bookshelf.   Then there are

$$n = 3 + 5 + 4 = 12$$

ways to choose one of the novels.

Product Rule Principle

The second counting principle follows:

> **Product Rule Principle**: Suppose an event E can occur in m ways and, independent of this event, an event F can occur in n ways.   Then combinations of events E and F can occur in mn ways.

This principle can also be stated in terms of sets, and it is simply a restatement of Theorem 1.11.

> **Product Rule Principle**: Suppose A and B are finite sets.   Then:
> $$n(A \times B) = n(A) \cdot n(B)$$

Clearly, this principle can also be extended to three or more events.   That is, suppose an event E_1 can occur in n_1 ways, then a second event E_2 can occur in n_2 ways, then a third event E_3 can occur in n_3 ways, and so on.   Then all of the events can occur in $n_1 \cdot n_2 \cdot n_3 \cdot \ldots$ ways.

EXAMPLE 2.2

(a) Suppose a restaurant has 3 different appetizers and 4 different entrees.   Then there are

$$n = 3(4) = 12$$

different ways to order an appetizer and an entree.

(b) Suppose airline A has 3 daily flights between Boston and Chicago, and airline B has 2 daily flights between Boston and Chicago.

   (1) There are $n = 3(2) = 6$ ways to fly airline A from Boston to Chicago, and then airline B from Chicago back to Boston.

   (2) There are $m = 3 + 2 = 5$ ways to fly from Boston to Chicago; and hence $n = 5(5) = 25$ ways to fly from Boston to Chicago and then back again.

(c) Suppose a college has 3 different history courses, 4 different literature courses, and 2 different science courses (with no prerequisites).

   (1) Suppose a student has to choose one of each of the courses.   The number of ways to do this is:

$$n = 3(4)(2) = 24$$

(2) Suppose a student only needs to choose one of the courses. Clearly, there are

$$m = 3 + 4 + 2 = 9$$

courses, and so the student will have 9 choices. In other words, here the sum rule is used rather than the multiplication rule since only one of the courses is chosen.

2.3 FACTORIAL NOTATION

The product of the positive integers from 1 to n inclusive occurs very often in mathematics and hence it is denoted by the special symbol $n!$, read "n factorial". That is,

$$n! = 1 \cdot 2 \cdot 3 \cdots (n - 2)(n - 1)n = n(n - 1)(n - 2) \cdots 3 \cdot 2 \cdot 1$$

In other words, $n!$ may be defined by

$$1! = 1 \quad \text{and} \quad n! = n \cdot (n - 1)!$$

It is also convenient to define $0! = 1$.

EXAMPLE 2.3

(a) $2! = 2 \cdot 1 = 2; \quad 3! = 3 \cdot 2 \cdot 1 = 6; \quad 4! = 4 \cdot 3 \cdot 2 \cdot 1 = 24; \quad 5! = 5 \cdot 4! = 5 \cdot 24 = 120$

(b) $\dfrac{8!}{6!} = \dfrac{8 \cdot 7 \cdot 6!}{6!} = 8 \cdot 7 = 56; \quad 12 \cdot 11 \cdot 10 = \dfrac{12 \cdot 11 \cdot 10 \cdot 9!}{9!} = \dfrac{12!}{9!}$

(c) $\dfrac{12 \cdot 11 \cdot 10}{3 \cdot 2 \cdot 1} = 12 \cdot 11 \cdot 10 \cdot \dfrac{1}{3!} = \dfrac{12!}{3! \, 9!}$

(d) $n(n - 1) \cdots (n - r + 1) = \dfrac{n(n - 1) \cdots (n - r + 1)(n - r)(n - r - 1) \cdots 3 \cdot 2 \cdot 1}{(n - r)(n - r - 1) \cdots 3 \cdot 2 \cdot 1} = \dfrac{n!}{(n - r)!}$

(e) Using (d), we get:

$$\frac{n(n - 1) \cdots (n - r + 1)}{r(r - 1) \cdots 3 \cdot 2 \cdot 1} = n(n - 1) \cdots (n - r + 1) \cdot \frac{1}{r!} = \frac{n!}{(n - r)!} \cdot \frac{1}{r!} = \frac{n!}{r!(n - r)!}$$

Stirling's Approximation to $n!$

A direct evaluation of $n!$ when n is very large is impossible, even with modern-day computers. Accordingly, one frequently uses the approximation formula

$$n! \sim \sqrt{2\pi n}\, n^n e^{-n}$$

(Here $e = 2.718\,28.\ \ldots$) The symbol $\sim$ means that as n gets larger and larger (that is, as $n \to \infty$), the ratio of both sides approaches 1.

2.4 BINOMIAL COEFFICIENTS

The symbol $\dbinom{n}{r}$, where r and n are positive integers with $r \le n$ [read: "nCr" or "n choose r"], is defined as follows:

$$\binom{n}{r} = \frac{n(n - 1)(n - 2) \cdots (n - r + 1)}{r(r - 1) \cdots 3 \cdot 2 \cdot 1} \quad \text{or} \quad \binom{n}{r} = \frac{n!}{r!(n - r)!}$$

The equivalence of the two formulas is shown in Example 2.3(e).

Note that $n - (n - r) = r$. This yields the following important relation:

Lemma 2.1: $\dbinom{n}{n-r} = \dbinom{n}{r}$ or, equivalently, $\dbinom{n}{a} = \dbinom{n}{b}$ where $a + b = n$.

Remark: Motivated by the second formula for $\dbinom{n}{r}$ and the fact that $0! = 1$, we define:

$$\binom{n}{0} = \frac{n!}{0!\,n!} = 1 \qquad \text{and, in particular,} \qquad \binom{0}{0} = \frac{0!}{0!\,0!} = 1$$

EXAMPLE 2.4

(a) $\dbinom{8}{2} = \dfrac{8 \cdot 7}{2 \cdot 1} = 28;$ $\dbinom{9}{4} = \dfrac{9 \cdot 8 \cdot 7 \cdot 6}{4 \cdot 3 \cdot 2 \cdot 1} = 126;$ $\dbinom{12}{5} = \dfrac{12 \cdot 11 \cdot 10 \cdot 9 \cdot 8}{5 \cdot 4 \cdot 3 \cdot 2 \cdot 1} = 792;$

$\dbinom{10}{3} = \dfrac{10 \cdot 9 \cdot 8}{3 \cdot 2 \cdot 1} = 120;$ $\dbinom{13}{1} = \dfrac{13}{1} = 13$

Note that $\dbinom{n}{r}$ has exactly r factors in both the numerator and the denominator.

(b) Suppose we want to compute $\dbinom{10}{7}$. By definition,

$$\binom{10}{7} = \frac{10 \cdot 9 \cdot 8 \cdot 7 \cdot 6 \cdot 5 \cdot 4}{7 \cdot 6 \cdot 5 \cdot 4 \cdot 3 \cdot 2 \cdot 1} = 120$$

On the other hand, $10 - 7 = 3$; hence using Lemma 2.1 we get:

$$\binom{10}{7} = \binom{10}{3} = \frac{10 \cdot 9 \cdot 8}{3 \cdot 2 \cdot 1} = 120$$

Observe that the second method saves both space and time.

Binomial Coefficients and Pascal's Triangle

The numbers $\dbinom{n}{r}$ are called the *binomial coefficients* since they appear as the coefficients in the expansion of $(a + b)^n$. Specifically, the following Binomial Theorem gives the general expression for the expansion of $(a + b)^n$:

Theorem 2.2 (Bionomial Theorem): $(a + b)^n = \displaystyle\sum_{k=0}^{n} \binom{n}{k} a^{n-k_b k}.$

This theorem is proved in Problem 2.34 using mathematical induction.

The coefficients of the successive powers of $a + b$ can be arranged in a triangular array of numbers, called *Pascal's triangle*, as pictured in Fig. 2-1. The numbers in Pascal's triangle have the following interesting properties:

(i) The first and last number in each row is 1.

(ii) Every other number in the array can be obtained by adding the two numbers appearing directly above it. For example, $10 = 4 + 6$, $15 = 5 + 10$, $20 = 10 + 10$.

Since the numbers appearing in Pascal's triangle are the binomial coefficients, property (ii) of Pascal's triangle comes from the following theorem (proved in Problem 2.7):

Theorem 2.3: $\dbinom{n+1}{r} = \dbinom{n}{r-1} + \dbinom{n}{r}$

$(a+b)^0 = 1$	1
$(a+b)^1 = a + b$	1 1
$(a+b)^2 = a^2 + 2ab + b^2$	1 2 1
$(a+b)^3 = a^3 + 3a^2b + 3ab^2 + b^3$	1 3 3 1
$(a+b)^4 = a^4 + 4a^3b + 6a^2b^2 + 4ab^3 + b^4$	1 4 6 4 1
$(a+b)^5 = a^5 + 5a^4b + 10a^3b^2 + 10a^2b^3 + 5ab^4 + b^5$	1 5 10 10 5 1
$(a+b)^6 = a^6 + 6a^5b + 15a^4b^2 + 20a^3b^3 + 15a^2b^4 + 6ab^5 + b^6$	1 6 15 20 15 6 1

Fig. 2-1. Pascal's triangle.

2.5 PERMUTATIONS

Any arrangement of a set of n objects in a given order is called a *permutation* of the objects (taken all at a time).   Any arrangement of any $r \le n$ of these objects in a given order is called an *r permutation* or a *permutation of the n objects taken r at a time*.   Consider, for example, the set of letters a, b, c, d.   Then:

   (i)   *bdca, dcba, acdb* are permutations of the four letters (taken all at a time).

   (ii)   *bad, adb, cbd, bca* are permutations of the four letters taken three at a time.

   (iii)   *ad, cb, da, bd* are permutations of the four letters taken two at a time.

The number of permutations of n objects taken r at a time will be denoted by

$$P(n, r)$$

Before we derive the general formula for $P(n, r)$ we consider a particular case.

EXAMPLE 2.5   Find the number of permutations of six objects, say A, B, C, D, E, F, taken three at a time.   In other words, find the number of "three-letter words" using only the given six letters without repetitions.
   Let the general three-letter word be represented by the following three boxes:

The first letter can be chosen in 6 different ways; following this, the second letter can be chosen in 5 different ways; and, following this, the last letter can be chosen in 4 different ways.   Write each number in the appropriate box as follows:

6	5	4

Accordingly, by the product rule principle, there are $6 \cdot 5 \cdot 4 = 120$ possible three-letter words without repetitions from the six letters, or there are 120 permutations of six objects taken three at a time.   Thus, we have shown that

$$P(6, 3) = 120$$

Derivation of the Formula for $P(n, r)$

The derivation of the formula for the number of permutations of n objects taken r at a time, or the number of r permutations of n objects, $P(n, r)$, follows the procedure in the preceding example. The first element in an r permutation of n objects can be chosen in n different ways; following this, the second element in the permutation can be chosen in $n - 1$ ways; and, following this, the third element in the permutation can be chosen in $n - 2$ ways.   Continuing in this manner, we have that the rth (last) element in the r permutation can be chosen in $n - (r - 1) = n - r + 1$ ways.   Thus, by the fundamental principle of counting, we have

$$P(n, r) = n(n - 1)(n - 2) \cdots (n - r + 1)$$

By Example 2.3(e), we see that

$$n(n - 1)(n - 2) \cdots (n - r + 1) = \frac{n(n - 1)(n - 2) \cdots (n - r + 1) \cdot (n - r)!}{(n - r)!} = \frac{n!}{(n - r)!}$$

Thus, we have proven the following theorem.

Theorem 2.4: $P(n, r) = \dfrac{n!}{(n - r)!}$.

Consider the case that $r = n$.   We get

$$P(n, n) = n(n - 1)(n - 2) \cdots 3 \cdot 2 \cdot 1 = n!$$

Accordingly,

Corollary 2.5:   There are $n!$ permutations of n objects (taken all at a time).

For example, there are $3! = 1 \cdot 2 \cdot 3 = 6$ permutations of the three letters a, b, c.   These are

$$abc, acb, bac, bca, cab, cba$$

Permutations with Repetitions

Frequently we want to know the number of permutations of a *multiset*, that is, a set of objects some of which are alike.   We will let

$$P(n; n_1, n_2, \ldots, n_r)$$

denote the number of permutations of n objects of which n_1 are alike, n_2 are alike, ..., n_r are alike.   The general formula follows:

Theorem 2.6: $P(n; n_1, n_2, \ldots, n_r) = \dfrac{n!}{n_1! n_2! \cdots n_r!}$

We indicate the proof of the above theorem by a particular example.   Suppose we want to form all possible five-letter "words" using the letters from the word "*BABBY*".   Now there are $5! = 120$ permutations of the objects B_1, A, B_2, B_3, Y, where we have distinguished the three B's.   Observe that the following 6 permutations produce the same word when the subscripts are removed:

$$B_1 B_2 B_3 AY, \quad B_1 B_3 B_2 AY, \quad B_2 B_1 B_3 AY, \quad B_2 B_3 B_1 AY, \quad B_3 B_1 B_2 AY, \quad B_3 B_2 B_1 AY$$

The 6 comes from the fact that there are $3! = 3 \cdot 2 \cdot 1 = 6$ different ways of placing the three B's in the first three positions in the permutation.   This is true for each set of three positions in which the three B's can appear.   Accordingly, there are

$$P(5; 3) = \frac{5!}{3!} = \frac{120}{6} = 20$$

different five-letter words that can be formed using the letters from the word "*BABBY*".

EXAMPLE 2.6

(a) Find the number m of seven-letter words that can be formed using the letters of the word "*BENZENE*".

   We seek the number of permutations of seven objects of which three are alike, the three E's, and two are alike, the two N's.   By Theorem 2.6,

$$m = P(7; 3, 2) = \frac{7!}{3!2!} = \frac{7 \cdot 6 \cdot 5 \cdot 4 \cdot 3 \cdot 2 \cdot 1}{3 \cdot 2 \cdot 1 \cdot 2 \cdot 1} = 420$$

(b) Find the number m of different signals, each consisting of eight flags in a vertical line, that can be formed from four indistinguishable red flags, three indistinguishable white flags, and a blue flag.

   We seek the number of permutations of eight objects of which four are alike, the red flags, and three are alike, the white flags.   By Theorem 2.6,

$$m = P(8; 4, 3) = \frac{8!}{4!3!} = \frac{8 \cdot 7 \cdot 6 \cdot 5 \cdot 4 \cdot 3 \cdot 2 \cdot 1}{4 \cdot 3 \cdot 2 \cdot 1 \cdot 3 \cdot 2 \cdot 1} = 280$$

Ordered Samples

   Many problems in combinatorial analysis and, in particular, probability are concerned with choosing an element from a set S containing n elements (or a card from a deck or a person from a population).   When we choose one element after another from the set S, say r times, we call the choice an ordered sample of size r.   We consider two cases:

(i) *Sampling with Replacement*: Here the element is replaced in the set S before the next element is chosen.   Since there are n different ways to choose each element (repetitions are allowed), the product rule principle tells us that there are

$$\overbrace{n \cdot n \cdot n \cdots n}^{r \text{ times}} = n^r$$

different ordered samples with replacement of size r.

(ii) *Sampling without Replacement*: Here the element is not replaced in the set S before the next element is chosen.   Thus, there are no repetitions in the ordered sample.   Accordingly, an ordered sample of size r without replacement is simply an r permutation of the elements in the set S with n elements.   Thus, there are

$$P(n, r) = n(n - 1)(n - 2) \cdots (n - r + 1) = \frac{n!}{(n - r)!}$$

different ordered samples without replacement of size r from a population (set) with n elements.   In other words, by the product rule, the first element can be chosen in n ways, the second in $n - 1$ ways, and so on.

EXAMPLE 2.7   Three cards are chosen in succession from a deck with 52 cards.   Find the number of ways this can be done: (a) with replacement, (b) without replacement.

(a) Since each card is replaced before the next card is chosen, each card can be chosen in 52 ways.   Thus,

$$52(52)(52) = 52^3 = 140,608$$

is the number of different ordered samples of size $r = 3$ with replacement.

(b) Since there is no replacement, the first card can be chosen in 52 ways, the second card in 51 ways, and the last card in 50 ways.   Thus,

$$P(52, 3) = 52(51)(50) = 132,600$$

is the number of different ordered samples of size $r = 3$ without replacement.

2.6 COMBINATIONS

Suppose we have a collection of n objects.   A *combination* of these n objects taken r at a time is any selection of r of the objects where order doesn't count.   In other words, an r *combination* of a set of n objects is any subset of r elements.   For example, the combinations of the letters a, b, c, d taken three at a time are

$$\{a, b, c\}, \{a, b, d\}, \{a, c, d\}, \{b, c, d\} \qquad \text{or simply} \qquad abc, abd, acd, bcd$$

Observe that the following combinations are equal:

$$abc, acb, bac, bca, cab, cba$$

That is, each denotes the same set $\{a, b, c\}$.

The number of combinations of n objects taken r at a time will be denoted by

$$C(n, r)$$

Before we derive the general formula for $C(n, r)$, we consider a particular case.

EXAMPLE 2.8   Find the number of combinations of four objects, a, b, c, d, taken three at a time.

Each combination consisting of three objects determines $3! = 6$ permutations of the objects in the combination as pictured in Fig. 2-2.   Thus, the number of combinations multiplied by 3! equals the number of permutations.   That is,

$$C(4, 3) \cdot 3! = P(4, 3) \qquad \text{or} \qquad C(4, 3) = \frac{P(4, 3)}{3!}$$

But $P(4, 3) = 4 \cdot 3 \cdot 2 = 24$ and $3! = 6$.   Thus, $C(4, 3) = 4$, which is noted in Fig. 2-2.

Combinations	Permutations
abc	abc, acb, bac, bca, cab, cba
abd	abd, adb, bad, bda, dab, dba
acd	acd, adc, cad, cda, dac, dca
bcd	bcd, bdc, cbd, cdb, dbc, dcb

Fig. 2-2

Formula for $C(n, r)$

Since any combination of n objects taken r at a time determines $r!$ permutations of the objects in the combination, we can conclude that

$$P(n, r) = r! \, C(n, r)$$

Thus, we obtain the following formula for $C(n, r)$:

Theorem 2.7: $C(n, r) = \dfrac{P(n, r)}{r!} = \dfrac{n!}{r!(n - r)!}.$

Recall that the binomial coefficient $\binom{n}{r}$ was defined to be $\dfrac{n!}{r!(n - r)!}$.   Accordingly,

$$\boxed{C(n, r) = \binom{n}{r}}$$

We shall use $C(n, r)$ and $\binom{n}{r}$ interchangeably.

EXAMPLE 2.9

(a)   Find the number m of committees of 3 that can be formed from 8 people.

Each committee is, essentially, a combination of the 8 people taken 3 at a time.   Thus

$$m = C(8, 3) = \binom{8}{3} = \frac{8 \cdot 7 \cdot 6}{3 \cdot 2 \cdot 1} = 56$$

(b)   A farmer buys 3 cows, 2 pigs, and 4 hens from a person who has 6 cows, 5 pigs, and 8 hens.   How many choices does the farmer have?

The farmer can choose the cows in $\binom{6}{3}$ ways, the pigs in $\binom{5}{2}$ ways, and the hens in $\binom{8}{4}$ ways.

Accordingly, altogether the farmer can choose the animals in

$$\binom{6}{3}\binom{5}{2}\binom{8}{4} = \frac{6 \cdot 5 \cdot 4}{3 \cdot 2 \cdot 1} \cdot \frac{5 \cdot 4}{2 \cdot 1} \cdot \frac{8 \cdot 7 \cdot 6 \cdot 5}{4 \cdot 3 \cdot 2 \cdot 1} = 20 \cdot 10 \cdot 70 = 14,000 \text{ ways}$$

EXAMPLE 2.10   Find the number m of ways that 9 toys can be divided between 4 children if the youngest is to receive 3 toys and each of the others 2 toys.

There are $C(9, 3) = 84$ ways to first choose 3 toys for the youngest.   Then there are $C(6, 2) = 15$ ways to choose 2 of the remaining 6 toys for the oldest.   Next, there are $C(4, 2) = 6$ ways to choose 2 of the remaining 4 toys for the second oldest.   The third oldest receives the remaining 2 toys.   Thus, by the product rule,

$$m = 84(15)(6) = 7560$$

Alternately, by Problem 2.37,

$$m = \frac{9!}{3!2!2!2!} = 7560$$

EXAMPLE 2.11   Find the number m of ways that 12 students can be partitioned into 3 teams, T_1, T_2, T_3, so that each team contains 4 students.

Method 1:   Let A be one of the students.   Then there are $C(11, 3)$ ways to choose 3 other students to be on the same team as A.   Now let B denote a student who is not on the same team as A; then there are $C(7, 3)$ ways to choose 3 students out of the remaining students to be on the same team as B.   The remaining 4 students constitute the third team.   Thus, altogether, the number m of ways to partition the students is as follows:

$$m = C(11, 3) \cdot C(7, 3) = \binom{11}{3} \cdot \binom{7}{3} = 165 \cdot 35 = 5775$$

Method 2: Each partition $[T_1, T_2, T_3]$ of the students can be arranged in $3! = 6$ ways as an ordered partition. By Problem 2.37 (or using the method in Example 2.10), there are

$$\frac{12!}{4!\,4!\,4!} = 34{,}650$$

such ordered partitions. Thus, there are $m = 34{,}650/6 = 5775$ (unordered) partitions.

2.7 TREE DIAGRAMS

A *tree diagram* is a device used to enumerate all the possible outcomes of a sequence of experiments or events where each event can occur in a finite number of ways. The construction of tree diagrams is illustrated in the following examples.

EXAMPLE 2.12 Find the product set $A \times B \times C$ where

$$A = \{1, 2\},\ B = \{a, b, c\},\ C = \{3, 4\}$$

The tree diagram for the $A \times B \times C$ appears in Fig. 2-3. Observe that the tree is constructed from left to right and that the number of branches at each point corresponds to the number of possible outcomes of the next event. Each endpoint of the tree is labeled by the corresponding element of $A \times B \times C$. As expected from Theorem 1.11, $A \times B \times C$ contains $n = 2 \cdot 3 \cdot 2 = 12$ elements.

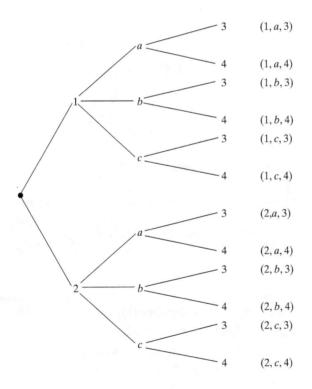

Fig. 2-3

EXAMPLE 2.13 Marc and Erik are to play a tennis tournament. The first person to win 2 games in a row or who wins a total of 3 games wins the tournament. Find the number of ways the tournament can occur.

The tree diagram showing the possible outcomes of the tournament appears in Fig. 2-4. Specifically, there are 10 endpoints which correspond to the following 10 ways that the tournament can occur:

MM, MEMM, MEMEM, MEMEE, MEE, EMM, EMEMM, EMEME, EMEE, EE

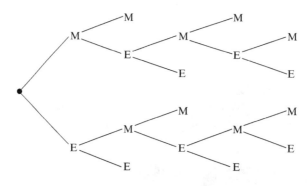

Fig. 2-4

The path from the beginning of the tree to the endpoint describes who won which game in the individual tournament.

Solved Problems

FACTORIAL NOTATION AND BINOMIAL COEFFICIENTS

2.1.   Compute: (*a*) 4!, 5!, 6!, 7!, 8!, 9!, 10!; (*b*) 50!

   (*a*) Use $(n + 1)! = (n + 1)n!$ after calculating 4! and 5!:

$$4! = 1 \cdot 2 \cdot 3 \cdot 4 = 24, \qquad\qquad 7! = 7(6!) = 7(720) = 5040$$
$$5! = 1 \cdot 2 \cdot 3 \cdot 4 \cdot 5 = 5(24) = 120, \qquad 8! = 8(7!) = 8(5040) = 40{,}320$$
$$6! = 6(5!) = 6(120) = 720, \qquad\qquad 9! = 9(8!) = 9(40{,}320) = 362{,}880$$
$$10! = 10(9!) = 10(362{,}880) = 3{,}628{,}800$$

   (*b*) Since *n* is very large, we use Stirling's approximation that $n! \sim \sqrt{2\pi n}\, n^n e^{-n}$ (where $e = 2.718$). Let

$$N = \sqrt{100\pi}\, 50^{50} e^{-50} \sim 50!$$

Evaluating N using a calculator, we get $N = 3.04 \times 10^{64}$ (which has 65 digits).
   Alternately, using (base 10) logarithms, we get

$$\begin{aligned}
\log N &= \log(\sqrt{100\pi}\, 50^{50} e^{-50}) \\
&= \tfrac{1}{2}\log 100 + \tfrac{1}{2}\log \pi + 50 \log 50 - 50 \log e \\
&= \tfrac{1}{2}(2) + \tfrac{1}{2}(0.497\,2) + 50(1.699\,0) - 50(0.434\,3) \\
&= 64.483\,6
\end{aligned}$$

The antilog yields $N = 3.04 \times 10^{64}$.

2.2. Compute: $(a)\ \dfrac{11!}{9!}$, $(b)\ \dfrac{6!}{9!}$.

$(a)\quad \dfrac{11!}{9!} = \dfrac{11 \cdot 10 \cdot 9 \cdot 8 \cdot 7 \cdot 6 \cdot 5 \cdot 4 \cdot 3 \cdot 2 \cdot 1}{9 \cdot 8 \cdot 7 \cdot 6 \cdot 5 \cdot 4 \cdot 3 \cdot 2 \cdot 1} = 11 \cdot 10 = 110$

Alternately, this could be solved as follows:

$$\frac{11!}{9!} = \frac{11 \cdot 10 \cdot 9!}{9!} = 11 \cdot 10 = 110$$

$(b)\quad \dfrac{6!}{9!} = \dfrac{6!}{9 \cdot 8 \cdot 7 \cdot 6!} = \dfrac{1}{9 \cdot 8 \cdot 7} = \dfrac{1}{504}$

2.3. Simplify: $(a)\ \dfrac{n!}{(n-1)!}$, $(b)\ \dfrac{(n+2)!}{n!}$.

$(a)\quad \dfrac{n!}{(n-1)!} = \dfrac{n(n-1)(n-2)\cdots 3 \cdot 2 \cdot 1}{(n-1)(n-2)\cdots 3 \cdot 2 \cdot 1} = n$

or simply $\dfrac{n!}{(n-1)!} = \dfrac{n(n-1)!}{(n-1)!} = n$

$(b)\quad \dfrac{(n+2)!}{n!} = \dfrac{(n+2)(n+1)n(n-1)(n-2)\cdots 3 \cdot 2 \cdot 1}{n(n-1)(n-2)\cdots 3 \cdot 2 \cdot 1} = (n+2)(n+1) = n^2 + 3n + 2$

or simply $\dfrac{(n+2)!}{n!} = \dfrac{(n+2)(n+1)n!}{n!} = (n+2)(n+1) = n^2 + 3n + 2$

2.4. Compute: $(a)\ \dbinom{14}{3}$, $(b)\ \dbinom{11}{4}$.

Recall that there are as many factors in the numerator as in the denominator.

$(a)\quad \dbinom{14}{3} = \dfrac{14 \cdot 13 \cdot 12}{3 \cdot 2 \cdot 1} = 364,$ $\qquad (b)\quad \dbinom{11}{4} = \dfrac{11 \cdot 10 \cdot 9 \cdot 8}{4 \cdot 3 \cdot 2 \cdot 1} = 330$

2.5. Compute: $(a)\ \dbinom{8}{6}$, $(b)\ \dbinom{10}{7}$.

$(a)\quad \dbinom{8}{6} = \dfrac{8 \cdot 7 \cdot 6 \cdot 5 \cdot 4 \cdot 3}{6 \cdot 5 \cdot 4 \cdot 3 \cdot 2 \cdot 1} = 28$

or, since $8 - 6 = 2$, we can use Lemma 2.1 to obtain:

$$\binom{8}{6} = \binom{8}{2} = \frac{8 \cdot 7}{2 \cdot 1} = 28$$

(b) Since $10 - 7 = 3$, Lemma 2.1 tells us that

$$\binom{10}{7} = \binom{10}{3} = \frac{10 \cdot 9 \cdot 8}{3 \cdot 2 \cdot 1} = 120$$

2.6. Prove: $\binom{17}{6} = \binom{16}{5} + \binom{16}{6}$

Now $\binom{16}{5} + \binom{16}{6} = \dfrac{16!}{5!\,11!} + \dfrac{16!}{6!\,10!}$. Multiply the first fraction by $\dfrac{6}{6}$ and the second by $\dfrac{11}{11}$ to obtain the same denominator in both fractions; and then add:

$$\binom{16}{5} + \binom{16}{6} = \frac{6 \cdot 16!}{6 \cdot 5! \cdot 11!} + \frac{11 \cdot 16!}{6! \cdot 11 \cdot 10!} = \frac{6 \cdot 16!}{6! \cdot 11!} + \frac{11 \cdot 16!}{6! \cdot 11!}$$

$$= \frac{6 \cdot 16! + 11 \cdot 16!}{6! \cdot 11!} = \frac{(6 + 11) \cdot 16!}{6! \cdot 11!} = \frac{17 \cdot 16!}{6! \cdot 11!} = \frac{17!}{6! \cdot 11!} = \binom{17}{6}$$

2.7. Prove Theorem 2.3: $\binom{n+1}{r} = \binom{n}{r-1} + \binom{n}{r}$.

(The technique in this proof is similar to that of the preceding problem.)

Now $\binom{n}{r-1} + \binom{n}{r} = \dfrac{n!}{(r-1)! \cdot (n-r+1)!} + \dfrac{n!}{r! \cdot (n-r)!}$. To obtain the same denominator in both fractions, multiply the first fraction by $\dfrac{r}{r}$ and the second fraction by $\dfrac{n-r+1}{n-r+1}$. Hence

$$\binom{n}{r-1} + \binom{n}{r} = \frac{r \cdot n!}{r \cdot (r-1)! \cdot (n-r+1)!} + \frac{(n-r+1) \cdot n!}{r! \cdot (n-r+1) \cdot (n-r)!}$$

$$= \frac{r \cdot n!}{r!(n-r+1)!} + \frac{(n-r+1) \cdot n!}{r!(n-r+1)!}$$

$$= \frac{r \cdot n! + (n-r+1) \cdot n!}{r!(n-r+1)!} = \frac{[r + (n-r+1)] \cdot n!}{r!(n-r+1)!}$$

$$= \frac{(n+1)n!}{r!(n-r+1)!} = \frac{(n+1)!}{r!(n-r+1)!} = \binom{n+1}{r}$$

COUNTING PRINCIPLES

2.8. Suppose a bookcase shelf has 5 history texts, 3 sociology texts, 6 anthropology texts, and 4 psychology texts. Find the number n of ways a student can choose: (a) one of the texts; (b) one of each type of text.

(a) Here the sum rule applies; hence $n = 5 + 3 + 6 + 4 = 18$.

(b) Here the product rule applies; hence $n = 5 \cdot 3 \cdot 6 \cdot 4 = 360$.

2.9. A restaurant has a menu with 4 appetizers, 5 entrees, and 2 desserts. Find the number n of ways a customer can order an appetizer, entree, and dessert.

Here the product rule applies since the customer orders one of each. Thus $n = 4 \cdot 5 \cdot 2 = 40$.

2.10. A history class contains 8 male students and 6 female students. Find the number n of ways that the class can elect: (a) 1 class representative; (b) 2 class representatives, 1 male and 1 female; (c) 1 president and 1 vice-president.

(a) Here the sum rule is used; hence $n = 8 + 6 = 14$.

(b) Here the product rule is used; hence $n = 8 \cdot 6 = 48$.

(c) There are 14 ways to elect the president, and then 13 ways to elect the vice-president. Thus, $n = 14 \cdot 13 = 182$.

2.11. There are 5 bus lines from city A to city B and 4 bus lines from city B to city C. Find the number n of ways a person can travel by bus:

(a) from A to C by way of B, (b) round-trip from A to C by way of B,
(c) round-trip from A to C by way of B, without using a bus line more than once.

(a) There are 5 ways to go from A to B and 4 ways to go from B to C; hence, by the product rule, $n = 5 \cdot 4 = 20$.

(b) There are 20 ways to go from A to C by way of B and 20 ways to return. Thus, by the product rule, $n = 20 \cdot 20 = 400$.

(c) The person will travel from A to B to C to B to A. Enter these letters with connecting arrows as follows:

$$A \to B \to C \to B \to A$$

There are 5 ways to go from A to B and 4 ways to go from B to C. Since a bus line is not to be used more than once, there are only 3 ways to go from C back to B and only 4 ways to go from B back to A. Enter these numbers above the corresponding arrows as follows:

$$A \xrightarrow{5} B \xrightarrow{4} C \xrightarrow{3} B \xrightarrow{4} A$$

Thus, by the product rule, $n = 5 \cdot 4 \cdot 3 \cdot 4 = 240$.

2.12. Suppose there are 12 married couples at a party. Find the number n of ways of choosing a man and a woman from the party such that the two are: (a) married to each other, (b) not married to each other.

(a) There are 12 married couples and hence there are $n = 12$ ways to choose one of the couples.

(b) There are 12 ways to choose, say, one of the men. Once the man is chosen, there are 11 ways to choose the women, anyone other than his wife. Thus, $n = 12(11) = 132$.

2.13. Suppose a password consists of 4 characters, the first 2 being letters in the (English) alphabet and the last 2 being digits. Find the number n of:

(a) passwords, (b) passwords beginning with a vowel

(a) There are 26 ways to choose each of the first 2 characters and 10 ways to choose each of the last 2 characters. Thus, by the product rule,
$$n = 26 \cdot 26 \cdot 10 \cdot 10 = 67,600$$

(b) Here there are only 5 ways to choose the first character. Hence $n = 5 \cdot 26 \cdot 10 \cdot 10 = 13,000$.

PERMUTATIONS AND ORDERED SAMPLES

2.14. State the essential difference between permutations and combinations, with examples.

Order counts with permutations, such as words, sitting in a row, and electing a president, vice-president, and treasurer. Order does not count with combinations, such as committees and teams (without counting positions). The product rule is usually used with permutations since the choice for each of the ordered positions may be viewed as a sequence of events.

2.15. Find the number n of ways that 4 people can sit in a row of 4 seats.

The 4 empty seats may be pictured by

_____, _____, _____, _____ .

The first seat can be occupied by any one of the 4 people, that is, there are 4 ways to fill the first seat. After the first person sits down, there are only 3 people left and so there are 3 ways to fill the second seat. Similarly, the third seat can be filled in 2 ways, and the last seat in 1 way. This is pictured by

$$\underline{\;4\;}\,,\,\underline{\;3\;}\,,\,\underline{\;2\;}\,,\,\underline{\;1\;}$$

Thus, by the product rule, $n = 4 \cdot 3 \cdot 2 \cdot 1 = 4! = 24$.

Alternately, n is the number of permutations of 4 things taken 4 at a time, and so

$$n = P(4, 4) = 4! = 24$$

2.16. A family has 3 boys and 2 girls. (a) Find the number of ways they can sit in a row. (b) How many ways are there if the boys and girls are each to sit together?

(a)   The 5 children can sit in a row in $5 \cdot 4 \cdot 3 \cdot 2 \cdot 1 = 5! = 120$ ways.

(b)   There are 2 ways to distribute them according to sex: *BBBGG* or *GGBBB*. In each case, the boys can sit in $3 \cdot 2 \cdot 1 = 3! = 6$ ways and the girls can sit in $2 \cdot 1 = 2! = 2$ ways. Thus, altogether, there are $2 \cdot 3! \cdot 2! = 2 \cdot 6 \cdot 2 = 24$ ways.

2.17. Find the number n of distinct permutations that can be formed from all the letters of each word: (a) THOSE, (b) UNUSUAL, (c) SOCIOLOGICAL.

This problem concerns permutations with repetitions.

(a) $n = 5! = 120$, since there are 5 letters and no repetitions.

(b) $n = \dfrac{7!}{3!} = 840$, since there are 7 letters of which 3 are U and no other letter is repeated.

(c) $n = \dfrac{12!}{3!\,2!\,2!\,2!}$, since there are 12 letters of which 3 are O, 2 are C, 2 are I, and 2 are L.

2.18. Find the number n of different signals, each consisting of 6 flags hung in a vertical line, that can be formed from 4 identical red flags and 2 identical blue flags.

This problem concerns permutations with repetitions. Thus, $n = \dfrac{6!}{4!\,2!} = 15$ since there are 6 flags of which 4 are red and 2 are blue.

2.19. Find the number n of ways that 7 people can arrange themselves: (a) in a row of 7 chairs, (b) around a circular table.

(a)   The 7 people can arrange themselves in a row in $n = 7 \cdot 6 \cdot 5 \cdot 4 \cdot 3 \cdot 2 \cdot 1 = 7!$ ways.

(b)   One person can sit at any place at the circular table. The other 6 people can then arrange themselves in $n = 6 \cdot 5 \cdot 4 \cdot 3 \cdot 2 \cdot 1 = 6!$ ways around the table.
This is an example of a *circular permutation*. In general, n objects can be arranged in a circle in $(n - 1)(n - 2) \cdots 3 \cdot 2 \cdot 1 = (n - 1)!$ ways.

2.20. Suppose repetitions are not allowed. (a) Find the number n of three-digit numbers that can be formed from the six digits: 2, 3, 5, 6, 7, 9. (b) How many of them are even? (c) How many of them exceed 400?

There are 6 digits, and the three-digit number may be pictured by

$$\underline{\qquad}\,,\,\underline{\qquad}\,,\,\underline{\qquad}\,.$$

In each case, write down the number of ways that one can fill each of the positions.

(a)   There are 6 ways to fill the first position, 5 ways for the second position, and 3 ways for the third position. This may be pictured by: $\underline{\;6\;}\,,\,\underline{\;5\;}\,,\,\underline{\;4\;}$. Thus, $n = 6 \cdot 5 \cdot 4 = 120$.
Alternately, n is the number of permutations of 6 things taken 3 at a time, and so

$$n = P(6, 3) = 6 \cdot 5 \cdot 4 = 120$$

(b)   Since the numbers must be even, the last digit must be either 2 or 4.   Thus, the third position is filled first and it can be done in 2 ways.   Then there are now 5 ways to fill the middle position and 4 ways to fill the first position.   This may be pictured by: __4__ , __5__ , __2__ .   Thus, $4 \cdot 5 \cdot 2 = 120$ of the numbers are even.

(c)   Since the numbers must exceed 400, they must begin with 5, 6, 7, or 9.   Thus, we first fill the first position and it can be done in 4 ways.   Then there are 5 ways to fill the second position and 4 ways to fill the third position.   This may be pictured by: __4__ , __5__ , __4__ .   Thus, $4 \cdot 5 \cdot 4 = 80$ of the numbers exceed 400.

2.21.   A class contains 8 students.   Find the number of ordered samples of size 3:

(a) with replacement, (b) without replacement.

(a)   Each student in the ordered sample can be chosen in 8 ways; hence there are $8 \cdot 8 \cdot 8 = 8^3 = 512$ samples of size 3 with replacement.

(b)   The first student in the sample can be chosen in 8 ways, the second in 7 ways, and the last in 6 ways.   Thus, there are $8 \cdot 7 \cdot 6 = 336$ samples of size 3 without replacement.

2.22.   Find n if: (a) $P(n, 2) = 72$,   (b) $2P(n, 2) + 50 = P(2n, 2)$.

(a) $P(n, 2) = n(n - 1) = n^2 - n$; hence

$$n^2 - n = 72 \quad \text{or} \quad n^2 - n - 72 = 0 \quad \text{or} \quad (n - 9)(n + 8) = 0$$

Since n must be positive, the only answer is $n = 9$.

(b) $P(n, 2) = n(n - 1) = n^2 - n$ and $P(2n, 2) = 2n(2n - 1) = 4n^2 - 2n$.   Hence:

$$2(n^2 - n) + 50 = 4n^2 - 2n \quad \text{or} \quad 2n^2 - 2n + 50 = 4n^2 - 2n$$
$$\text{or} \quad 50 = 2n^2 \quad \text{or} \quad n^2 = 25$$

Since n must be positive, the only answer is $n = 5$.

COMBINATIONS AND PARTITIONS

2.23.   There are 12 students who are eligible to attend the National Student Association annual meeting.   Find the number n of ways a delegation of 4 students can be selected from the 12 eligible students.

This concerns combinations, *not* permutations, since order does not count in a delegation.   There are "12 choose 4" such delegations.   That is,

$$n = C(12, 4) = \binom{12}{4} = \frac{12 \cdot 11 \cdot 10 \cdot 9}{4 \cdot 3 \cdot 2 \cdot 1} = 495$$

2.24.   A student is to answer 8 out of 10 questions on an exam.

(a)   Find the number n of ways the student can choose the eight questions.

(b)   Find n if the student must answer the first three questions.

(a)   The 8 questions can be selected "10 choose 8" ways.   That is,

$$n = C(10, 8) = \binom{10}{8} = \binom{10}{2} = \frac{10 \cdot 9}{2 \cdot 1} = 45$$

(b)   If the first 3 questions are answered, then the student must choose the other 5 questions from the remaining 7 questions.   Hence

$$n = C(7, 5) = \binom{7}{5} = \binom{7}{2} = \frac{7 \cdot 6}{2 \cdot 1} = 21$$

2.25. A class contains 10 students with 6 men and 4 women.   Find the number n of ways:

(a)   A 4-member committee can be selected from the students.

(b)   A 4-member committee with 2 men and 2 women.

(c)   The class can elect a president, vice-president, treasurer, and secretary.

(a)   This concerns combinations, *not* permutations, since order does not count in a committee.   There are "10 choose 4" such committees.   That is,

$$n = C(10, 4) = \binom{10}{4} = \frac{10 \cdot 9 \cdot 8 \cdot 7}{4 \cdot 3 \cdot 2 \cdot 1} = 210$$

(b)   The 2 men can be chosen from the 6 men in $\binom{6}{2}$ ways and the 2 women can be chosen from the 4 women in $\binom{4}{2}$ ways.   Thus, by the product rule,

$$n = \binom{6}{2}\binom{4}{2} = \frac{6 \cdot 5}{2 \cdot 1} \cdot \frac{4 \cdot 3}{2 \cdot 1} = 15(6) = 90 \text{ ways}$$

(c)   This concerns permutations, *not* combinations, since order does count.   Thus,

$$n = P(6, 4) = 6 \cdot 5 \cdot 4 \cdot 3 = 360$$

2.26. A box contains 7 blue socks and 5 red socks.   Find the number n of ways two socks can be drawn from the box if: (a) They can be any color; (b) They must be the same color.

(a)   There are "12 choose 2" ways to select 2 of the 12 socks.   That is,

$$n = C(12, 2) = \binom{12}{2} = \frac{12 \cdot 11}{2 \cdot 1} = 66$$

(b)   There are $C(7, 2) = 21$ ways to choose 2 of the 7 blue socks and $C(5, 2) = 10$ ways to choose 2 of the 5 red socks.   By the sum rule, $n = 21 + 10 = 31$.

2.27. Let $A, B, \ldots, L$ be 12 given points in the plane $\mathbf{R}^2$ such that no 3 of the points lie on the same line.   Find the number n of:

(a)   Lines in $\mathbf{R}^2$ where each line contains two of the points.

(b)   Lines in $\mathbf{R}^2$ containing A and one of the other points.

(c)   Triangles whose vertices come from the given points.

(d)   Triangles whose vertices are A and two of the other points.

Since order does not count, this problem involves combinations.

(a)   Each pair of points determines a line; hence

$$n = \text{"12 choose 2"} = C(12, 2) = \binom{12}{2} = \frac{12 \cdot 11}{2 \cdot 1} = 66$$

(b)   We need only choose one of the 11 remaining points; hence $n = 11$.

(c)   Each triple of points determines a triangle; hence

$$n = \text{"12 choose 3"} = C(12, 3) = \binom{12}{3} = 220$$

(d) We need only choose two of the 11 remaining points; hence $n = C(11, 2) = 55$. (Alternately, there are $C(11, 3) = 165$ triangles without A as a vertex; hence $220 - 165 = 55$ of the triangles do have A as a vertex.)

2.28. There are 12 students in a class. Find the number n of ways that 12 students can take 3 different tests if 4 students are to take each test.

There are $C(12, 4) = 495$ ways to choose 4 students to take the first test; following this, there are $C(8, 4) = 70$ ways to choose 4 students to take the second test. The remaining students take the third test. Thus

$$n = 70(495) = 34{,}650$$

2.29. Find the number n of ways 12 students can be partitioned into 3 teams A_1, A_2, A_3, so that each team contains 4 students. (Compare with the preceding Problem 2.28.)

Let A denote one of the students. There are $C(11, 3) = 165$ ways to choose 3 other students to be on the same team as A. Now let B be a student who is not on the same team as A. Then there are $C(7, 3) = 35$ ways to choose 3 from the remaining students to be on the same team as B. The remaining 4 students form the third team. Thus, $n = 35(165) = 5925$.

Alternately, each partition $[A_1, A_2, A_3]$ can be arranged in $3! = 6$ ways as an ordered partition. By the preceding Problem 2.28, there are 34,650 such ordered partitions. Thus, $n = 34{,}650/6 = 5925$.

2.30. Find the number n of committees of 5 with a given chairperson that can be selected from 12 persons.

Method 1: The chairperson can be chosen in 12 ways and, following this, the other 4 on the committee can be chosen from the remaining 11 people in $C(11, 4) = 330$ ways. Thus,

$$n = 12(330) = 3960$$

Method 2: The 5-member committee can be chosen from the 12 persons in $C(12, 5) = 792$ ways. Each committee can then select a chairman in 5 ways. Thus,

$$n = 5(792) = 3960$$

2.31. There are n married couples at a party. (a) Find the number N of (unordered) pairs at the party. (b) Suppose every person shakes hands with every other person other than his or her spouse. Find the number M of handshakes.

(a) There are $2n$ people at the party, and so there are "$2n$ choose 2" pairs. That is,

$$N = C(2n, 2) = \frac{2n(2n - 1)}{2} = n(2n - 1) = 2n^2 - n$$

(b) M is equal to the number of pairs who are not married. There are n married pairs. Thus, using (a),

$$M = 2n^2 - n - n = 2n^2 - 2n = 2n(n - 1)$$

TREE DIAGRAMS

2.32. Construct the tree diagram that gives the permutations of $\{a, b, c\}$.

The tree diagram, drawn downward with the "root" on the top, appears in Fig. 2-5. Each path from the root to an endpoint ("leaf") of the tree represents a permutation. There are 6 such paths which yield the following 6 permutations:

abc, acb, bac, bca, cab, cba

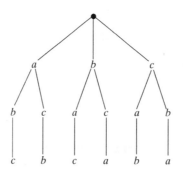

Fig. 2-5

2.33. Audrey has time to play roulette at most 5 times. At each play she wins or loses $1. She begins with $1 and will stop playing before 5 plays if she loses all her money.

(*a*) Find the number of ways the betting can occur.

(*b*) How many cases will she stop before playing 5 times?

(*c*) How many cases will she leave without any money?

Construct the appropriate tree diagram as shown in Fig. 2-6. Each number in the diagram denotes the number of dollars she has at that moment in time. Thus, the root, which is circled, is labeled with the number 1.

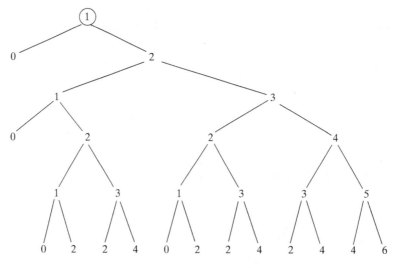

Fig. 2-6

(a) There are 14 paths from the root of the tree to an endpoint ("leaf"), so the betting can occur in 14 different ways.

(b) There are only 2 paths with less than 5 edges, so Audrey will not play 5 times in only 2 of the cases.

(c) There are 4 paths which end in 0 or, in other words, only 4 of the leaves are labeled with 0. Thus, Audrey will leave without any money in 4 of the cases.

MISCELLANEOUS PROBLEMS

2.34. Prove Theorem 2.2 (binomial theorem): $(a + b)^n = \sum_{k=0}^{n} \binom{n}{k} a^{n-k} b^k$.

The theorem is true for $n = 1$, since

$$\sum_{r=0}^{1} \binom{1}{r} a^{1-r} b^r = \binom{1}{0} a^1 b^0 + \binom{1}{1} a^0 b^1 = a + b = (a + b)^1$$

We assume the theorem holds for $(a + b)^n$ and prove it is true for $(a + b)^{n+1}$

$(a + b)^{n+1} = (a + b)(a + b)^n$

$$= (a + b)\left[a^n + \binom{n}{1} a^{n-1} b + \cdots + \binom{n}{r-1} a^{n-r+1} b^{r-1} + \binom{n}{r} a^{n-r} b^r + \cdots + \binom{n}{1} ab^{n-1} + b^n \right]$$

Now the term in the product which contains b^r is obtained from

$$b\left[\binom{n}{r-1} a^{n-r+1} b^{r-1} \right] + a\left[\binom{n}{r} a^{n-r} b^r \right] = \binom{n}{r-1} a^{n-r+1} b^r + \binom{n}{r} a^{n-r+1} b^r$$

$$= \left[\binom{n}{r-1} + \binom{n}{r} \right] a^{n-r+1} b^r$$

But, by Theorem 2.3 $\binom{n}{r-1} + \binom{n}{r} = \binom{n+1}{r}$. Thus, the term containing b^r is

$$\binom{n+1}{r} a^{n-r+1} b^r$$

Note that $(a + b)(a + b)^n$ is a polynomial of degree $n + 1$ in b. Consequently,

$$(a + b)^{n+1} = (a + b)(a + b)^n = \sum_{r=0}^{n+1} \binom{n+1}{r} a^{n-r+1} b^r$$

which was to be proved.

2.35. Prove: $\binom{4}{0} + \binom{4}{1} + \binom{4}{2} + \binom{4}{3} + \binom{4}{4} = 16$.

Note that $16 = 2^4 = (1 + 1)^4$. Expanding $(1 + 1)^4$, using the binomial theorem, yields:

$$16 = (1 + 1)^4 = \binom{4}{0} 1^4 + \binom{4}{1} 1^3 1^1 + \binom{4}{2} 1^2 1^2 + \binom{4}{3} 1^1 1^3 + \binom{4}{4} 1^4$$

$$= \binom{4}{0} + \binom{4}{1} + \binom{4}{2} + \binom{4}{3} + \binom{4}{4}$$

2.36. Let n and $n_1, n_2, \ldots, n_r$ be nonnegative integers such that $n_1 + n_2 + \cdots + n_r = n$. The *multinomial coefficients* are denoted and defined by

$$\binom{n}{n_1, n_2, \ldots, n_r} = \frac{n!}{n_1! n_2! \cdots n_r!}$$

Compute the following multinomial coefficients:

(a) $\binom{6}{3, 2, 1}$,   (b) $\binom{8}{4, 2, 2, 0}$,   (c) $\binom{10}{5, 3, 2, 2}$

Use the above formula to obtain:

(a) $\binom{6}{3, 2, 1} = \frac{6!}{3! 2! 1!} = \frac{6 \cdot 5 \cdot 4 \cdot 3 \cdot 2 \cdot 1}{3 \cdot 2 \cdot 1 \cdot 2 \cdot 1 \cdot 1} = 60$

(b) $\binom{8}{4, 2, 2, 0} = \frac{8!}{4! 2! 2! 0!} = \frac{8 \cdot 7 \cdot 6 \cdot 5 \cdot 4 \cdot 3 \cdot 2 \cdot 1}{4 \cdot 3 \cdot 2 \cdot 1 \cdot 2 \cdot 1 \cdot 2 \cdot 1 \cdot 1} = 420$

(Here we use the act that $0! = 1$.)

(c) The expression $\binom{10}{5, 3, 2, 2}$ has no meaning, since $5 + 3 + 2 + 2 \neq 10$.

2.37. Suppose S contains n elements, and let $n_1, n_2, \ldots, n_r$ be positive integers such that

$$n_1 + n_2 + \cdots + n_r = n$$

Prove there exists

$$\binom{n}{n_1, n_2, \ldots, n_r} = \frac{n!}{n_1! n_2! \cdots n_r!}$$

different ordered partitions of S of the form $[A_1, A_2, \ldots, A_r]$ where A_1 contains n_1 elements, A_2 contains n_2 elements, $\ldots$, A_r contains n_r elements.

We begin with n elements in S; hence there are $\binom{n}{n_1}$ ways of selecting the cell A_1. Following this, there are $n - n_1$ elements left in S, that is, in $S \setminus A_1$; hence there are $\binom{n - n_1}{n_2}$ ways of selecting the cell A_2. Similarly, for $i = 3, 4, \ldots, r$, there are $\binom{n - n_1 - \cdots - n_{i-1}}{n_i}$ ways of selecting the cell A_i.

Accordingly, there are

$$\binom{n}{n_1}\binom{n - n_1}{n_2}\binom{n - n_1 - n_2}{n_2} \cdots \binom{n - n_1 - \cdots - n_{r-1}}{n_r} \tag{*}$$

different ordered partitions of S. Now (*) is equal to

$$\frac{n!}{n_1!(n - n_1)!} \cdot \frac{(n - n_1)!}{n_2!(n - n_1 - n_2)!} \cdot \frac{(n - n_1 - n_2)!}{n_3!(n - n_1 - n_2 - n_3)!} \cdot \ldots \cdot \frac{(n - n_1 - \cdots - n_{r-1})!}{n_r!(n - n_1 - n_2 - \cdots - n_r)!}$$

But this is equal to

$$\binom{n}{n_1, n_2, \ldots, n_r} = \frac{n!}{n_1! n_2! \cdots n_r!}$$

since each numerator after the first is cancelled by the second term in the preceding denominator and since $(n - n_1 - \cdots - n_r)! = 0! = 1$. Thus, the theorem is proved.

Supplementary Problems

FACTORIAL NOTATION AND BINOMIAL COEFFICIENTS

2.38. Find: (a) 10!, 11!, 12! (b) 60! (*Hint*: Use Stirling's approximation to $n!$.)

2.39. Compute: (a) $\dfrac{16!}{14!}$, (b) $\dfrac{14!}{11!}$, (c) $\dfrac{8!}{10!}$, (d) $\dfrac{10!}{13!}$.

2.40. Simplify: (a) $\dfrac{(n+1)!}{n!}$, (b) $\dfrac{n!}{(n-2)!}$, (c) $\dfrac{(n-1)!}{(n+2)!}$, (d) $\dfrac{(n-r+1)!}{(n-r-1)!}$.

2.41. Compute: (a) $\dbinom{5}{2}$, (b) $\dbinom{7}{3}$, (c) $\dbinom{14}{2}$, (d) $\dbinom{6}{4}$, (e) $\dbinom{20}{17}$, (f) $\dbinom{18}{15}$.

2.42. Show that: (a) $\dbinom{n}{0} + \dbinom{n}{1} + \dbinom{n}{2} + \dbinom{n}{3} + \cdots + \dbinom{n}{n} = 2^n$,

(b) $\dbinom{n}{0} - \dbinom{n}{1} + \dbinom{n}{2} - \dbinom{n}{3} + \cdots \pm \dbinom{n}{n} = 0$.

2.43. Evaluate the following multinomial coefficients (defined in Problem 2.36):

(a) $\dbinom{6}{2,3,1}$, (b) $\dbinom{7}{3,2,2,0}$, (c) $\dbinom{9}{3,5,1}$, (d) $\dbinom{8}{4,3,2}$.

2.44. Find the (a) ninth and (b) tenth rows of Pascal's triangle, assuming the following is the eighth row:

$$1 \quad 8 \quad 28 \quad 70 \quad 56 \quad 28 \quad 8 \quad 1$$

COUNTING PRINCIPLES, SUM AND PRODUCT RULES

2.45. A store sells clothes for men. It has 3 different kinds of jackets, 7 different kinds of shirts, and 5 different kinds of pants. Find the number of ways a person can buy:

(a) one of the items for a present, (b) one of each of the items for a present.

2.46. A restaurant has, on its dessert menu, 4 kinds of cakes, 2 kinds of cookies, and 3 kinds of ice cream. Find the number of ways a person can select: (a) one of the desserts, (b) one of each kind of dessert.

2.47. A class contains 8 male students and 6 female students. Find the number of ways that the class can elect: (a) a class representative; (b) 2 class representatives, 1 male and 1 female; (c) a president and a vice-president.

2.48. Suppose a password consists of 4 characters where the first character must be a letter of the (English) alphabet, but each of the other characters may be a letter or a digit. Find the number of:

(a) passwords, (b) passwords beginning with one of the 5 vowels.

2.49. Suppose a code consists of 2 letters followed by 3 digits. Find the number of:

(a) codes, (b) codes with distinct letters, (c) codes with the same letters.

2.50. There are 6 roads between A and B and 4 roads between B and C. Find the number n of ways a person can drive: (a) from A to C by way of B, (b) round-trip from A to C by way of B, (c) round-trip from A to C by way of B without using the same road more than once.

PERMUTATIONS AND ORDERED SAMPLES

2.51. Find the number n of ways a judge can award first, second, and third places in a contest with 18 contestants.

2.52. Find the number n of ways 6 people can ride a toboggan where: (a) anyone can drive, (b) one of 3 must drive.

2.53. A debating team consists of 3 boys and 3 girls. Find the number n of ways they can sit in a row where: (a) there are no restrictions, (b) the boys and girls are each to sit together, (c) just the girls are to sit together.

2.54. Find the number n of permutations that can be formed from all the letters of each word: (a) QUEUE, (b) COMMITTEE, (c) PROPOSITION, (d) BASEBALL.

2.55. Find the number n of different signals, each consisting of 8 flags hung in a vertical line, that can be formed from 4 identical red flags, 2 identical blue flags, and 2 identical green flags.

2.56. Find the number n of ways 5 large books, 4 medium-size books, and 3 small books can be placed on a shelf so that all books of the same size are together.

2.57. A box contains 12 light bulbs. Find the number n of ordered samples of size 3:

(a) with replacement, (b) without replacement.

2.58. A class contains 10 students. Find the number n of ordered samples of size 4:

(a) with replacement, (b) without replacement.

COMBINATIONS

2.59. A restaurant has 6 different desserts. Find the number of ways a customer can choose 2 of the desserts.

2.60. A store has 8 different mystery books. Find the number of ways a customer can buy 3 of the books.

2.61. A box contains 6 blue socks and 4 white socks. Find the number of ways two socks can be drawn from the box where: (a) there are no restrictions, (b) they are different colors, (c) they are to be the same color.

2.62. A class contains 9 boys and 3 girls. Find the number of ways a teacher can select a committee of 4.

2.63. Repeat Problem 2.62, but where: (a) there are to be 2 boys and 2 girls, (b) there is to be exactly 1 girl, (c) there is to be at least 1 girl.

2.64. A woman has 11 close friends. Find the number of ways she can invite 5 of them to dinner.

2.65. Repeat Problem 2.64, but where 2 of the friends are married and will not attend separately.

2.66. Repeat Problem 2.64, but where 2 of the friends are not on speaking terms and will not attend together.

2.67. A person is dealt a poker hand (5 cards) from an ordinary deck with 52 cards. Find the number of ways the person can be dealt: (a) four of a kind, (b) a flush.

2.68. A student must answer 10 out of 13 questions. (a) How many choices are there? (b) How many if the student must answer the first 2 questions? (c) How many if the student must answer the first or second question but not both?

PARTITIONS

2.69. Find the number of ways 6 toys may be divided evenly among 3 children.

2.70. Find the number of ways 6 students can be partitioned into 3 teams containing 2 students each. (Compare with Problem 2.69.)

2.71. Find the number of ways 6 students can be partitioned into 2 teams where each team contains 2 or more students.

2.72. Find the number of ways 9 toys may be divided among 4 children if the youngest is to receive 3 toys and each of the others 2 toys.

2.73. There are 9 students in a class. Find the number of ways the students can take 3 tests if 3 students are to take each test.

2.74. There are 9 students in a class. Find the number of ways the students can be partitioned into 3 teams containing 3 students each. (Compare with Problem 2.73.)

TREE DIAGRAMS

2.75. Teams A and B play in the world series of baseball where the team that first wins 4 games wins the series. Suppose A wins the first game and that the team that wins the second game also wins the fourth game. (a) Find the number n of ways the series can occur, and list the n ways the series can occur. (b) How many ways will B win the series? (c) How many ways will the series last 7 games?

2.76. Suppose $A, B, \ldots, F$ in Fig. 2-7 denote islands, and the lines connecting them bridges. A person begins at A and walks from island to island. The person stops for lunch when he or she cannot continue to walk without crossing the same bridge twice. (a) Construct the appropriate tree diagram, and find the number of ways the person can walk before eating lunch. (b) At which islands can he or she eat lunch?

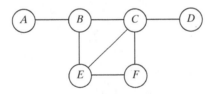

Fig. 2-7

Answers to Supplementary Problems

2.38. (a) 3,628,800; 39,916,800; 479,001,600. (b) log(60!) = 81.92, so 60! ≈ 6.59 × 10^{81}.

2.39. (a) 240; (b) 2184; (c) 1/90; (d) 1/1716.

2.40. (a) $n + 1$; (b) $n(n - 1) = n^2 - n$; (c) $1/[n(n + 1)(n + 2)]$; (d) $(n - r)(n - r + 1)$.

2.41. (a) 10; (b) 35; (c) 91; (d) 15; (e) 1140; (f) 816.

2.42. *Hint*: Expand (*a*) $(1 + 1)^n$; (*b*) $(1 - 1)^n$.

2.43. (*a*) 60; (*b*) 210; (*c*) 504; (*d*) Not defined.

2.44. (*a*) 1, 9, 36, 84, 126, 126, 84, 36, 9, 1; (*b*) 1, 10, 45, 120, 210, 252, 210, 120, 45, 10, 1.

2.45. (*a*) 15; (*b*) 105.

2.46. (*a*) 9; (*b*) 24.

2.47. (*a*) 14; (*b*) 48; (*c*) 182.

2.48. (*a*) $26 \cdot 36^3$; (*b*) $5 \cdot 36^3$.

2.49. (*a*) $26^2 \cdot 10^3 = 676{,}000$; (*b*) $26 \cdot 25 \cdot 10^3 = 650{,}000$; (*c*) $26 \cdot 10^3 = 26{,}000$.

2.50. (*a*) 24; (*b*) $24^2 = 576$; (*c*) 360.

2.51. $n = 18 \cdot 17 \cdot 16 = 4896$.

2.52. (*a*) $6! = 720$; (*b*) $3 \cdot 5! = 360$.

2.53. (*a*) $6! = 720$; (*b*) $2 \cdot 3! \cdot 3! = 72$; (*c*) $4 \cdot 3! \cdot 3! = 144$.

2.54. (*a*) 30; (*b*) $\dfrac{9!}{2!2!2!} = 45{,}360$; (*c*) $\dfrac{11!}{2!3!2!} = 1{,}663{,}200$; (*d*) $\dfrac{8!}{2!2!2!} = 5040$.

2.55. $n = \dfrac{8!}{4!2!2!} = 420$.

2.56. $3!5!4!3! = 103{,}680$.

2.57. (*a*) $12^3 = 1728$; (*b*) 1320.

2.58. (*a*) $10^4 = 10{,}000$; (*b*) $10 \cdot 9 \cdot 8 \cdot 7 = 5040$.

2.59. $C(6, 2) = 15$.

2.60. $C(8, 3) = 56$.

2.61. (*a*) $C(10, 2) = 45$; (*b*) $6 \cdot 4 = 24$; (*c*) $C(6, 2) + C(4, 2) = 21$ or $45 - 24 = 21$.

2.62. $C(12, 4) = 495$.

2.63. (*a*) $C(9, 2) \cdot C(3, 2) = 108$; (*b*) $C(9, 3) \cdot 3 = 252$;
(*c*) $9 + 108 + 252 = 369$ or $C(12, 4) - C(9, 4) = 495 - 126 = 369$.

2.64. $C(11, 5) = 462$.

2.65. 210.

2.66. 252.

2.67. (a) $13 \cdot 48 = 624$; (b) $4 \cdot C(13, 5) = 5148$.

2.68. (a) $C(13, 10) = C(13, 3) = 286$; (b) $2 \cdot C(11, 9) = 2 \cdot C(11, 2) = 110$.

2.69. 90.

2.70. 15.

2.71. (*Hint*: The number of subsets excluding $\varnothing$ and the 6 singleton subsets.) $2^5 - 1 - 6 = 25$.

2.72. $\dfrac{9!}{3!2!2!2!} = 7560$.

2.73. $\dfrac{9!}{3!3!3!} = 1680$.

2.74. $\dfrac{1680}{3!} = 280$.

2.75. Construct the appropriate tree diagram as in Fig. 2-8. Note that the tree begins at A, the winner of the first game, and that there is only one choice in the fourth game, the winner of the second game. (a) The diagram shows that $n = 15$ and that the series can occur in the following 15 ways:

> *AAAA, AABAA, AABABA, AABABBA, AABABBB, ABABAA, ABABABA, ABABABB,*
> *ABABBAA, ABABBAB, ABABBB, ABBBAAA, ABBBAAB, ABBBAB, ABBBB*

(b) 6; (c) 8.

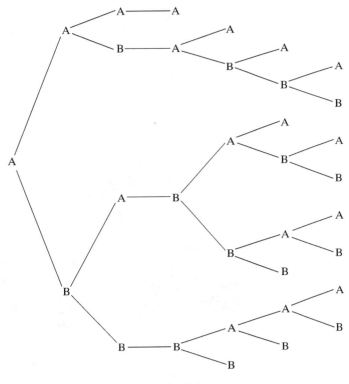

Fig. 2-8

2.76. (a) See Fig. 2-9. There are 11 ways to take his walk. (b) *B*, *D*, or *E*.

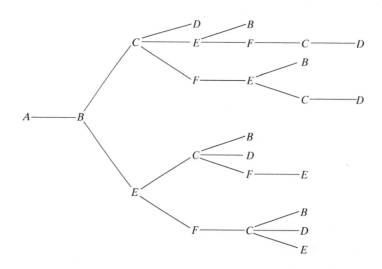

Fig. 2-9

CHAPTER 3

Introduction to Probability

3.1 INTRODUCTION

Probability theory is a mathematical modeling of the phenomenon of chance or randomness. If a coin is tossed in the air, it can land heads or tails, but we do not know which of these will occur in a single toss. However, suppose we repeat this experiment of tossing a coin; let s be the number of successes, that is, that a head appears, and let n be the number of tosses. Then it has been empirically observed that the ratio $f = s/n$, called the *relative frequency* of the outcome, becomes stable in the long run, that is, the ratio $f = s/n$ approaches a limit. If the coin is perfectly balanced, then we expect that the coin will land heads approximately 50 percent of the time or, in other words, the relative frequency will approach 1/2. Alternately, assuming the coin is perfectly balanced, we can arrive at the value 1/2 deductively. That is, one side of the coin is as likely to occur as the other; hence the chances of getting a head is one in two which means the probability of getting a head is 1/2. Although the specific outcome on any one toss is unknown, the behavior over the long run is determined. This stable long-run behavior of random phenomena forms the basis of probability theory.

Consider another experiment, the tossing of a six-sided die (Fig. 3-1) and observing the number of dots, or pips, that appear on the top face. Suppose the experiment is repeated n times and let s be the number of times 4 dots appear on top. Again, as n increases, the relative frequency $f = s/n$ of the outcome 4 becomes more stable. Assuming the die is perfectly balanced, we would expect that the

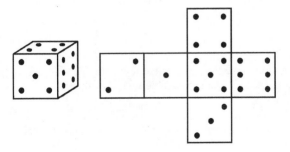

Fig. 3-1

stable or long-run value of this ratio is 1/6, and we say the probability of getting a 4 is 1/6.   Alternately, we can arrive at the value 1/6 deductively.   That is, with a perfectly balanced die, any one side of the die is as likely as any other to occur on top.   Thus, the chances of getting a 4 is one in six or, in other words, the probability of getting a 4 is 1/6.   Again, although the specific outcome on any one toss is unknown, the behavior over the long run is determined.

The historical development of probability theory is similar to the above discussion.   That is, letting E denote the outcome of an experiment, called an *event*, there were two ways to obtain the probability p of E:

(a) **Classical (A Priori) Definition:** Suppose an event E can occur in s ways out of a total of n equally likely possible ways.   Then $p = s/n$.

(b) **Frequency (A Posteriori) Definition:** Suppose after n repetitions, where n is very large, an event E occurs s times.   Then $p = s/n$.

Both of the above definitions have serious flaws.   The classical definition is essentially circular since the idea of "equally likely" is the same as that of "with equal probability" which has not been defined.   The frequency definition is not well defined since "very large" has not been defined.

The modern treatment of probability theory is axiomatic using set theory.   Specifically, a mathematical model of an experiment is obtained by arbitrarily assigning probabilities to all the events, except that the assignments must satisfy certain axioms listed below.   Naturally, the reliability of our mathematical model for a given experiment depends upon the closeness of the assigned probabilities to the actual limiting relative frequencies.   This then gives rise to problems of testing and reliability, which form the subject matter of statistics.

3.2 SAMPLE SPACE AND EVENTS

The set S of all possible outcomes of some experiment is called the *sample space*.   A particular outcome, that is, an element of S, is called a *sample point*.   An *event A* is a set of outcomes or, in other words, a subset of the sample space S.   The event $\{a\}$ consisting of a single point $a \in S$ is called an *elementary event*.   The empty set $\varnothing$ and S are subsets of S and hence they are events; $\varnothing$ is sometimes called the *impossible* or *null* event, and S is sometimes called the *certain* or *sure* event.

Events can be combined to form new events using the various set operations:

(i) $A \cup B$ is the event that occurs iff A occurs or B occurs (or both).

(ii) $A \cap B$ is the event that occurs iff A occurs and B occurs.

(iii) A^c, the complement of A, is the event that occurs iff A does not occur.

(Here "iff" is an abbreviation of "if and only if".)

Events A and B are called *mutually exclusive* if they are disjoint, that is, if $A \cap B = \varnothing$.   In other words, A and B are mutually exclusive if they cannot occur simultaneously.   Three or more events are *mutually exclusive* if every two of them are mutually exclusive.

EXAMPLE 3.1

(a) **Experiment:** Toss a die and observe the number (of dots) that appears on top face.
     The sample space S consists of the six possible numbers, that is,

$$S = \{1, 2, 3, 4, 5, 6\}$$

Let A be the event that an even number occurs, B that an odd number occurs, and C that a number greater than 3 occurs, that is, let

$$A = \{2, 4, 6\}, \qquad B = \{1, 3, 5\}, \qquad C = \{4, 5, 6\}$$

Then:

$A \cup C = \{2, 4, 5, 6\}$ = the event that an even number or a number exceeding 3 occurs

$A \cap C = \{4, 6\}$ = the event that an even number and a number exceeding 3 occurs

$C^c = \{1, 2, 3\}$ = the event that a number exceeding 3 does not occur.

Note that A and B are mutually exclusive, that is, that $A \cap B = \emptyset$. In other words, an even number and an odd number cannot occur simultaneously.

(b) **Experiment:** Toss a coin three times and observe the sequence of heads (H) and tails (T) that appears. The sample space S consists of the following eight elements:

$$S = \{HHH, HHT, HTH, HTT, THH, THT, TTH, TTT\}$$

Let A be the event that two or more heads appear consecutively, and B that all the tosses are the same, that is,

$$A = \{HHH, HHT, THH\} \quad \text{and} \quad B = \{HHH, TTT\}$$

Then $A \cap B = \{HHH\}$ is the elementary event in which only heads appear. The event that five heads appear is the empty set $\emptyset$.

(c) **Experiment:** Toss a coin until a head appears, and then count the number of times the coin is tossed.

The sample space of this experiment is $S = \{1, 2, 3, \ldots, \infty\}$. Here ∞ refers to the case when a head never appears, and so the coin is tossed an infinite number of times. Since every positive integer is an element of S, the sample space is infinite. In fact, this is an example of a sample space which is *countably infinite*.

(d) **Experiment:** Let a pencil drop, head first, into a rectangular box and note the point at the bottom of the box that the pencil first touches. Here S consists of all the points on the bottom of the box. Let the rectangular area in Fig. 3-2 represent these points. Let A and B be the events that the pencil drops into the corresponding areas illustrated in Fig. 3-2. Then $A \cap B$ is the event that the pencil drops in the shaded region in Fig. 3-2.

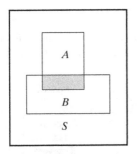

Fig. 3-2

Remark: The sample space S in Example 3.1(d) is an example of a continuous sample space. (A sample space S is *continuous* if it is an interval or a product of intervals.) In such a case, only special subsets (called *measurable* sets) will be events. On the other hand, if the sample space S is *discrete*, that is, if S is finite or countably infinite, then every subset of S is an event.

EXAMPLE 3.2 Toss of a pair of dice A pair of dice is tossed and the two numbers appearing on the top faces are recorded. There are six possible numbers, $1, 2, \ldots, 6$, on each die. Thus, S consists of the pairs of numbers from 1 to 6, and hence $n(S) = 6 \cdot 6 = 36$. Figure 3-3 shows these 36 pairs of numbers arranged in an array where the rows are labeled by the first die and the columns by the second die. Let A be the event that the sum of the two numbers is 6, and let B be the event that the largest of the two numbers is 4. That is, let

$$A = \{(1, 5), (2, 4), (3, 3), (4, 2), (5, 1)\}$$
$$B = \{(1, 4), (2, 4), (3, 4), (4, 4), (4, 3), (4, 2), (4, 1)\}$$

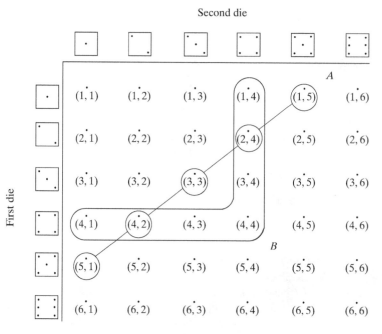

Fig. 3-3

These events are pictured in Fig. 3-3. Then the event "A and B" consists of those pairs of integers whose sum is 6 and whose largest number is 4 or, in other words, the intersection of A and B. Thus

$$A \cap B = \{(2, 4), (4, 2)\}$$

Similarly, "A or B", the sum is 6 or the largest is 4, is the union $A \cup B$, and "not A", the sum is not 6, is the complement A^c.

EXAMPLE 3.3 Deck of cards A card is drawn from an ordinary deck of 52 cards which is pictured in Fig. 3-4(a). The sample space S consists of the four *suits*, clubs (C), diamonds (D), hearts (H), and spades (S),

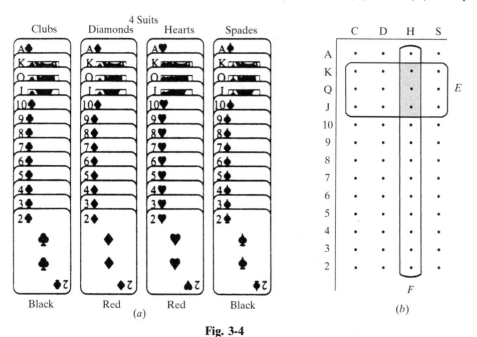

Fig. 3-4

where each suit contains 13 cards which are numbered 2 to 10, and jack (J), queen (Q), king (K), and ace (A). The hearts (H) and diamonds (D) are red cards, and the spades (S) and clubs (C) are black cards. Figure 3-4(b) pictures 52 points which represent the deck S of cards in the obvious way. Let E be the event of a *picture card*, that is, a jack (J), queen (Q), or king (K), and let F be the event of a heart. Then

$$E \cap F = \{JH, QH, KH\}$$

is the event of a heart and a picture card, as shaded in Fig. 3-4(b).

3.3 AXIOMS OF PROBABILITY

Let S be a sample space, let $\mathscr{C}$ be the class of all events, and let P be a real-valued function defined on $\mathscr{C}$. Then P is called a *probability function*, and $P(A)$ is called the *probability* of the event A, when the following axioms hold:

[$\mathbf{P_1}$]   For any event A, we have $P(A) \geq 0$.

[$\mathbf{P_2}$]   For the certain event S, we have $P(S) = 1$.

[$\mathbf{P_3}$]   For any two disjoint events A and B, we have

$$P(A \cup B) = P(A) + P(B)$$

[$\mathbf{P_3'}$]   For any infinite sequence of mutually disjoint events $A_1, A_2, A_3, \ldots$, we have

$$P(A_1 \cup A_2 \cup A_3 \cup \cdots) = P(A_1) + P(A_2) + P(A_3) + \cdots$$

Furthermore, when P does satisfy the above axioms, the sample space S will be called a *probability space*.

The first axiom states that the probability of any event is nonnegative, and the second axiom states that the certain or sure event S has probability 1. The next remarks concern the two axioms [$\mathbf{P_3}$] and [$\mathbf{P_3'}$]. The axiom [$\mathbf{P_3}$] formalizes the natural assumption that if A and B are two disjoint events, then the probability of either of them occurring is the sum of their individual probabilities. Using mathematical induction, we can then extend this *additive property* for two sets to any finite number of disjoint events, that is, for any mutually disjoint sets $A_1, A_2, \ldots, A_n$, we have

$$P(A_1 \cup A_2 \cup \cdots \cup A_n) = P(A_1) + P(A_2) + \cdots + P(A_n) \qquad (*)$$

We emphasize that [$\mathbf{P_3'}$] does not follow from [$\mathbf{P_3}$], even though (*) is true for every positive integer n. However, if the sample space S is finite, then only [$\mathbf{P_3}$] is needed, that is, [$\mathbf{P_3'}$] is superfluous.

Theorems on Probability Spaces

The following theorems follow directly from our axioms, and will be proved here. We use $\square$ to indicate the end of a proof.

Theorem 3.1:   The impossible event or, in other words, the empty set $\varnothing$ has probability zero, that is, $P(\varnothing) = 0$.

   Proof: For any event A, we have $A \cup \varnothing = A$ where A and $\varnothing$ are disjoint. By [$\mathbf{P_3}$],

$$P(A) = P(A \cup \varnothing) = P(A) + P(\varnothing)$$

Adding $-P(A)$ to both sides gives $P(\varnothing) = 0$. $\square$

The next theorem, called the *complement rule*, formalizes our intuition that if we hit a target, say, $p = 1/3$ of the times, then we miss the target $q = 1 - p = 2/3$ of the times. [Recall that A^c denotes the complement of the set A.]

Theorem 3.2 (Complement Rule):   For any event A, we have

$$P(A^c) = 1 - P(A)$$

Proof: $S = A \cup A^c$ where A and A^c are disjoint.   By $[\mathbf{P}_2]$, $P(S) = 1$.   Thus, by $[\mathbf{P}_3]$,

$$1 = P(S) = P(A \cup A^c) = P(A) + P(A^c)$$

Adding $-P(A)$ to both sides gives us $P(A^c) = 1 - P(A)$. $\square$

The next theorem tells us that the probability of any event must lie between 0 and 1.   That is,

Theorem 3.3:   For any event A, we have $0 \le P(A) \le 1$.

Proof: By $[\mathbf{P}_1]$, $P(A) \ge 0$.   Hence we need only show that $P(A) \le 1$.   Since $S = A \cup A^c$ where A and A^c are disjoint, we get

$$1 = P(S) = P(A \cup A^c) = P(A) + P(A^c)$$

Adding $-P(A^c)$ to both sides gives us $P(A) = 1 - P(A^c)$.   Since $P(A^c) \ge 0$, we get $P(A) \le 1$, as required. $\square$

The following theorem applies to the case that one event is a subset of another event.

Theorem 3.4:   If $A \subseteq B$, then $P(A) \le P(B)$.

Proof: If $A \subseteq B$, then, as indicated by Fig. 3-5(a), $B = A \cup (B \setminus A)$ where A and $B \setminus A$ are disjoint.   Hence

$$P(B) = P(A) + P(B \setminus A)$$

By $[\mathbf{P}_1]$, we have $P(B \setminus A) \ge 0$; hence $P(A) \le P(B)$. $\square$

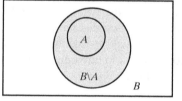
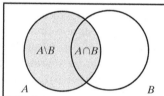
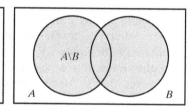

(a) B is shaded.   (b) A is shaded.   (c) $A \cup B$ is shaded.

Fig. 3-5

The following theorem concerns two arbitrary events.

Theorem 3.5:   For any two events A and B, we have

$$P(A \setminus B) = P(A) - P(A \cap B)$$

Proof: As indicated by Fig. 3-5(b), $A = (A \setminus B) \cup (A \cap B)$ where $A \setminus B$ and $A \cap B$ are disjoint.   Accordingly, by $[\mathbf{P}_3]$,

$$P(A) = P(A \setminus B) + P(A \cap B)$$

from which our result follows. $\square$

The next theorem, called the *general addition rule*, or simply *addition rule*, is similar to the inclusion-exclusion principle for sets.

Theorem (Addition Rule) 3.6: For any two events A and B,

$$P(A \cup B) = P(A) + P(B) - P(A \cap B)$$

Proof: As indicated by Fig. 3-5(c), $A \cup B = (A \setminus B) \cup B$ where $A \setminus B$ and B are disjoint sets. Thus, using Theorem 3.5,

$$P(A \cup B) = P(A \setminus B) + P(B) = P(A) - P(A \cap B) + P(B)$$
$$= P(A) + P(B) - P(A \cap B)$$

which is our result. $\square$

Applying the above theorem twice (Problem 3.34), we obtain:

Corollary 3.7: For any events, A, B, C, we have

$$P(A \cup B \cup C) = P(A) + P(B) + P(C) - P(A \cap B) - P(A \cap C) - P(B \cap C) + P(A \cap B \cap C)$$

Clearly, like the analogous inclusion-exclusion principle for sets, the addition rule can be extended by induction to any finite number of sets.

3.4 FINITE PROBABILITY SPACES

Consider a finite sample space S where we assume, unless otherwise stated, that the class $\mathscr{C}$ of all events consists of all subsets of S. As noted above, S becomes a probability space by assigning probabilities to the events in $\mathscr{C}$ so they satisfy the probability axioms. This section shows how this is usually done when the sample space S is finite. The next section discusses infinite sample spaces.

Finite Equiprobable Spaces

Suppose S is a finite sample space, and suppose the physical characteristics of the experiment suggest that the various outcomes of the experiment be assigned equal probabilities. Such a probability space S, where each point is assigned the same probability, is called a *finite equiprobable space*. Specifically, if S has n elements, then each point in S is assigned the probability $1/n$ and each event A containing r points is assigned the probability r/n. In other words,

$$P(A) = \frac{\text{number of elements in } A}{\text{number of elements in } S} = \frac{n(A)}{n(S)}$$

or

$$P(A) = \frac{\text{number of ways that the event } A \text{ can occur}}{\text{number of ways that the sample space } S \text{ can occur}}$$

We emphasize that the above formula for $P(A)$ can only be used with respect to an equiprobable space, and cannot be used in general.

We state the above result formally.

Theorem 3.8: Let S be a finite sample space and, for any subset A of S, let $P(A) = n(A)/n(S)$. Then P satisfies axioms $[\mathbf{P}_1]$, $[\mathbf{P}_2]$, and $[\mathbf{P}_3]$.

The expression "at random" will be used only with respect to an equiprobable space; formally, the statement "choose a point at random from a set S" shall mean that S is an equiprobable space where each point in S has the same probability.

EXAMPLE 3.4 A card is selected at random from an ordinary deck of 52 playing cards. (See Fig. 3-4.) Consider the following events [where a face card is a jack (J), queen (Q), or king (K)]:

$$A = \{\text{heart}\} \quad \text{and} \quad B = \{\text{face card}\}$$

(a) Find $P(A)$, $P(B)$, and $P(A \cap B)$. (b) Find $P(A \cup B)$.

(a) Since we have an equiprobable space,

$$P(A) = \frac{\text{number of hearts}}{\text{number of cards}} = \frac{13}{52} = \frac{1}{4}, \quad P(B) = \frac{\text{number of face cards}}{\text{number of cards}} = \frac{12}{52} = \frac{3}{13},$$

$$P(A \cap B) = \frac{\text{number of heart face cards}}{\text{number of cards}} = \frac{3}{52}$$

(b) Since we want $P(A \cup B)$, the probability that the card is a heart or a face card, we can count the number of such cards and use Theorem 3.8. Alternately, we can use (a) and the Addition Rule Theorem 3.6 to obtain

$$P(A \cup B) = P(A) + P(B) - P(A \cap B) = \frac{1}{4} + \frac{3}{13} - \frac{3}{52} = \frac{22}{52} = \frac{11}{26}$$

EXAMPLE 3.5 Suppose a student is selected at random from 80 students where 30 are taking mathematics, 20 are taking chemistry, and 10 are taking mathematics and chemistry. Find the probability p that the student is taking mathematics (M) or chemistry (C).

Since the space is equiprobable, we have:

$$P(M) = \frac{30}{80} = \frac{3}{8}, \; P(C) = \frac{20}{80} = \frac{1}{4}, \; P(M \text{ and } C) = P(M \cap C) = \frac{10}{80} = \frac{1}{8}$$

Thus, by the Addition Rule (Theorem 3.6),

$$p = P(M \text{ or } C) = P(M \cup C) = P(M) + P(C) - P(M \cap C) = \frac{3}{8} + \frac{1}{4} - \frac{1}{8} = \frac{1}{2}$$

Finite Probability Spaces

Let S be a finite sample space, say $S = \{a_1, a_2, \ldots, a_n\}$. A *finite probability space*, or *finite probability model*, is obtained by assigning to each point a_i in S a real number p_i, called the *probability* of a_i, satisfying the following properties:

(i) Each p_i is nonnegative, that is, $p_i \geq 0$.

(ii) The sum of the p_i is 1, that is,

$$\sum p_i = p_1 + p_2 + \cdots + p_n = 1$$

The *probability* $P(A)$ of an event A is defined as the sum of the probabilities of the points in A, that is,

$$P(A) = \sum_{a_i \in A} P(a_i) = \sum_{a_i \in A} p_i$$

For notational convenience, we write $P(a_i)$ instead of $P(\{a_i\})$.

Sometimes the points in a finite sample space S and their assigned probabilities are given in the form of a table as follows:

Outcome	a_1	a_2	$\cdots$	a_n
Probability	p_1	p_2	$\cdots$	p_n

Such a table is called a *probability distribution*.

The fact that $P(A)$, the sum of the probabilities of the points in A, does define a probability space is stated formally below (and proved in Problem 3.32).

Theorem 3.9:   The above function $P(A)$ satisfies the axioms

$$[\mathbf{P}_1], [\mathbf{P}_2], \text{ and } [\mathbf{P}_3].$$

EXAMPLE 3.6   Experiment Let three coins be tossed and the number of heads observed. [Compare with Example 3.1(b).]   Then the sample space is $S = \{0, 1, 2, 3\}$.   The following assignments on the elements of S define a probability space:

Outcome	0	1	2	3
Probability	1/8	3/8	3/8	1/8

That is, each probability is nonnegative, and the sum of the probabilities is 1.   Let A be the event that at least one head appears, and let B be the event that all heads or all tails appear, that is, let

$$A = \{1, 2, 3\} \quad \text{and} \quad B = \{0, 3\}$$

Then, by definition,

$$P(A) = P(1) + P(2) + P(3) = \frac{3}{8} + \frac{3}{8} + \frac{1}{8} = \frac{7}{8}$$

and

$$P(B) = P(0) + P(3) = \frac{1}{8} + \frac{1}{8} = \frac{1}{4}$$

EXAMPLE 3.7   Three horses A, B, C are in a race; A is twice as likely to win as B, and B is twice as likely to win as C.

(a)   Find their respective probabilities of winning, that is, find $P(A)$, $P(B)$, $P(C)$.

(b)   Find the probability that B or C wins.

(a)   Let $P(C) = p$.   Since B is twice as likely to win as C, $P(B) = 2p$; and since A is twice as likely to win as B, $P(A) = 2P(B) = 2(2p) = 4p$.   Now the sum of the probabilities must be 1; hence

$$p + 2p + 4p = 1 \quad \text{or} \quad 7p = 1 \quad \text{or} \quad p = \frac{1}{7}$$

Accordingly, $P(A) = 4p = \frac{4}{7}$, $\quad P(B) = 2p = \frac{2}{7}$, $\quad P(C) = p = \frac{1}{7}$

(b)   Note $\{B, C\}$ is the event that B or C wins, so we want $P(\{B, C\})$.   By definition, we simply add up the probabilities of the points in $\{B, C\}$.   Thus

$$P(\{B, C\}) = P(B) + P(C) = \frac{2}{7} + \frac{1}{7} = \frac{3}{7}$$

3.5   INFINITE SAMPLE SPACES

This section considers infinite sample spaces S.   There are two cases, the case where S is countably infinite and the case where S is uncountable.   We note that a finite or a countably infinite probability space S is said to be *discrete*, whereas an uncountable space S is said to be *nondiscrete*.   Moreover, an uncountable space S which consists of a continuum of points, such as an interval or product of intervals, is said to be *continuous*.

Countably Infinite Sample Spaces

Suppose S is a countably infinite sample space; say

$$S = \{a_1, a_2, a_3, \ldots\}$$

Then, as in the finite case, we obtain a probability space by assigning each $a_i \in S$ a real number p_i, called its probability, such that:

(i)   Each p_i is nonnegative, that is, $p_i \geq 0$.

(ii)   The sum of the p_i is equal to 1, that is,

$$p_1 + p_2 + p_3 + \cdots = \sum_{i=1}^{\infty} p_i = 1$$

The probability $P(A)$ of an event A is then the sum of the probabilities of its points.

EXAMPLE 3.8   Consider the sample space $S = \{1, 2, 3, \ldots, \infty\}$ of the experiment of tossing a coin until a head appears; here n denotes the number of times the coin is tossed.   A probability space is obtained by setting

$$p(1) = \frac{1}{2}, \, p(2) = \frac{1}{4}, \, p(3) = \frac{1}{8}, \, \ldots, \, p(n) = \frac{1}{2^n}, \, \ldots, \, p(\infty) = 0$$

Consider the events:

$$A = \{n \text{ is at most } 3\} = \{1, 2, 3\} \qquad \text{and} \qquad B = \{n \text{ is even}\} = \{2, 4, 6, \ldots\}$$

Find $P(A)$ and $P(B)$.

Adding the probabilities of the points in the sets (events) yields:

$$P(A) = P(1, 2, 3) = \frac{1}{2} + \frac{1}{4} + \frac{1}{8} = \frac{7}{8}$$

$$P(B) = P(2, 4, 6, 8, \ldots) = \frac{1}{4} + \frac{1}{4^2} + \frac{1}{4^3} + \cdots$$

Note that $P(B)$ is a geometric series with $a = 1/4$ and $r = 1/4$; hence

$$P(B) = \frac{a}{1 - r} = \frac{1/4}{3/4} = \frac{1}{3}$$

Uncountable Spaces

The only uncountable sample spaces S which we will consider here are those with some finite geometrical measurement $m(S)$, such as length, area, or volume, and where a point in S is selected at random.   The probability of an event A, that is, that the selected point belongs to A, is then the ratio of $m(A)$ to $m(S)$.   Thus

$$P(A) = \frac{\text{length of } A}{\text{length of } S} \qquad \text{or} \qquad P(A) = \frac{\text{area of } A}{\text{area of } S} \qquad \text{or} \qquad P(A) = \frac{\text{volume of } A}{\text{volume of } S}$$

Such a probability space S is said to be *uniform*.

EXAMPLE 3.9   A point is chosen at random inside a rectangle measuring 3 by 5 in.   Find the probability p that the point is at least 1 in from the edge.

Let S denote the set of points inside the rectangle and let A denote the set of points at least 1 in from the edge. S and A are pictured in Fig. 3-6.   Note that A is a rectangular area measuring 1 in by 3 in.   Thus

$$p = \frac{\text{area of } A}{\text{area of } S} = \frac{1 \cdot 3}{3 \cdot 5} = \frac{1}{5}$$

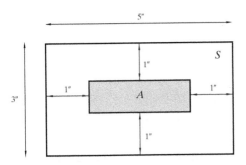

Fig. 3-6. A is shaded.

3.6 CLASSICAL BIRTHDAY PROBLEM

The classical birthday problem concerns the probability that n people have distinct birthdays where $n \leq 365$. Here we ignore leap years and assume that a person's birthday can fall on any day with equal probability.

Since there are n people and 365 different days, there are 365^n ways in which the n people can have their birthdays. On the other hand, if the n persons are to have distinct birthdays, then:

(i)   The first person can be born on any of the 365 days.

(ii)   The second person can be born on the remaining 364 days.

(iii)   The third person can be born on the remaining 363 days, and so on.

Thus there are:

$$365 \cdot 364 \cdot 363 \cdots (365 - n + 1)$$

ways that n persons can have distinct birthdays.   Therefore

$$P(n \text{ people have distinct birthdays}) = \frac{365 \cdot 364 \cdot 363 \cdots (365 - n + 1)}{365^n}$$

Accordingly, the probability p that two or more people have the same birthday is as follows:

$$p = 1 - [\text{probability that no two people have the same birthday}]$$

$$= 1 - \frac{365 \cdot 364 \cdot 363 \cdots (365 - n + 1)}{365^n}$$

The value of p where n is a multiple of 10 up to 60 follows:

n	10	20	30	40	50	60
p	0.117	0.411	0.706	0.891	0.970	0.994

We note that $p = 0.476$ for $n = 22$ and that $p = 0.507$ for $n = 23$.   Accordingly:

> In a group of 23 people, it is more likely
> that at least two of them have the same birthday
> than that they all have distinct birthdays.

The above table also tells us that, in a group of 60 or more people, the probability that two or more of them have the same birthday exceeds 99 percent.

Solved Problems

SAMPLE SPACES AND EVENTS

3.1. Let A and B be events. Find an expression and exhibit the Venn diagram for the event:

(a) A but not B, (b) neither A nor B, (c) either A or B, but not both.

(a) Since A but not B occurs, shade the area of A outside of B, as in Fig. 3-7(a). Note that B^c, the complement of B, occurs, since B does not occur; hence A and B^c occur. In other words, the event is $A \cap B^c$.

(b) "Neither A nor B" means "not A and not B" or $A^c \cap B^c$. By DeMorgan's law, this is also the set $(A \cup B)^c$; hence shade the area outside of A and outside of B, that is, outside $A \cup B$, as in Fig. 3-7(b).

(c) Since A or B, but not both, occurs, shade the area of A and B, except where they intersect, as in Fig. 3-7(c). The event is equivalent to the occurrence of A but not B or B but not A. Thus, the event is $(A \cap B^c) \cup (B \cap A^c)$. Alternately, the event is $A \oplus B$, the symmetric difference of A and B.

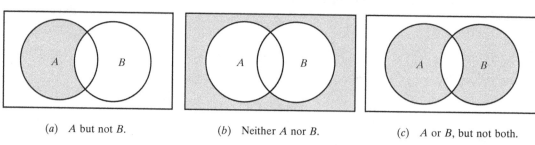

(a) A but not B.   (b)   Neither A nor B.   (c) A or B, but not both.

Fig. 3-7

3.2. Let A, B, C be events. Find an expression and exhibit the Venn diagram for the event:

(a) A and B but not C occurs, (b) only A occurs.

(a) Since A and B but not C occurs, shade the intersection of A and B which lies outside of C, as in Fig. 3-8(a). The event consists of the elements in A, in B, and in C^c (not in C), that is, the event is the intersection $A \cap B \cap C^c$.

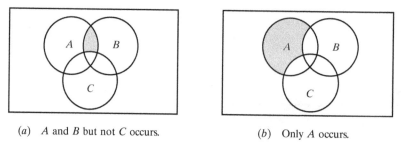

(a) A and B but not C occurs.   (b)   Only A occurs.

Fig. 3-8

(b) Since only A is to occur, shade the area of A which lies outside of B and C, as in Fig. 3-8(b). The event consists of the elements in A, in B^c (not in B), and in C^c (not in C), that is, the event is the intersection $A \cap B^c \cap C^c$.

3.3. Let a coin and a die be tossed; and let the sample space S consist of the 12 elements:

$$S = \{H1, H2, H3, H4, H5, H6, T1, T2, T3, T4, T5, T6\}$$

Express explicitly the following events:

(a) $A = \{$heads and an even number$\}$, (b) $B = \{$a number less than 3$\}$,
(c) $C = \{$tails and an odd number$\}$.

(a) The elements of A are those elements of S which consist of an H and an even number; hence

$$A = \{H2, H4, H6\}$$

(b) The elements of B are those elements of S whose second component is less than 3, that is, 1 or 2; hence

$$B = \{H1, H2, T1, T2\}$$

(c) The elements of C are those elements of S which consist of a T and an odd number; hence

$$C = \{T1, T3, T5\}$$

3.4. Consider the events A, B, C in the preceding Problem 3.3. Express explicitly the event that:

(a) A or B occurs.   (b) B and C occur.   (c)   Only B occurs.

Which pair of the events A, B, C are mutually exclusive?

(a) "A or B" is the set union $A \cup B$; hence

$$A \cup B = \{H2, H4, H6, H1, T1, T2\}$$

(b) "B and C" is the set intersection $B \cap C$; hence

$$B \cap C = \{T1\}$$

(c) "Only B" consists of the elements of B which are not in A and not in B, that is, the set intersection $B \cap A^c \cap C^c$; hence

$$B \cap A^c \cap C^c = \{H1, T2\}$$

Only A and C are mutually exclusive, that is, $A \cap C = \varnothing$.

3.5. A pair of dice is tossed and the two numbers appearing on the top are recorded. Recall that S consists of 36 pairs of numbers which are pictured in Fig. 3-3. Find the number of elements in each of the following events:

(a) $A = \{$two numbers are equal$\}$ (c) $C = \{$5 appears on first die$\}$
(b) $B = \{$sum is 10 or more$\}$ (d) $D = \{$5 appears on at least one die$\}$

   Use Fig. 3-3 to help count the number of elements in each of the events:

(a) $A = \{(1, 1), (2, 2), \ldots, (6, 6)\}$,   so $n(A) = 6$.
(b) $B = \{(6, 4), (5, 5), (4, 6), (6, 5), (5, 6), (6, 6)\}$,   so $n(B) = 6$.
(c) $C = \{(5, 1), (5, 2), \ldots, (5, 6)\}$,   so $n(C) = 6$.
(d) There are six pairs with 5 as the first element, and six pairs with 5 as the second element. However, $(5, 5)$ appears in both places. Hence

$$n(D) = 6 + 6 - 1 = 11$$

Alternately, count the pairs in Fig. 3-3 which are in D to get $n(D) = 11$.

FINITE EQUIPROBABLE SPACES

3.6. Determine the probability p of each event:

(a) An even number appears in the toss of a fair die.

(b) At least one tail appears in the toss of 3 fair coins.

(c) A white marble appears in the random drawing of 1 marble from a box containing 4 white, 3 red, and 5 blue marbles.

Each sample space S is an equiprobable space. Hence, for each event E, use

$$P(E) = \frac{\text{number of elements in } E}{\text{number of elements in } S} = \frac{n(E)}{n(S)}$$

(a) The event can occur in three ways (a 2, 4, or 6) out of 6 equally like cases; hence $p = 3/6 = 1/2$.

(b) Assuming the coins are distinguished, there are 8 cases:

$$HHH, \ HHT, \ HTH, \ HTT, \ THH, \ THT, \ TTH, \ TTT$$

Only the first case is not favorable; hence $p = 7/8$.

(c) There are $4 + 3 + 5 = 12$ marbles of which 4 are white; hence $p = 4/12 = 1/3$.

3.7. A single card is drawn from an ordinary deck S of 52 cards. (See Fig. 3-4.) Find the probability p that the card is a: (a) king, (b) face card (jack, queen, or king), (c) red card (heart or diamond), (d) red face card.

Here $n(S) = 52$.

(a) There are 4 kings; hence $p = 4/52 = 1/13$.

(b) There are $4(3) = 12$ face cards; hence $p = 12/52 = 3/13$.

(c) There are 13 hearts and 13 diamonds; hence $p = 26/52 = 1/2$.

(d) There are 6 face cards which are red; hence $p = 6/52 = 3/26$.

3.8. Consider the sample space S and events A, B, C in Problem 3.3 where a coin and a die are tossed. Suppose the coin and die are fair; hence S is an equiprobable space. Find:

(a) $P(A)$, $P(B)$, $P(C)$,   (b) $P(A \cup B)$, $P(B \cap C)$, $P(B \cap A^c \cap C^c)$

Since S is an equiprobable space, use $P(E) = n(E)/n(S)$. Here $n(S) = 12$. We need only count the number of elements in each given set, and then divide by 12.

(a) By Problem 3.3, $P(A) = 3/12$, $P(B) = 4/12$, $P(C) = 3/12$.

(b) By Problem 3.4, $P(A \cup B) = 6/12$, $P(B \cap C) = 1/12$, $P(B \cap A^c \cap C^c) = 2/12$.

3.9. A box contains 15 billiard balls which are numbered from 1 to 15. A ball is drawn at random and the number recorded. Find the probability p that the number is:

(a) even, (b) less than 5, (c) even and less than 5, (d) even or less than 5.

(a) There are 7 numbers, 2, 4, 6, 8, 10, 12, 14, which are even; hence $p = 7/15$.

(b) There are 4 numbers, 1, 2, 3, 4, which are less than 5, hence $p = 4/15$.

(c) There are 2 numbers, 2 and 4, which are even and less than 5; hence $p = 2/15$.

(d) By the addition rule (Theorem 3.6),

$$p = \frac{7}{15} + \frac{4}{15} - \frac{2}{15} = \frac{9}{15}$$

Alternately, there are 9 numbers, 1, 2, 3, 4, 6, 8, 10, 12, 14, which are even or less than 5; hence $p = 9/15$.

3.10. A box contains 2 white sox and 2 blue sox. Two sox are drawn at random. Find the probability p they are a match (same color).

There are $C(4, 2) = \binom{4}{2} = 6$ ways to draw 2 of the sox. Only two pairs will yield a match. Thus $p = 2/6 = 1/3$.

3.11. Five horses are in a race. Audrey picks 2 of the horses at random and bets on them. Find the probability p that Audrey picked the winner.

There are $C(5, 2) = \binom{5}{2} = 10$ ways to pick 2 of the horses. Four of the pairs will contain the winner. Thus, $p = 4/10 = 2/5$.

3.12. A class contains 10 men and 20 women of which half the men and half the women have brown eyes. Find the probability p that a person chosen at random is a man or has brown eyes.

Let $A = \{\text{men}\}$, $B = \{\text{brown eyes}\}$. We seek $P(A \cup B)$. First find:

$$P(A) = \frac{10}{30} = \frac{1}{3}, \; P(B) = \frac{15}{30} = \frac{1}{2}, \; P(A \cap B) = \frac{5}{30} = \frac{1}{6}$$

Thus, by the addition rule (Theorem 3.6),

$$P(A \cup B) = P(A) + P(B) - P(A \cap B) = \frac{1}{3} + \frac{1}{2} - \frac{1}{6} = \frac{2}{3}$$

3.13. Six married people are standing in a room. Two people are chosen at random. Find the probability p that: (a) they are married; (b) one is male and one is female.

There are $C(12, 2) = 66$ ways to choose 2 people from the 12 people.

(a) There are 6 married couples; hence $p = 6/66 = 1/11$.

(b) There are 6 ways to choose the male and 6 ways to choose the female; hence $p = (6 \cdot 6)/66 = 36/66 = 6/11$.

3.14. Suppose 5 marbles are placed in 5 boxes at random. Find the probability p that exactly 1 of the boxes is empty.

There are exactly 5^5 ways to place the 5 marbles in the 5 boxes. If exactly 1 box is empty, then 1 box contains 2 marbles and each of the remaining boxes contains 1 marble. There are 5 ways to select the empty box, then 4 way to select the box containing 2 marbles, and $C(5, 2) = 10$ ways to select 2 marbles to go into this box. Finally, there are 3! ways to distribute the remaining 3 marbles among the remaining 3 boxes. Thus

$$p = \frac{5 \cdot 4 \cdot 10 \cdot 3!}{5^5} = \frac{48}{125}$$

3.15. Two cards are drawn at random from an ordinary deck of 52 cards. (See Fig. 3-4.) Find the probability p that: (a) both are hearts, (b) one is a heart and one is a spade.

There are $C(52, 2) = 1326$ ways to choose 2 cards from the 52-card deck. In other words, $n(S) = 1326$.

(a) There are $C(13, 2) = 78$ ways to draw 2 hearts from the 13 hearts; hence

$$p = \frac{\text{number of ways 2 hearts can be drawn}}{\text{number of ways 2 cards can be drawn}} = \frac{78}{1326} = \frac{3}{51}$$

(b) There are 13 hearts and 13 spades, so there are $13 \cdot 13 = 169$ ways to draw a heart and a spade. Thus, $p = 169/1326 = 13/102$.

FINITE PROBABILITY SPACES

3.16. A sample space S consists of four elements, that is, $S = \{a_1, a_2, a_3, a_4\}$. Under which of the following functions P does S become a probability space?

(a) $P(a_1) = 0.4$, $P(a_2) = 0.3$, $P(a_3) = 0.2$, $P(a_4) = 0.3$.
(b) $P(a_1) = 0.4$, $P(a_2) = -0.2$, $P(a_3) = 0.7$, $P(a_4) = 0.1$.
(c) $P(a_1) = 0.4$, $P(a_2) = 0.2$, $P(a_3) = 0.1$, $P(a_4) = 0.3$.
(d) $P(a_1) = 0.4$, $P(a_2) = 0$, $P(a_3) = 0.5$, $P(a_4) = 0.1$.

(a) The sum of the values on the points in S exceeds one; hence P does not define S to be a probability space.
(b) Since $P(a_2)$ is negative, P does not define S to be a probability space.
(c) Each value is nonnegative and their sum is one; hence P does define S to be a probability space.
(d) Although $P(a_2) = 0$, each value is still nonnegative and their sum does equal. Thus, P does define S to be a probability space.

3.17. A coin is weighted so that heads is twice as likely to appear as tails. Find $P(T)$ and $P(H)$.

Let $P(T) = p$; then $P(H) = 2p$. Now set the sum of the probabilities equal to one, that is, set $p + 2p = 1$. Then $p = 1/3$. Thus $P(H) = 1/3$ and $P(B) = 2/3$.

3.18. Suppose A and B are events with $P(A) = 0.6$, $P(B) = 0.3$, and $P(A \cap B) = 0.2$. Find the probability that:

(a) A does not occur.   (c) A or B occurs.
(b) B does not occur.   (d) Neither A nor B occurs.

(a) By the complement rule, $P(\text{not } A) = P(A^c) = 1 - P(A) = 0.4$.
(b) By the complement rule, $P(\text{not } B) = P(B^c) = 1 - P(B) = 0.7$.
(c) By the addition rule,

$$P(A \text{ or } B) = P(A \cup B) = P(A) + P(B) - P(A \cap B)$$
$$= 0.6 + 0.3 - 0.2 = 0.7$$

(d) Recall [Fig. 3-7(b)] that neither A nor B is the complement of $A \cup B$. Therefore

$$P(\text{neither } A \text{ nor } B) = P((A \cup B)^c) = 1 - P(A \cup B) = 1 - 0.7 = 0.3$$

3.19. A die is weighted so that the outcomes produce the following probability distribution:

Outcome	1	2	3	4	5	6
Probability	0.1	0.3	0.2	0.1	0.1	0.2

Consider the events:

$$A = \{\text{even number}\}, \quad B = \{2, 3, 4, 5\}, \quad C = \{x : x < 3\}, \quad D = \{x : x > 7\}$$

Find the following probabilities:

$$(a) \quad P(A) \qquad (b) \quad P(B) \qquad (c) \quad P(C) \qquad (d) \quad P(D)$$

For any event E, find $P(E)$ by summing the probabilities of the elements in E.

(a) $A = \{2, 4, 6\}$, so $P(A) = 0.3 + 0.1 + 0.2 = 0.6$.

(b) $P(B) = 0.3 + 0.2 + 0.1 + 0.1 = 0.7$.

(c) $C = \{1, 2\}$, so $P(C) = 0.1 + 0.3 = 0.4$.

(d) $D = \varnothing$, the empty set.   Hence $P(D) = 0$.

3.20. For the data in Problem 3.19, find: (a) $P(A \cap B)$, (b) $P(A \cup C)$, (c) $P(B \cap C)$.

First find the elements in the event, and then add the probabilities of the elements.

(a) $A \cap B = \{2, 4\}$,   so $P(A \cap B) = 0.3 + 0.1 = 0.4$.

(b) $A \cup C = \{1, 2, 3, 4, 5\} = \{6\}^c$,   so $P(A \cup C) = 1 - 0.2 = 0.8$.

(c) $B \cap C = \{2\}$,   so $P(B \cap C) = 0.3$.

3.21. Let A and B be events such that $P(A \cup B) = 0.8$, $P(A) = 0.4$, and $P(A \cap B) = 0.3$.   Find:
(a) $P(A^c)$;   (b) $P(B)$;   (c) $P(A \cap B^c)$;   (d) $P(A^c \cap B^c)$.

(a) By the complement rule, $P(A^c) = 1 - P(A) = 1 - 0.4 = 0.6$.

(b) By the addition rule, $P(A \cup B) = P(A) + P(B) - P(A \cap B)$.   Substitute in this formula to obtain:

$$0.8 = 0.4 + P(B) + 0.3 \qquad \text{or} \qquad P(B) = 0.1$$

(c) $P(A \cap B^c) = P(A \setminus B) = P(A) - P(A \cap B) = 0.4 - 0.3 = 0.1$.

(d) By DeMorgan's law, $(A \cup B)^c = A^c \cap B^c$.   Thus

$$P(A^c \cap B^c) = P((A \cup B)^c) = 1 - P(A \cup B) = 1 - 0.8 = 0.2$$

3.22. Suppose $S = \{a_1, a_2, a_3, a_4\}$, and suppose P is a probability function defined on S.

(a) Find $P(a_1)$ if $P(a_2) = 0.4$, $P(a_3) = 0.2$, $P(a_3) = 0.1$.

(b) Find $P(a_1)$ and $P(a_2)$ if $P(a_3) = P(a_4) = 0.2$ and $P(a_1) = 3P(a_2)$.

(a) Let $P(a_1) = p$.   For P to be a probability function, the sum of the probabilities on the sample points must equal one.   Thus, we have

$$p + 0.4 + 0.2 + 0.1 = 1 \qquad \text{or} \qquad p = 0.3$$

(b) Let $P(a_2) = p$ so $P(a_1) = 3p$.   Thus

$$3p + p + 0.2 + 0.2 = 1 \qquad \text{or} \qquad p = 0.15$$

Hence $P(a_2) = 0.15$ and $P(a_1) = 0.45$.

ODDS

3.23. Suppose $P(E) = p$. The *odds* that E occurs is defined to be the ratio $p : (1 - p)$. Find p if the odds that E occurs are a to b.

Set the ratio $p : (1 - p)$ to $a : b$ to obtain

$$\frac{p}{1-p} = \frac{a}{b} \quad \text{or} \quad bp = a - ap \quad \text{or} \quad ap + bp = a \quad \text{or} \quad p = \frac{a}{a+b}$$

3.24. The odds that an event E occurs is 3 to 2. Find the probability of E.

Let $p = P(E)$. Set the odds equal to $p : (1 - p)$ to obtain

$$\frac{p}{1-p} = \frac{3}{2} \quad \text{or} \quad 2p = 3 - 3p \quad \text{or} \quad 5p = 3 \quad \text{or} \quad p = \frac{3}{5}$$

Alternately, use the formula in Problem 3.21 to directly obtain

$$p = \frac{a}{a+b} = \frac{3}{3+2} = \frac{3}{5}$$

3.25. Suppose $P(E) = 5/12$. Express the odds that E occurs in terms of positive integers.

First compute $1 - P(E) = 7/12$. The odds that E occurs are

$$\frac{P(E)}{1 - P(E)} = \frac{5/12}{7/12} = \frac{5}{7}$$

Thus, the odds are 5 to 7.

UNCOUNTABLE UNIFORM SPACES

3.26. A point is chosen at random inside a circle. Find the probability p that the point is closer to the center of the circle than to its circumference.

Let S denote the set of points inside the circle with radius r, and let A denote the set of points inside the concentric circle with radius $\frac{1}{2}r$, as pictured in Fig. 3-9(a). Thus, A consists precisely of those points of S which are closer to the center than to its circumference. Therefore

$$p = p(A) = \frac{\text{area of } A}{\text{area of } S} = \frac{\pi(r/2)^2}{\pi r^2} = \frac{1}{4}$$

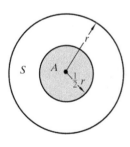

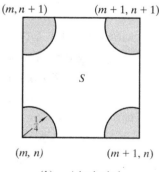

(a)   *A* is shaded.   (b)   *A* is shaded.

Fig. 3-9

3.27. Consider the plane $\mathbf{R}^2$, and let X denote the subset of points with integer coordinates. A coin of radius 1/4 is tossed randomly on the plane. Find the probability that the coin covers a point of X.

Let S denote the set of points inside a square with corners

$$(m, n), \qquad (m, n+1), \qquad (m+1, n), \qquad (m+1, n+1)$$

where m and n are integers. Let A denote the set of points in S with distance less than 1/4 from any corner point, as pictured in Fig. 3-9(b). Note that the area of A is equal to the area inside a circle of radius 1/4. Suppose the center of the coin falls in S. Then the coin will cover a point in X if and only if its center falls in A. Accordingly,

$$p = \frac{\text{area of } A}{\text{area of } S} = \frac{\pi(1/4)^2}{1} = \frac{\pi}{16} \approx 0.2$$

(Note: We cannot take S to be all of $\mathbf{R}^2$ since the area of $\mathbf{R}^2$ is infinite.)

3.28. On the real line $\mathbf{R}$, points a and b are selected at random such that $0 \le a \le 3$ and $-2 \le b \le 0$, as shown in Fig. 3-10(a). Find the probability p that the distance between a and b is greater than 3.

The sample space S consists of the ordered pairs (a, b) and so forms a rectangular region shown in Fig. 3-10(b). On the other hand, the set A of points (a, b) for which $d = a - b > 3$ consists of those points which lie below the line $x - y = 3$, and hence form the shaded region in Fig. 3-10(b). Thus

$$p = P(A) = \frac{\text{area of } A}{\text{area of } S} = \frac{2}{6} = \frac{1}{3}$$

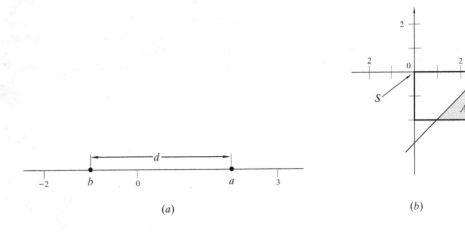

(a)   (b)

Fig. 3-10

3.29. Three points are selected at random from the circumference of a circle. Find the probability p that the three points lie on a semicircle.

Suppose the length of the circumference is $2s$. Let x denote the clockwise arc length from a to b, and let y denote the clockwise arc length from a to c, as pictured in Fig. 3-11(a). Thus

$$0 < x < 2s \quad \text{and} \quad 0 < y < 2s \qquad (*)$$

Let S denote the set of points in $\mathbf{R}^2$ for which the condition $(*)$ holds. Let A denote the set of points of S for which any of the following conditions hold:

(i) $x, y < s$,   (iii) $x < s$ and $y - x > s$,

(ii) $x, y > s$,   (iv) $y < s$ and $x - y > s$.

Then A, shaded in Fig. 3-11(b), consists of those points for which a, b, c lie on a semicircle. Thus

$$p = \frac{\text{area of } A}{\text{area of } S} = \frac{3s^2}{4s^2} = \frac{3}{4}$$

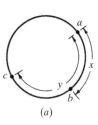
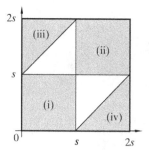

(b) A is shaded.

Fig. 3-11

3.30.   A coin of diameter 1/2 is tossed randomly onto the plane $\mathbf{R}^2$. Find the probability p that the coin does not intersect any line of the form $x = k$ where k is an integer.

The lines are all vertical, and the distance between adjacent lines is one. Let S denote a horizontal line segment between adjacent lines, say, $x = k$ and $x = k + 1$; and let A denote the points of S which are at least 1/4 from either end, as pictured in Fig. 3-12. Note that the length of S is 1 and the length of A is 1/2. Suppose the center of the coin falls in S. Then the coin will not intersect the lines if and only if its center falls in A. Accordingly

$$p = \frac{\text{length of } A}{\text{length of } S} = \frac{1/2}{1} = \frac{1}{2}$$

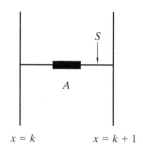

Fig. 3-12

MISCELLANEOUS PROBLEMS

3.31. Show that axiom $[\mathbf{P}_3]$ follows from axiom $[\mathbf{P}_3']$.

First we show that $P(\varnothing) = 0$ using $[\mathbf{P}_3']$ instead of $[\mathbf{P}_3]$. We have $\varnothing = \varnothing + \varnothing + \varnothing + \cdots$ where the empty sets are disjoint. Say $P(\varnothing) = a$. Then, by $[\mathbf{P}_3']$,

$$P(\varnothing) = P(\varnothing + \varnothing + \varnothing + \cdots) = P(\varnothing) + P(\varnothing) + P(\varnothing) + \cdots$$

However, zero is the only real number a satisfying $a = a + a + a + \cdots$. Therefore, $P(\varnothing) = 0$.

Suppose A and B are disjoint. Then $A, B, \varnothing, \varnothing, \ldots$ are disjoint, and

$$A \cup B = A \cup B \cup \varnothing \cup \varnothing \cup \ldots$$

Hence, by $[\mathbf{P}_3']$,

$$P(A \cup B) = P(A \cup B \cup \varnothing \cup \varnothing \cup \cdots) = P(A) + P(B) + P(\varnothing) + P(\varnothing) + \cdots$$
$$= P(A) + P(B) + 0 + 0 + \cdots = P(A) + P(B)$$

which is $[\mathbf{P}_3]$.

3.32. Prove Theorem 3.9. Suppose $S = \{a_1, a_2, \ldots, a_n\}$ and each a_i is assigned the probability p_i where: (i) $p_i \geq 0$, and (ii) $\Sigma p_i = 1$. For any event A, let

$$P(A) = \Sigma(p_j : a_j \in A)$$

Then P satisfies: (a) $[\mathbf{P}_1]$, (b) $[\mathbf{P}_2]$, (c) $[\mathbf{P}_3]$.

(a) Each $p_j \geq 0$; hence $P(A) = \Sigma p_j \geq 0$.

(b) Every $a_j \in S$; hence $P(S) = p_1 + p_2 + \cdots + p_n = 1$.

(c) Suppose A and B are disjoint, and

$$P(A) = \Sigma(p_j : a_j \in A), \text{ and } P(B) = \Sigma(p_k : a_k \in B)$$

Then the a_j's and a_k's are distinct. Therefore

$$P(A \cup B) = \Sigma(p_t : p_t \in A \cup B) = \Sigma(p_t : a_t \in A) + \Sigma(p_t : a_t \in B) = P(A) + P(B)$$

3.33. Let $S = \{a_1, a_2, \ldots, a_s\}$ and $T = \{b_1, b_2, \ldots, b_t\}$ be finite probability spaces. Let the number $p_{ij} = P(a_i)P(b_j)$ be assigned to the ordered pair (a_i, b_j) in the product set

$$S \times T = \{(s, t) : s \in S, t \in T\}$$

Show that the p_{ij} define a probability space on $S \times T$, that is, show that:

(i) The p_{ij} are nonnegative.   (ii) The sum of the p_{ij} equals one.

(This is called the *product probability space*. We emphasize that this is not the only probability function that can be defined on the product set $S \times T$.)

Since $P(a_i), P(b_j) \geq 0$, for each i and each j, we have $p_{ij} = P(a_i)P(b_j) \geq 0$. Hence (i) is true. Also, we have

$$p_{11} + p_{12} + \cdots + p_{1t} + p_{21} + p_{22} + \cdots + p_{2t} + \cdots + p_{s1} + p_{s2} + \cdots + p_{st}$$
$$= P(a_1)P(b_1) + \cdots + P(a_1)P(b_t) + \cdots + P(a_s)P(b_1) + \cdots + P(a_s)P(b_t)$$
$$= P(a_1)[P(b_1) + \cdots + P(b_t)] + \cdots + P(a_s)[P(b_1) + \cdots + P(b_t)]$$
$$= P(a_1) \cdot 1 + \cdots + P(a_s) \cdot 1 = P(a_1) + \cdots + P(a_s) = 1$$

That is,

$$\sum_{i,j} a_{ij} = \sum_{i,j} P(a_i)P(b_j) = \sum_i P(a_i) \sum_j P(b_j) = \sum_i P(a_i) \cdot 1 = \sum_i P(a_i) = 1$$

Thus, (ii) is true.

3.34. Prove Corollary 3.7: For any events A, B, C, we have:

$$P(A \cup B \cup C) = P(A) + P(B) + P(C) - P(A \cap B) - P(A \cap C) - P(B \cap C) + P(A \cap B \cap C)$$

Let $D = B \cup C$. Then $A \cap D = A \cap (B \cup C) = (A \cap B) \cup (A \cap C)$. Using the addition rule (Theorem 3.6), we get

$$P(A \cap D) = P[(A \cap B) \cup (A \cap C)] = P(A \cap B) + P(A \cap C) - P(A \cap B \cap A \cap C)$$
$$= P(A \cap B) + P(A \cap C) - P(A \cap B \cap C)$$

Using the addition rule (Theorem 3.6) again, we get

$$P(A \cup B \cup C) = P(A \cup D) = P(A) + P(D) - P(A \cap D) = P(A) + P(B \cup C) - P(A \cap D)$$
$$= P(A) + [P(B) + P(C) - P(B \cap C)] - [P(A \cap B) + P(A \cap C) - P(A \cap B \cap C)]$$
$$= P(A) + P(B) + P(C) - P(A \cap B) - P(A \cap C) - P(B \cap C) + P(A \cap B \cap C)$$

3.35. A die is tossed 100 times. The following table lists the six numbers and the frequency with which each number appeared:

Number	1	2	3	4	5	6
Frequency	14	17	20	18	15	16

(a) Find the relative frequency f of each of the following events:

$$A = \{3 \text{ appears}\}, \quad B = \{5 \text{ appears}\}, \quad C = \{\text{even number appears}\}$$

(b) Find a probability model of the data.

(a) The relative frequency $f = \dfrac{\text{number of successes}}{\text{total number of trials}}$. Thus

$$f_A = \frac{20}{100} = 0.20, \qquad f_B = \frac{15}{100} = 0.15, \qquad f_C = \frac{17 + 18 + 16}{100} = 0.52$$

(b) The geometric symmetry of the die indicates that we first assume an equal probability space. Statistics is then used to decide whether or not the given data supports the assumption of a fair die.

Supplementary Problems

SAMPLE SPACES AND EVENTS

3.36. Let A and B be events. Find an expression and exhibit the Venn diagram for the event that:

(a) A or not B occurs, (b) only A occurs.

3.37. Let A, B, and C be events. Find an expression and exhibit the Venn diagram for the event that:

(a) A or C, but not B occurs,   (c) none of the events occurs,
(b) exactly one of the three events occurs,   (d) at least two of the events occur.

3.38. A penny, a dime, and a die are tossed. Describe a suitable sample space S, and find $n(S)$.

3.39. For the space S in Problem 3.38, express explicitly the following events:

$$A = \{\text{two heads and an even number}\}, \quad B = \{2 \text{ appears}\}$$
$$C = \{\text{exactly one head and an odd number}\}$$

3.40. For the events A, B, C in Problem 3.39, express explicitly the event:

(a) A and B,   (b) only B,   (c) B and C,   (d) A but not B.

FINITE EQUIPROBABLE SPACES

3.41. Determine the probability of each event:

(a) An odd number appears in the toss of a fair die.

(b) 1 or more heads appear in the toss of 4 fair coins.

(c) Both numbers exceed 4 in the toss of 2 fair dice.

(d) Exactly one 6 appears in the toss of 2 fair dice.

(e) A red or a face card appears when a card is randomly selected from a 52-card deck.

3.42. A student is chosen at random to represent a class with 5 freshmen, 4 sophomores, 8 juniors, and 3 seniors. Find the probability that the student is

(a) a sophomore   (b) a senior   (c) a junior or a senior

3.43. One card is selected at random from 25 cards numbered 1 to 25. Find the probability that the number on the card is: (a) even, (b) divisible by 3, (c) even and divisible by 3, (d) even or divisible by 3, (e) ends in the digit 2.

3.44. Three bolts and three nuts are in a box. Two parts are chosen at random. Find the probability that one is a bolt and one is a nut.

3.45. A box contains 2 white sox, 2 blue sox, and 2 red sox. Two sox are drawn at random. Find the probability they are a match (same color).

3.46. Of 120 students, 60 are studying French, 50 are studying Spanish, and 20 are studying both French and Spanish. A student is chosen at random. Find the probability that the student is studying:

(a) French and Spanish   (d) only French

(b) French or Spanish   (e) exactly one of the two languages.

(c) neither French nor Spanish

3.47. Of 10 girls in a class, 3 have blue eyes. Two of the girls are chosen at random. Find the probability that:

(a) both have blue eyes   (c) at least one has blue eyes

(b) neither has blue eyes   (d) exactly one has blue eyes.

3.48. Ten students A, B, ... are in a class. A committee of 3 is chosen from the class. Find the probability that

(a) A belongs to the committee.   (c) A and B belong to the committee.

(b) B belongs to the committee.   (d) A or B belongs to the committee.

FINITE PROBABILITY SPACES

3.49. Under which of the following functions does $S = \{a_1, a_2, a_3\}$ become a probability space?

(a) $P(a_1) = 0.3, P(a_2) = 0.4, P(a_3) = 0.5$ (c) $P(a_1) = 0.3, P(a_2) = 0.2, P(a_3) = 0.5$

(b) $P(a_1) = 0.7, P(a_2) = -0.2, P(a_3) = 0.5$ (d) $P(a_1) = 0.3, P(a_2) = 0, P(a_3) = 0.7$

3.50. A coin is weighted so that heads is three times as likely to appear as tails. Find $P(H)$ and $P(T)$.

3.51. Suppose A and B are events with $P(A) = 0.7$, $P(B) = 0.5$, and $P(A \cap B) = 0.4$. Find the probability that

(a) A does not occur.   (c) A but not B occurs.

(b) A or B occurs.   (d) Neither A nor B occurs.

3.52. Consider the following probability distribution:

Outcome	1	2	3	4	5	6
Probability	0.1	0.3	0.1	0.2	0.2	0.1

Consider the following events:

$$A = \{\text{even number}\}, \qquad B = \{2, 3, 4, 5\}, \qquad C = \{1, 2\}$$

Find: (a) $P(A)$, (b) $P(B)$, (c) $P(C)$, (d) $P(\varnothing)$, (e) $P(S)$.

3.53. For the events A, B, C in Problem 3.52, find:

(a) $P(A \cap B)$,   (b) $P(A \cup C)$,   (c) $P(B \cap C)$,   (d) $P(A^c)$,   (e) $P(B \cap C^c)$.

3.54. Three students A, B, and C are in a swimming race. A and B have the same probability of winning and each is twice as likely to win as C. Find the probability that

(a) B wins   (b) C wins   (c) B or C wins

3.55. Let P be a probability function on $S = \{a_1, a_2, a_3\}$. Find $P(a_1)$ if

(a) $P(a_2) = 0.3, P(a_3) = 0.5$;   (c) $P(\{a_2, a_3\}) = 2P(a_1)$;

(b) $P(a_1) = 2P(a_2)$ and $P(a_3) = 0.7$;   (d) $P(a_3) = 2P(a_2)$ and $P(a_2) = 3P(a_1)$.

ODDS

3.56. Find the probability of an event E if the odds that it will occur are: (a) 2 to 1, (b) 5 to 11.

3.57. Find the odds that an event E occurs if: (a) $P(E) = 2/7$, (b) $P(E) = 0.4$.

3.58. In a swimming race, the odds that A will win are 2 to 3 and the odds that B will win are 1 to 4. Find the probability p and the odds that: (a) A will lose, (b) A or B will win, (c) neither A nor B will win.

NONCOUNTABLE UNIFORM SPACES

3.59. A point is chosen at random inside a circle with radius r. Find the probability p that the point is at most $\frac{1}{3}r$ from the center.

3.60. A point A is selected at random inside an equilateral triangle whose side length is 3. Find the probability p that the distance of A from any corner is greater than 1.

3.61. A coin of diameter 1/2 is tossed randomly onto the plane $\mathbf{R}^2$. Find the probability p that the coin does not intersect any line of the form: (a) $x = k$ or $y = k$ where k is an integer, (b) $x + y = k$ where k is an integer.

3.62. A point X is selected at random from a line segment AB with midpoint O. Find the probability p that the line segments AX, XB, and AO can form a triangle.

MISCELLANEOUS PROBLEMS

3.63. A die is tossed 50 times. The following table gives the 6 numbers and their frequency of occurrence:

Number	1	2	3	4	5	6
Frequency	7	9	8	7	9	10

Find the relative frequency of each event: (a) 4 appears, (b) an odd number appears, (c) a number greater than 4 appears.

3.64. Use mathematical induction to prove: For any events $A_1, A_2, \ldots, A_n$,

$$P(A_1 \cup \cdots \cup A_n) = \sum_i P(A_i) - \sum_{i<j} P(A_i \cap A_j) + \sum_{i<j<k} P(A_i \cap A_j \cap A_k) - \cdots \pm P(A_1 \cap \cdots \cap A_n)$$

Remark: This result generalizes Theorem 3.6 (addition rule) for two sets and Corollary 3.7 for three sets.

3.65. Consider the countably infinite sample space $S = \{a_1, a_2, a_3, \ldots\}$. Suppose $P(a_1) = 1/4$ and suppose $P(a_{k+1}) = rP(a_k)$ for $k = 1, 2, \ldots$. Find r and $P(a_3)$.

Answers to Supplementary Problems

3.36. (a) $A \cup B^c$; (b) $A \cap B^c$.

3.37. (a) $(A \cup C) \cap B$; (b) $(A \cap B^c \cap C^c) \cup (A^c \cap B \cap C^c) \cup (A^c \cap B^c \cap C)$; (c) $(A \cup B \cup B)^c = A^c \cap B^c \cap C^c$; (d) $(A \cap B) \cup (A \cap C) \cup (B \cap C)$.

3.38. $n(S) = 24$; $S = \{H, T\} \times \{H, T\} \times \{1, 2, , \ldots, 6\} = \{HH1, \ldots, HH6, HT1, \ldots, TT6\}$.

3.39. $A = \{HH2, HH4, HH6\}$; $B = \{HH2, HT2, TH2, TT2\}$; $C = \{HT1, HT3, HT5, TH1, TH3, TH5\}$.

3.40. (a) $\{HH2\}$; (b) $\{HT2, TH2, TT2\}$; (c) $\varnothing$; (d) $\{HH4, HH6\}$.

3.41. (a) 3/6; (b) 15/16; (c) 4/36; (d) 10/36; (e) 32/52.

3.42. (a) 4/20; (b) 3/20; (c) 11/20.

3.43. (a) 12/25; (b) 8/25; (c) 4/25; (d) 16/25; (e) 3/25.

3.44. $9/15 = 3/5$.

3.45. $3/15 = 1/5$.

3.46. (*a*) 1/6; (*b*) 3/4; (*c*) 1/4; (*d*) 1/3; (*e*) 7/12.

3.47. (*a*) 1/15; (*b*) 7/15; (*c*) 8/15; (*d*) 7/15.

3.48. (*a*) 3/10; (*b*) 3/10; (*c*) 1/15; (*d*) 8/15.

3.49. (*c*) and (*d*).

3.50. $P(H) = 3/4; P(T) = 1/4$.

3.51. (*a*) 0.3; (*b*) 0.8; (*c*) 0.2; (*d*) 0.2.

3.52. (*a*) 0.6; (*b*) 0.8; (*c*) 0.4; (*d*) 0; (*e*) 1.

3.53. (*a*) 0.5; (*b*) 0.7; (*c*) 0.3; (*d*) 0.4; (*e*) 0.5.

3.54. (*a*) 2/5; (*b*) 1/5; (*c*) 3/5.

3.55. (*a*) 0.2; (*b*) 0.2; (*c*) 1/3; (*d*) 0.1.

3.56. (*a*) 2/3; (*b*) 5/16.

3.57. (*a*) 2 to 5; (*b*) 2 to 3.

3.58. (*a*) $p = 3/5$, odds 3 to 2; (*b*) $p = 3/5$, odds 3 to 2; (*c*) $p = 2/5$, odds 2 to 3.

3.59. 1/9.

3.60. $1 - 2\pi/(9\sqrt{3}) = 1 - 2\sqrt{3}\,\pi/27$.

3.61. (*a*) 1/4; (*b*) $1 - \sqrt{2}/2$.

3.62. 1/2.

3.63. (*a*) 7/50; (*b*) 24/50; (*c*) 19/50.

3.65. $r = 3/4; P(a_3) = 9/64$.

CHAPTER 4

Conditional Probability and Independence

4.1 INTRODUCTION

The notions of conditional probability and independence will be motivated by two well-known examples.

(a) **Gender Gap:** Suppose candidate A receives 54 percent of the entire vote, but only 48 percent of the female vote. Let $P(A)$ denote the probability that a random person voted for A, but let $P(A|W)$ denote the probability that a random woman voted for A. Then

$$P(A) = 0.54 \qquad \text{but} \qquad P(A|W) = 0.48$$

$P(A|W)$ is called the *condition probability* of A given W. Note that $P(A|W)$ only looks at the reduced sample space consisting of women. The fact that $P(A) \neq P(A|W)$ is called the *gender gap* in politics. On the other hand, suppose $P(A) = P(A|W)$. We then say there is no gender gap, that is, the probability that a person voted for A is "independent" of the gender of the voter.

(b) **Insurance Rates:** Auto insurance rates usually depend on the probability that a random person will be involved in an accident. It is well known that male drivers under 25 years old get into more accidents than the general public. That is, letting $P(A)$ denote the probability of an accident and letting E denote male drivers under 25 years old, the data tell us that

$$P(A) < P(A|E)$$

Again we use the notation $P(A|E)$ to denote the probability of an accident given that the driver is male and under 25 years old.

This chapter formally defines conditional probability and independence. We also cover finite stochastic processes, Bayes' theorem, and independent repeated trials.

4.2 CONDITIONAL PROBABILITY

Suppose E is an event in a sample space S with $P(E) > 0$. The probability that an event A occurs once E has occurred or, specifically, the *conditional probability of A given E*, written $P(A \mid E)$, is defined as follows:

$$P(A \mid E) = \frac{P(A \cap E)}{P(E)}$$

As pictured in the Venn diagram in Fig. 4-1, $P(A \mid E)$ measures, in a certain sense, the relative probability of A with respect to the reduced space E.

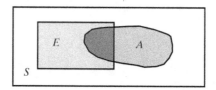

Fig. 4-1

Now suppose S is an equiprobable space, and we let $n(A)$ denote the number of elements in the event A. Then

$$P(A \cap E) = \frac{n(A \cap E)}{n(S)}, \quad P(E) = \frac{n(E)}{n(S)}, \quad \text{and so} \quad P(A \mid E) = \frac{P(A \cap E)}{P(E)} = \frac{n(A \cap E)}{n(E)}$$

We state this result formally.

Theorem 4.1: Suppose S is an equiprobable space and A and B are events. Then

$$P(A \mid E) = \frac{\text{number of elements in } A \cap E}{\text{number of elements in } E} = \frac{n(A \cap E)}{n(E)}$$

EXAMPLE 4.1 A pair of fair dice is tossed. The sample space S consists of the 36 ordered pairs (a, b) where a and b can be any of the integers from 1 to 6. (See Fig. 3-3.) Thus, the probability of any point is 1/36. Find the probability that one of the dice is 2 if the sum is 6. That is, find $P(A \mid E)$ where

$$E = \{\text{sum is 6}\} \quad \text{and} \quad A = \{2 \text{ appears on at least one die}\}$$

Also find $P(A)$.

Now E consists of five elements, specifically

$$E = \{(1, 5), (2, 4), (3, 3), (4, 2), (5, 1)\}$$

Two of them, $(2, 4)$ and $(4, 2)$, belong to A, that is, $A \cap E = \{(2, 4), (4, 2)\}$. By Theorem 4.1, $P(A \mid E) = 2/5$.

On the other hand, A consists of 11 elements, specifically:

$$A = \{(2, 1), (2, 2), (2, 3), (2, 4), (2, 5), (2, 6), (1, 2), (3, 2), (4, 2), (5, 2), (6, 2)\}$$

and S consists of 36 elements; hence $P(A) = 11/36$.

EXAMPLE 4.2 Suppose a couple has two children. The sample space is $S = \{dd, bg, gb, gg\}$ where we assume an equiprobable space, that is, we assume probability 1/4 for each point. Find the probability p that both children are boys if it is known that: (*a*) At least one of the children is a boy. (*b*) The older child is a boy.

(*a*) Here the reduced sample space consists of three elements $\{bb, bg, gb\}$; hence $p = 1/3$.

(*b*) Here the reduced sample space consists of two elements $\{bb, bg\}$; hence $p = 1/2$.

Multiplication Theorem for Conditional Probability

Suppose A and B are events in a sample space S with $P(A) > 0$. By definition of conditional probability and using $A \cap B = B \cap A$, we obtain:

$$P(B|A) = \frac{P(A \cap B)}{P(A)}$$

Multiplying both sides by $P(A)$ gives us the following useful formula:

Theorem 4.2 (Multiplication Theorem for Conditional Probability):

$$P(A \cap B) = P(A)P(B|A)$$

The multiplication theorem gives us a formula for the probability that events A and B both occur. It can be extended to three or more events. For three events, we get:

Corollary 4.3: $P(A \cap B \cap C) = P(A)P(B|A)P(C|A \cap B)$

That is, the probability that A, B, and C occur is equal to the product of the following:

(i)   The probability that A occurs.
(ii) The probability that B occurs, assuming that A occurred.
(iii) The probability that C occurs, assuming that A and B have occurred.

We apply this result in the following example.

EXAMPLE 4.3 A lot contains 12 items of which 4 are defective. Three items are drawn at random from the lot one after the other. Find the probability p that all 3 are nondefective.

We compute the following 3 probabilities:

(i)   The probability that the first item is nondefective is $\frac{8}{12}$ since 8 of 12 items are nondefective.
(ii) Assuming that the first item is nondefective, the probability that the second item is nondefective is $\frac{7}{11}$ since only 7 of the remaining 11 items are nondefective.
(iii) Assuming that the first and second items are nondefective, the probability that the third item is nondefective is $\frac{6}{10}$ since only 6 of the remaining 10 items are now nondefective.

Accordingly, by the multiplication theorem,

$$p = \frac{8}{12} \cdot \frac{7}{11} \cdot \frac{6}{10} = \frac{14}{55}$$

4.3 FINITE STOCHASTIC PROCESSES AND TREE DIAGRAMS

A *(finite) stochastic process* is a finite sequence of experiments where each experiment has a finite number of outcomes with given probabilities. A convenient way of describing such a process is by means of a *labeled tree diagram*, as illustrated below. The multiplication theorem (Theorem 4.2) can then be used to compute the probability of an event which is represented by a given path of the tree.

EXAMPLE 4.4 Suppose the following three boxes are given:

Box X has 10 lightbulbs of which 4 are defective.
Box Y has 6 lightbulbs of which 1 is defective.
Box Z has 8 lightbulbs of which 3 are defective.

A box is chosen at random, and then a bulb is randomly selected from the chosen box.

(a)   Find the probability p that the bulb is nondefective.

(b)   If the bulb is nondefective, find the probability that it came from box Z.

Here we perform a sequence of two experiments:

(i)   Select one of the three boxes.

(ii) Select a bulb which is either defective (D) or nondefective (N).

The tree diagram in Fig. 4-2 describes this process and gives the probability of each edge of the tree. The multiplication theorem tells us that the probability of a given path of the tree is the product of the probabilities of each edge of the path. For example, the probability of selecting box X and then a nondefective bulb N from box X is as follows:

$$\frac{1}{3} \cdot \frac{3}{5} = \frac{1}{5}$$

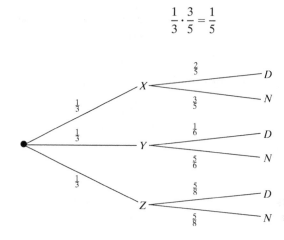

Fig. 4-2

(a)   Since there are three disjoint paths which lead to a nondefective bulb N, the sum of the probabilities of these paths gives us the required probability. Namely

$$p = P(N) = \frac{1}{3} \cdot \frac{3}{5} + \frac{1}{3} \cdot \frac{5}{6} + \frac{1}{3} \cdot \frac{5}{8} = \frac{247}{360} \approx 0.686$$

(b)   Here we want to compute $P(Z|N)$, the conditional probability of box Z given a nondefective bulb N.

Now box Z and a nondefective bulb N, that is, the event $Z \cap N$, can only occur on the bottom path. Therefore

$$P(Z \cap N) = \frac{1}{3} \cdot \frac{5}{8} = \frac{5}{24}$$

By part (a), we have $P(N) = 247/360$. Accordingly, by the definition of conditional probability,

$$P(Z|N) = \frac{P(Z \cap N)}{P(N)} = \frac{5/24}{247/360} = \frac{75}{247} = 0.304$$

In other words, we divide the probability of the successful path by the probability of the reduced sample space consisting of all the paths leading to N.

EXAMPLE 4.5   Suppose a coin, weighted so that $P(H) = 2/3$ and $P(T) = 1/3$, is tossed. If heads appears, then a number is selected at random from the numbers 1 through 9; if tails appears, then a number is selected at random from the numbers 1 through 5. Find the probability that an even number appears.

Note that the probability of selecting an even number from the numbers 1 through 9 is $\frac{4}{9}$ since there are 4 even numbers out of the 9 numbers, whereas the probability of selecting an even number from the numbers 1 through 5 is $\frac{2}{5}$ since there are 2 even numbers out of the 5 numbers. Thus, Fig. 4-3 is the tree diagram with the respective

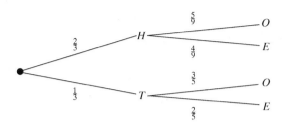

Fig. 4-3

probabilities which represents the above stochastic process. (Here E means an even number is selected and 0 means an odd number is selected.) There are two paths in the tree which lead to an even number, HE and TE. Thus

$$p = P(E) = \frac{2}{3} \cdot \frac{4}{9} + \frac{1}{3} \cdot \frac{2}{5} = \frac{58}{135} \approx 0.43$$

4.4 PARTITIONS, TOTAL PROBABILITY, AND BAYES' FORMULA

Suppose a set S is the union of mutually disjoint subsets $A_1, A_2, \ldots, A_n$, that is, suppose the sets $A_1, A_2, \ldots, A_n$ form a partition of the set S. Furthermore, suppose E is any subset of S. Then, as illustrated in Fig. 4-4 for the case $n = 3$,

$$E = E \cap S = E \cap (A_1 \cup A_2 \cup \ldots \cup A_n) = (E \cap A_1) \cup (E \cap A_2) \cup \ldots \cup (E \cap A_n)$$

Moreover, the n subsets on the right in the above equation, are also mutually disjoint, that is, form a partition of E.

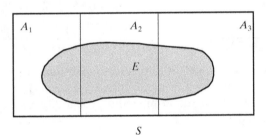

Fig. 4-4

Law of Total Probability

Now suppose S is a sample space and the above subsets $A_1, A_2, \ldots, A_n$, E are events. Since the $E \cap A_k$ are disjoint, we obtain

$$P(E) = P(E \cap A_1) + P(E \cap A_2) + \cdots + P(E \cap A_n)$$

Using the multiplication theorem for conditional probability, we also obtain

$$P(E \cap A_k) = P(A_k \cap E) = P(A_k)P(E|A_k)$$

Thus, we arrive at the following theorem:

Theorem 4.4 (Total Probability):   Let E be an event in a sample space S, and let $A_1, A_2, \ldots, A_n$, be mutually disjoint events whose union is S. Then

$$P(E) = P(A_1)P(E|A_1) + P(A_2)P(E|A_2) + \cdots + P(A_n)P(E|A_n)$$

The equation in Theorem 4.4 is called the *law of total probability*. We emphasize that the sets A_1, $A_2, \ldots, A_n$ are pairwise disjoint and their union is all of S, that is, that the A's form a *partition* of S.

EXAMPLE 4.6   A factory uses three machines X, Y, Z to produce certain items. Suppose:

    (1)   Machine X produces 50 percent of the items of which 3 percent are defective.

    (2)   Machine Y produces 30 percent of the items of which 4 percent are defective.

    (3)   Machine Z produces 20 percent of the items of which 5 percent are defective.

Find the probability p that a randomly selected item is defective.

Let D denote the event that an item is defective. Then, by the law of total probability,

$$P(D) = P(X)P(D|X) + P(Y)P(D|Y) + P(Z)P(D|Z)$$
$$= (0.50)(0.03) + (0.30)(0.04) + (0.20)(0.05) = 0.037 = 3.7\%$$

Bayes' Theorem

Suppose the events $A_1, A_2, \ldots, A_n$ do form a partition of the sample space S, and E is any event. Then, for $k = 1, 2, \ldots, n$, the multiplication theorem for conditional probability tells us that $P(A_k \cap E) = P(A_k)P(E|A_k)$. Therefore

$$P(A_k|E) = \frac{P(A_k \cap E)}{P(E)} = \frac{P(A_k)P(E|A_k)}{P(E)}$$

Using the law of total probability (Theorem 4.4) for the denominator $P(E)$, we arrive at the next theorem.

Theorem 4.5 (Bayes' Formula):   Let E be an event in a sample space S, and let $A_1, A_2, \ldots, A_n$ be disjoint events whose union is S. Then, for $k = 1, 2, \ldots, n$,

$$P(A_k|E) = \frac{P(A_k)P(E|A_k)}{P(A_1)P(E|A_1) + P(A_2)P(E|A_2) + \cdots + P(A_n)P(E|A_n)}$$

The above equation is called *Bayes' rule* or *Bayes' formula*, after the English mathematician Thomas Bayes (1702–1761). If we think of the events $A_1, A_2, \ldots, A_n$ as possible causes of the event E, then Bayes' formula enables us to determine the probability that a particular one of the A's occurred, given that E occurred.

EXAMPLE 4.7   Consider the factory in Example 4.6. Suppose a defective item is found among the output. Find the probability that it came from each of the machines, that is, find $P(X|D)$, $P(Y|D)$, and $P(Z|D)$.

Recall that $P(D) = P(X)P(D|X) + P(Y)P(D|Y) + P(Z)P(D|Z) = 0.037$. Therefore, by Bayes' formula,

$$P(X|D) = \frac{P(X)P(D|X)}{P(D)} = \frac{(0.50)(0.03)}{0.037} = \frac{15}{37} = 40.5\%$$

$$P(Y|D) = \frac{P(Y)P(D|Y)}{P(D)} = \frac{(0.30)(0.04)}{0.037} = \frac{12}{37} = 32.5\%$$

$$P(Z|D) = \frac{P(Z)P(D|Z)}{P(D)} = \frac{(0.20)(0.05)}{0.037} = \frac{10}{37} = 27.0\%$$

Stochastic Interpretation of Total Probability and Bayes' Formula

Frequently, problems involving the total probability law and Bayes' formula can be interpreted as a two-step stochastic process. Figure 4-5 gives the stochastic tree corresponding to Fig. 4-4 where the first step in the tree involves the events A_1, A_2, A_3 which partition S and the second step involves the arbitrary event E.

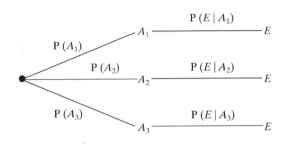

Fig. 4-5

Suppose we want $P(E)$.   Using the tree diagram, we obtain

$$P(E) = P(A_1)P(E|A_1) + P(A_2)P(E|A_2) + P(A_3)P(E|A_3)$$

Furthermore, for $k = 1, 2, 3,$

$$P(A_k|E) = \frac{P(A_k \cap E)}{P(E)} = \frac{P(A_k)P(E|A_k)}{P(E)}$$

$$= \frac{P(A_k)P(E|A_k)}{P(A_1)P(E|A_1) + P(A_2)P(E|A_2) + P(A_3)P(E|A_3)}$$

Observe that the above two formulas are simply the total probability law and Bayes' formula, for the case $n = 3$.   The stochastic approach also applies to any positive integer n.

EXAMPLE 4.8   Suppose a student dormitory in a college consists of:

(1)   30 percent freshmen of whom 10 percent own a car
(2)   40 percent sophomores of whom 20 percent own a car
(3)   20 percent juniors of whom 40 percent own a car
(4)   10 percent seniors of whom 60 percent own a car

(a)   Find the probability that a student in the dormitory owns a car.
(b)   If a student does own a car, find the probability that the student is a junior.

Let A, B, C, D denote, respectively, the set of freshmen, sophomores, juniors, and seniors, and let E denote the set of students owning a car.   Figure 4-6 is a stochastic tree describing the given data.

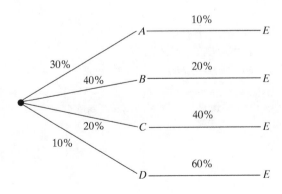

Fig. 4-6

(*a*)   We seek $P(E)$.   By the law of total probability or by using Fig. 4-6, we obtain

$$P(E) = (0.30)(0.10) + (0.40)(0.20) + (0.20)(0.40) + (0.10)(0.60)$$
$$= 0.03 + 0.08 + 0.08 + 0.06 = 0.25 = 25\%$$

(*b*)   We seek $P(C|E)$.   By Bayes' formula,

$$P(C|E) = \frac{P(C)P(E|C)}{P(E)} = \frac{(0.20)(0.40)}{0.25} = \frac{8}{25} = 32\%$$

4.5   INDEPENDENT EVENTS

Events A and B in a probability space S are said to be *independent* if the occurrence of one of them does not influence the occurrence of the other.   More specifically, B is independent of A if $P(B)$ is the same as $P(B|A)$.   Now suppose we substitute $P(B)$ for $P(B|A)$ in the multiplication theorem that $P(A \cap B) = P(A)P(B|A)$.   This yields

$$\boxed{P(A \cap B) = P(A)P(B)}$$

We formally use the above equation as our definition of independence.

Definition: Events A and B are *independent* if $P(A \cap B) = P(A)P(B)$; otherwise they are *dependent*.

We emphasize that independence is a symmetric relationship.   In particular

$$\boxed{P(A \cap B) = P(A)P(B) \text{ implies both } P(B|A) = P(B) \text{ and } P(A|B) = P(A)}$$

Note also that disjoint (mutually exclusive) events are not independent unless one of them has zero probability.   That is, suppose $A \cap B = \varnothing$ and A and B are independent.   Then

$$P(A)P(B) = P(A \cap B) = 0 \qquad \text{and so} \qquad P(A) = 0 \qquad \text{or} \qquad P(B) = 0$$

EXAMPLE 4.9   A fair coin is tossed three times yielding the equiprobable space

$$S = \{HHH, HHT, HTH, HTT, THH, THT, TTH, TTT\}$$

Consider the events:

   $A = \{$first toss is heads$\} = \{HHH, HHT, HTH, HTT\}$
   $B = \{$second toss is heads$\} = \{HHH, HHT, THH, THT\}$
   $C = \{$exactly two heads in a row$\} = \{HHT, THH\}$

Clearly A and B are independent events; this fact is verified below.   On the other hand, the relationship between A and C and between B and C is not obvious.   We claim that A and C are independent, but that B and C are dependent.   Note that

$$P(A) = \frac{4}{8} = \frac{1}{2}, \qquad P(B) = \frac{4}{8} = \frac{1}{2}, \qquad P(C) = \frac{2}{8} = \frac{1}{4}$$

Also

$$P(A \cap B) = P(\{HHH, HHT\}) = \frac{1}{4}, \qquad P(A \cap C) = P(\{HHT\}) = \frac{1}{8}, \qquad P(B \cap C) = P(\{HHT, THH\}) = \frac{1}{4}$$

Accordingly

$$P(A)P(B) = \frac{1}{2} \cdot \frac{1}{2} = \frac{1}{4} = P(A \cap B); \text{ hence } A \text{ and } B \text{ are independent.}$$

$$P(A)P(C) = \frac{1}{2} \cdot \frac{1}{4} = \frac{1}{8} = P(A \cap C); \text{ hence } A \text{ and } C \text{ are independent.}$$

$$P(B)P(C) = \frac{1}{2} \cdot \frac{1}{4} = \frac{1}{8} \neq P(B \cap C); \text{ hence } B \text{ and } C \text{ are not independent.}$$

Frequently, we will postulate that two events are independent, or the experiment itself will imply that two events are independent.

EXAMPLE 4.10   The probability that A hits a target is $\frac{1}{4}$, and the probability that B hits the target is $\frac{2}{5}$.   Both shoot at the target.   Find the probability that at least one of them hits the target, that is, find the probability that A or B (or both) hits the target.

$$\text{Here } P(A) = \frac{1}{2} \quad \text{and} \quad P(B) = \frac{2}{5}, \quad \text{and we seek} \quad P(A \cup B)$$

Furthermore, we assume that A and B are independent events, that is, that the probability that A or B hits the target is not influenced by what the other does.   Therefore

$$P(A \cap B) = P(A)P(B) = \frac{1}{4} \cdot \frac{2}{5} = \frac{1}{10}$$

Accordingly, by the addition rule in Theorem 3.6,

$$P(A \cup B) = P(A) + P(B) - P(A \cap B) = \frac{1}{4} + \frac{2}{5} - \frac{1}{10} = \frac{11}{20}$$

Independence of Three or More Events

Three events A, B, C are *independent* if the following two conditions hold:

(1)   They are pairwise independent, that is,

$$P(A \cap B) = P(A)P(B), \quad P(A \cap C) = P(A)P(C), \quad P(B \cap C) = P(B)P(C)$$

(2) $P(A \cap B \cap C) = P(A)P(B)P(C)$

Example 4.11 below shows that condition (2) does not follow from condition (1), that is, three events may be pairwise independent but not independent themselves.   [Problem 4.32 shows that condition (1) does not follow from condition (2).]

Independence of more than three events is defined analogously.   Namely, the events $A_1, A_2, \ldots, A_n$ are *independent* if any proper subset of them is independent and

$$P(A_1 \cap A_2 \cap \ldots \cap A_n) = P(A_1)P(A_2) \ldots P(A_n)$$

Observe that induction is used in this definition.

EXAMPLE 4.11   A pair of fair coins is tossed yielding the equiprobable space $S = \{HH, HT, TH, TT\}$.   Consider the events:

$$A = \{\text{head on first toss}\} = \{HH, HT\}, \quad B = \{\text{head on second toss}\} = \{HH, TH\},$$
$$C = \{\text{head on exactly one coin}\} = \{HT, TH\}$$

Then $P(A) = P(B) = P(C) = \frac{2}{4} = \frac{1}{2}$.   Also

$$P(A \cap B) = P(\{HH\}) = \frac{1}{4}, \quad P(A \cap C) = P(\{HT\}) = \frac{1}{4}, \quad P(B \cap C) = P(\{TH\}) = \frac{1}{4}$$

Thus, condition (1) is satisfied, that is, the events are pairwise independent. On the other hand,

$$A \cap B \cap C = \emptyset \quad \text{and so} \quad P(A \cap B \cap C) = P(\emptyset) = 0 \neq P(A)P(B)P(C)$$

Thus, condition (2) is not satisfied, and so the three events are not independent.

4.6 INDEPENDENT REPEATED TRIALS

Previously, we discussed probability spaces which were associated with an experiment repeated a finite number of times, such as the tossing of a coin three times. This concept of repetition is formalized as follows:

Definition: Let S be a finite probability space. The probability space of n *independent* or *repeated trials*, denoted by S_n, consists of ordered n-tuples of elements of S with the probability of an n-tuple defined to be the product of the probability of its components, that is,

$$P((s_1, s_2, \ldots, s_n)) = P(s_1)P(s_2) \cdots P(s_n)$$

EXAMPLE 4.12 Suppose that whenever three horses a, b, c race together, their respective probabilities of winning are 1/2, 1/3, and 1/6. In other words,

$$S = \{a, b, c\} \text{ with } P(a) = \frac{1}{2}, \quad P(b) = \frac{1}{3}, \quad \text{and} \quad P(c) = \frac{1}{6}$$

Suppose the horses race twice. Then the sample space S_2 of the two repeated trials follows:

$$S_2 = \{aa, ab, ac, ba, bb, bc, ca, cb, cc\}$$

For notational convenience, we have written ac for the ordered pair (a, c). The probability of each point of S_2 follows:

$$P(aa) = P(a)P(a) = \frac{1}{2} \cdot \frac{1}{2} = \frac{1}{4}, \qquad P(ba) = \frac{1}{6}, \qquad P(ca) = \frac{1}{12}$$

$$P(ab) = P(a)P(b) = \frac{1}{2} \cdot \frac{1}{3} = \frac{1}{6}, \qquad P(bb) = \frac{1}{9}, \qquad P(cb) = \frac{1}{18}$$

$$P(ac) = P(a)P(c) = \frac{1}{2} \cdot \frac{1}{6} = \frac{1}{12}, \qquad P(bc) = \frac{1}{18}, \qquad P(cc) = \frac{1}{36}$$

Thus, the probability that c wins the first race and a wins the second race is $P(ca) = \frac{1}{12}$.

Repeated Trials as a Stochastic Process

From another point of view, the probability space of a repeated-trials process may be viewed as a stochastic process whose tree diagram has the following properties:

(i)   Each branch point has the same outcomes.

(ii) All branches leading to the same outcome have the same probability.

For example, the tree diagram for the repeated-trials process in Example 4.12 appears in Fig. 4-7. Observe that

(i)   Each branch point has outcomes a, b, c.

(ii) All branches leading to outcome a have probability $\frac{1}{2}$, to outcome b have probability $\frac{1}{3}$, and to outcome c have probability $\frac{1}{6}$.

These two properties are expected as noted above.

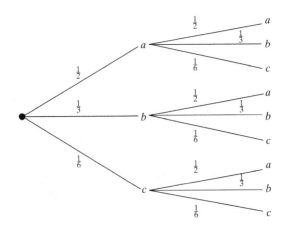

Fig. 4-7

Solved Problems

CONDITIONAL PROBABILITY

4.1. Three fair coins, a penny, a nickel, and a dime, are tossed.   Find the probability p that they are all heads if: (a) the penny is heads, (b) at least one of the coins is heads, (c) the dime is tails.

The sample space has eight elements:

$$S = \{HHH, HHT, HTH, HTT, THH, THT, TTH, TTT\}$$

(a)   If the penny is heads, the reduced sample space is $A = \{HHH, HHT, HTH, HTT\}$.   All coins are heads in only 1 of the 4 cases; hence $p = 1/4$.

(b)   If one or more of the coins is heads, the reduced sample space is

$$B = \{HHH, HHT, HTH, HTT, THH, THT, TTH\}$$

All coins are heads in only 1 of the 7 cases; hence $p = 1/7$.

(c)   If the dime (third coin) is tails, the reduced sample space is $C = \{HHT, HTT, THT, TTT\}$.   None contains all heads; hence $p = 0$.

4.2. A billiard ball is drawn at random from a box containing 15 billiard balls numbered 1 to 15, and the number n is recorded.

(a) Find the probability p that n exceeds 10.

(b) If n is even, find the probability p that n exceeds 10.

(a)   The n can be one of the 5 numbers, 11, 12, 13, 14, 15.   Hence $p = 5/15 = 1/3$.

(b)   The reduced sample space E consists of the 7 even numbers, that is, $E = \{2, 4, 6, 8, 10, 12, 14\}$.   Of these, only 2, 12, and 14, exceed 10.   Hence $p = 2/7$.

4.3. A pair of fair dice is thrown.   Find the probability p that the sum is 10 or greater if:

(a) 5 appears on the first die, (b) 5 appears on at least one die.

Figure 3-3 shows the 36 ways the pair of dice can be thrown.

(a) If 5 appears on the first die, the reduced sample space A has six elements:

$$A = \{(5,1), (5,2), (5,3), (5,4), (5,5), (5,6)\}$$

The sum is 10 or more on 2 of the 6 outcomes, $(5,5)$ and $(5,6)$. Thus, $p = 2/6 = 1/3$.

(b) If 5 appears on at least one die, the reduced sample space B has 11 elements:

$$B = \{(5,1), (5,2), (5,3), (5,4), (5,5), (5,6), (1,5), (2,5), (3,5), (4,5), (6,5)\}$$

The sum is 10 or more on 3 of the 11 outcomes, $(5,5)$, $(5,6)$, and $(6,5)$. Thus, $p = 3/11$.

4.4. In a certain college, 25 percent of the students failed mathematics, 15 percent failed chemistry, and 10 percent failed both mathematics and chemistry. A student is selected at random.

(a) If the student failed chemistry, what is the probability that he or she failed mathematics?
(b) If the student failed mathematics, what is the probability that he or she failed chemistry?
(c) What is the probability that the student failed mathematics or chemistry?
(d) What is the probability that the student failed neither mathematics nor chemistry?

(a) We seek $P(M|C)$, the probability that the student failed mathematics, given that he or she failed chemistry. By definition,

$$P(M|C) = \frac{P(M \cap C)}{P(C)} = \frac{0.10}{0.15} = \frac{10}{15} = \frac{2}{3}$$

(b) We seek $P(C|M)$, the probability that the student failed chemistry, given that he or she failed mathematics. By definition,

$$P(C|M) = \frac{P(M \cap C)}{P(M)} = \frac{0.10}{0.25} = \frac{10}{25} = \frac{2}{5}$$

(c) By the addition rule (Theorem 3.6),

$$P(M \cup C) = P(M) + P(C) - P(M \cap C) = 0.25 + 0.15 - 0.10 = 0.30$$

(d) Students who failed neither mathematics nor chemistry form the complement of the set $M \cup C$, that is, form the set $(M \cup C)^c$. Hence

$$P((M \cup C)^c) = 1 - P(M \cup C) = 1 - 0.30 = 0.70$$

4.5. A pair of fair dice is thrown. If the two numbers appearing are different, find the probability p that: (a) the sum is 6, (b) an ace appears, (c) the sum is 4 or less.

There are 36 ways the pair of dice can be thrown (Fig. 3-3) and 6 of them, $(1,1), (2,2), \ldots, (6,6)$, have the same numbers. Thus, the reduced sample space E will consist of $36 - 6 = 30$ elements.

(a) The sum 6 can appear in 4 ways: $(1,5), (2,4), (4,2), (5,1)$. [We cannot include $(3,3)$ since the numbers must be different.] Thus, $p = 4/30 = 2/15$.
(b) An ace can appear in 10 ways: $(1,2), (1,3), \ldots, (1,6)$ and $(2,1), (3,1), \ldots, (6,1)$. [We cannot include $(1,1)$ since the numbers must be different.] Thus, $p = 10/30 = 1/3$.
(c) The sum is 4 or less in 4 ways: $(3,1), (1,3), (2,1), (1,2)$. [We cannot include $(1,1)$ and $(2,2)$ since the numbers must be different.] Thus, $p = 4/30 = 2/15$.

4.6. Let A and B be events with $P(A) = 0.6$, $P(B) = 0.3$, and $P(A \cap B) = 0.2$. Find:

(a) $P(A|B)$ and $P(B|A)$, (b) $P(A \cup B)$, (c) $P(A^c)$ and $P(B^c)$.

(a) By definition of conditional probability,

$$P(A|B) = \frac{P(A \cap B)}{P(B)} = \frac{0.2}{0.3} = \frac{2}{3}, \quad P(B|A) = \frac{P(A \cap B)}{P(A)} = \frac{0.2}{0.6} = \frac{1}{3}$$

(*b*) By the addition rule (Theorem 3.6),

$$P(A \cup B) = P(A) + P(B) - P(A \cap B) = 0.6 + 0.3 - 0.2 = 0.7$$

(*c*) By the complement rule,

$$P(A^c) = 1 - P(A) = 1 - 0.6 = 0.4 \quad \text{and} \quad P(B^c) = 1 - P(B) = 1 - 0.3 = 0.7$$

4.7. Consider the data in Problem 4.6. Find: (*a*) $P(A^c|B^c)$, (*b*) $P(B^c|A^c)$.

First compute $P(A^c \cap B^c)$. By DeMorgan's law, $(A \cup B)^c = A^c \cap B^c$. Hence, by the complement rule,

$$P(A^c \cap B^c) = P((A \cup B)^c) = 1 - P(A \cup B) = 1 - 0.7 = 0.3$$

(*a*) $P(A^c|B^c) = \dfrac{P(A^c \cap B^c)}{P(B^c)} = \dfrac{0.3}{0.7} = \dfrac{3}{7}$

(*b*) $P(B^c|A^c) = \dfrac{P(A^c \cap B^c)}{P(A^c)} = \dfrac{0.3}{0.4} = \dfrac{3}{4}$

4.8. Let A and B be events with $P(A) = \frac{3}{8}$, $P(B) = \frac{5}{8}$, and $P(A \cup B) = \frac{3}{4}$. Find $P(A|B)$ and $P(B|A)$.

First find $P(A \cap B)$ using the addition rule that $P(A \cup B) = P(A) + P(B) - P(A \cap B)$. We have

$$\frac{3}{4} = \frac{3}{8} + \frac{5}{8} - P(A \cap B) \quad \text{or} \quad P(A \cap B) = \frac{1}{4}$$

Now use the definition of conditional probability to get

$$P(A|B) = \frac{P(A \cap B)}{P(B)} = \frac{1/4}{5/8} = \frac{2}{5} \quad \text{and} \quad P(B|A) = \frac{P(A \cap B)}{P(A)} = \frac{1/4}{3/8} = \frac{2}{3}$$

4.9. Find $P(B|A)$ if: (*a*) A is a subset of B, (*b*) A and B are mutually exclusive (disjoint). [Assume $P(A) > 0$.]

(*a*) If A is a subset of B [as pictured in Fig. 4-8(*a*)], then whenever A occurs, B must occur; hence $P(B|A) = 1$. Alternately, if A is a subset of B, then $A \cap B = A$; hence

$$P(B|A) = \frac{P(A \cap B)}{P(A)} = \frac{P(A)}{P(A)} = 1$$

(*b*) If A and B are mutually exclusive, that is, disjoint [as pictured in Fig. 4-8(*b*)], then whenever A occurs, B cannot occur; hence $P(B|A) = 0$. Alternately, if A and B are disjoint, then $A \cap B = \varnothing$; hence

$$P(B|A) = \frac{P(A \cap B)}{P(A)} = \frac{P(\varnothing)}{P(A)} = \frac{0}{P(A)} = 0$$

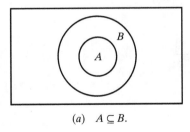

(*a*) $A \subseteq B$.

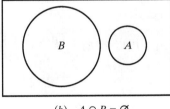

(*b*) $A \cap B = \varnothing$.

Fig. 4-8

4.10. Let E be an event for which $P(E) > 0$. Show that the conditional probability function $P(* \mid E)$ satisfies the axioms of a probability space, that is

[$\mathbf{P_1}$] For any event A, we have $P(A \mid E) \geq 0$.

[$\mathbf{P_2}$] For any certain event S, we have $P(S \mid E) = 1$.

[$\mathbf{P_3}$] For any two disjoint events A and B, we have

$$P(A \cup B \mid E) = P(A \mid E) + P(B \mid E)$$

[$\mathbf{P_3'}$] For any infinite sequence of mutually disjoint events $A_1, A_2, \ldots$, we have

$$P(A_1 \cup A_2 \cup \cdots \mid E) = P(A_1 \mid E) + P(A_2 \mid E) + \cdots$$

(a) We have $P(A \cap E) \geq 0$ and $P(E) > 0$; hence

$$P(A \mid E) = \frac{P(A \cap E)}{P(E)} \geq 0$$

Thus, [$\mathbf{P_1}$] holds.

(b) We have $S \cap E = E$; hence

$$P(S \mid E) = \frac{P(S \cap E)}{P(E)} = \frac{P(E)}{P(E)} = 1$$

Thus, [$\mathbf{P_2}$] holds.

(c) If A and B are disjoint events, then so are $A \cap E$ and $B \cap E$. Furthermore,

$$(A \cup B) \cap E = (A \cap E) \cup (B \cap E)$$

Hence,

$$P[(A \cup B) \cap E] = P[(A \cap E) \cup (B \cap E)] = P(A \cap E) + P(B \cap E)$$

Therefore

$$P(A \cup B \mid E) = \frac{P[(A \cup B) \cap E]}{P(E)} = \frac{P(A \cap E) + P(B \cap E)}{P(E)}$$

$$= \frac{P(A \cap E)}{P(E)} + \frac{P(B \cap E)}{P(E)} = P(A \mid E) + P(B \mid E).$$

Thus, [$\mathbf{P_3}$] holds.

(d) [Similar to (c).] If $A_1, A_2, \ldots$ are mutually disjoint events, then so are $A_1 \cap E, A_2 \cap E, \ldots$. Also, by the generalized distributive law,

$$(A_1 \cup A_2 \cup \cdots) \cap E = (A_1 \cap E) \cup (A_2 \cap E) \cup \cdots$$

Thus

$$P[(A_1 \cup A_2 \cup \cdots) \cap E] = P[(A_1 \cap E) \cup (A_2 \cap E) \cup \cdots]$$
$$= P(A_1 \cap E) + P(A_2 \cap E) + \cdots$$

Therefore

$$P(A_1 \cup P_2 \cup \cdots \mid E) = \frac{P[(A_1 \cup A_2 \cup \cdots) \cap E]}{P(E)}$$

$$= \frac{P(A_1 \cap E) + P(A_2 \cap E) + \cdots}{P(E)} = \frac{P(A_1 \cap E)}{P(E)} + \frac{P(A_2 \cap E)}{P(E)} + \cdots$$

$$= P(A_1 \mid E) + P(A_2 \mid E) + \cdots$$

Thus, [$\mathbf{P_3'}$] holds.

MULTIPLICATION THEOREM

4.11. A class has 12 men and 4 women. Suppose 3 students are selected at random from the class. Find the probability p that they are all men.

The probability that the first student is a man is 12/16 since there are 12 men out of the 16 students. If the first student is a man, then the probability that the second student is a man is 11/15 since there are 11 men left out of the 15 students left. Finally, if the first 2 students are men, then the probability that the third student is a man is 10/14 since there are now only 10 men out of the 14 students left. Accordingly, by the Multiplication Theorem 4.2, the probability that all 3 are men is

$$p = \frac{12}{16} \cdot \frac{11}{15} \cdot \frac{10}{14} = \frac{11}{28}$$

Another Method: There are $C(16,3) = 560$ ways to select 3 students out of 16 students, and $C(12,3) = 220$ ways to select 3 men from the 12 men. Thus

$$p = \frac{220}{560} = \frac{11}{28}$$

A Third Method: Suppose the students are selected one after the other. Then there are $16 \cdot 15 \cdot 14$ ways to select 3 students, and there are $12 \cdot 11 \cdot 10$ ways to select the 3 men. Thus

$$p = \frac{16 \cdot 15 \cdot 14}{12 \cdot 11 \cdot 10} = \frac{11}{28}$$

4.12. A person is dealt 5 cards from an ordinary 52-card deck (Fig. 3-4). Find the probability p that they are all spades.

The probability that the first card is a spade is 13/52, that the second is a spade is 12/51, that the third is a spade is 11/50, and that the fourth is a spade is 10/48. (We assume in each case that the previous cards were spades.) Thus, by the Multiplication Theorem 4.2,

$$p = \frac{13}{52} \cdot \frac{12}{51} \cdot \frac{11}{50} \cdot \frac{10}{48} = \frac{33}{66,640} = 0.000\,49$$

Another Method: There are $C(52,5)$ ways to select 5 cards from the 52-card deck, and $C(13,5)$ ways to select 5 spades from the 13 spades. Thus

$$p = \frac{C(13,5)}{C(52,5)} \approx 0.000\,49$$

4.13. A box contains 7 red marbles and 3 white marbles. Three marbles are drawn from the box one after the other. Find the probability p that the first 2 are red and the third is white.

The probability that the first marble is red is 7/10 since there are 7 red marbles out of the 10 marbles. If the first marble is red, then the probability that the second marble is red is 6/9 since there are 6 red marbles out of the remaining 9 marbles. Finally, if the first 2 marbles are red, then the probability that the third marble is white is 3/8 since there are 3 white marbles out of the remaining 8 marbles in the box. Accordingly, by the Multiplication Theorem 4.2,

$$p = \frac{7}{10} \cdot \frac{6}{9} \cdot \frac{3}{8} = \frac{7}{40} = 0.175 = 17.5\%$$

4.14. Students in a class are selected at random, one after the other, for an examination. Find the probability p that the men and women in the class alternate if:

(*a*) the class consists of 4 men and 3 women, (*b*) the class consists of 3 men and 3 women.

(a) If the men and women are to alternate, then the first student must be a man. The probability that the first is a man is 4/7. If the first is a man, the probability that the second is a woman is 3/6 since there are 3 women out of the 6 students left. Continuing in this manner, we obtain that the probability that the third is a man is 3/5, the fourth is a woman is 2/4, that the fifth is a man is 2/3, that the sixth is a woman is 1/2, and that the last is a man is 1/1. Thus

$$p = \frac{4}{7} \cdot \frac{3}{6} \cdot \frac{3}{5} \cdot \frac{2}{4} \cdot \frac{2}{3} \cdot \frac{1}{2} \cdot \frac{1}{1} = \frac{1}{35}$$

(b) There are two mutually exclusive cases: the first student is a man and the first is a woman. If the first student is a man, then, by the multiplication theorem, the probability p_1 that the students alternate is

$$p_1 = \frac{3}{6} \cdot \frac{3}{5} \cdot \frac{2}{4} \cdot \frac{2}{3} \cdot \frac{1}{2} \cdot \frac{1}{1} = \frac{1}{20}$$

If the first student is a woman, then, by the multiplication theorem, the probability p_2 that the students alternate is

$$p_2 = \frac{3}{6} \cdot \frac{3}{5} \cdot \frac{2}{4} \cdot \frac{2}{3} \cdot \frac{1}{2} \cdot \frac{1}{1} = \frac{1}{20}$$

Thus, $p = p_1 + p_2 = \frac{1}{20} + \frac{1}{20} = \frac{1}{10}$.

FINITE STOCHASTIC PROCESSES

4.15. Let X, Y, Z be three coins in a box. Suppose X is a fair coin, Y is two-headed, and Z is weighted so that the probability of heads is 1/3. A coin is selected at random and is tossed. (a) Find the probability that heads appears, that is, find $P(H)$. (b) If heads appears, find the probability that it is the fair coin X, that is, find $P(X|H)$. (c) If tails appears, find the probability it is the coin Z, that is, find $P(Z|T)$.

Construct the corresponding two-step stochastic tree diagram in Fig. 4-9(a).

(a) Heads appears along three of the paths; hence

$$P(H) = \frac{1}{3} \cdot \frac{1}{2} + \frac{1}{3} \cdot 1 + \frac{1}{3} \cdot \frac{1}{3} = \frac{11}{18}$$

(b) Note X and heads H appear only along the top path in Fig. 4-9(a); hence

$$P(X \cap H) = (1/3)(1/2) = 1/6 \quad \text{and so} \quad P(X|H) = \frac{P(X \cap H)}{P(H)} = \frac{1/6}{11/18} = \frac{3}{11}$$

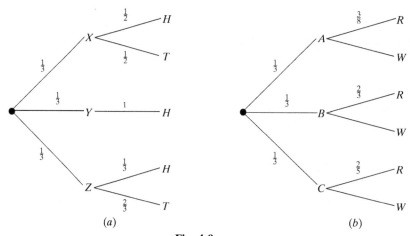

(a)   (b)

Fig. 4-9

(c) $P(T) = 1 - P(H) = 1 - 11/18 = 7/18$.   Alternately, tails appears along two of the paths and so

$$P(T) = \frac{1}{3} \cdot \frac{1}{2} + \frac{1}{3} \cdot \frac{2}{3} = \frac{7}{18}$$

Note Z and tails T appear only along the bottom path in Fig. 4-9(a); hence

$$P(Z \cap T) = (1/3)(2/3) = 2/9 \quad \text{and so} \quad P(Z \mid T) = \frac{P(Z \cap T)}{P(T)} = \frac{2/9}{7/18} = \frac{4}{7}$$

4.16.   Suppose the following three boxes are given:

Box A contains 3 red and 5 white marbles.

Box B contains 2 red and 1 white marbles.

Box C contains 2 red and 3 white marbles.

A box is selected at random, and a marble is randomly drawn from the box.   If the marble is red, find the probability that it came from box A.

Construct the corresponding stochastic tree diagram as in Fig. 4-9(b).   We seek $P(A \mid R)$, the probability that A was selected, given that the marble is red.   Thus, it is necessary to find $P(A \cap R)$ and $P(R)$.   Note that A and R only occur on the top path; hence $P(A \cap R) = (1/3)(3/8) = 1/8$.   There are three paths leading to a red marble R; hence

$$P(R) = \frac{1}{3} \cdot \frac{3}{8} + \frac{1}{3} \cdot \frac{2}{3} + \frac{1}{3} \cdot \frac{2}{5} = \frac{173}{360} \approx 0.48$$

Thus

$$P(A \mid R) = \frac{P(A \cap R)}{P(R)} = \frac{1/8}{173/360} = \frac{45}{173} \approx 0.26$$

4.17.   Box A contains 9 cards numbered 1 through 9, and box B contains 5 cards numbered 1 through 5.   A box is selected at random, and a card is randomly drawn from the box.   If the number is even, find the probability that the card came from box A.

Construct the corresponding stochastic tree diagram as in Fig. 4-10(a).   We seek $P(A \mid E)$, the probability that A was selected, given that the number is even.   Thus, it is necessary to find $P(A \cap E)$ and $P(E)$.   Note that A and E only occur on the top path; hence $P(A \cap E) = (1/2)(4/9) = 2/9$.   Two paths lead to an even number E; hence

$$P(E) = \frac{1}{2} \cdot \frac{4}{9} + \frac{1}{2} \cdot \frac{2}{5} = \frac{19}{45} \quad \text{and so} \quad P(A \mid E) = \frac{P(A \cap E)}{P(E)} = \frac{2/9}{19/45} = \frac{10}{19} \approx 0.53$$

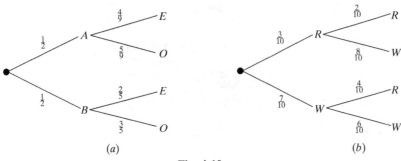

(a)   (b)

Fig. 4-10

4.18. A box contains 3 red marbles and 7 white marbles. A marble is drawn from the box and the marble is replaced by a marble of the other color. A second marble is drawn from the box.

(a) Find the probability p that the second marble is red.

(b) If both marbles were of the same color, find the probability p that they both were white.

Construct the corresponding stochastic tree diagram as in Fig. 4-10(b).

(a) Two paths lead to a red marble R; hence

$$p = \frac{3}{10} \cdot \frac{2}{10} + \frac{7}{10} \cdot \frac{4}{10} = \frac{17}{50} = 0.34$$

(b) Note that W appears twice only on the bottom path; hence $P(WW) = (7/10)(6/10) = 21/50$ is the probability that both were white. There are two paths, the top path and the bottom path, where the marbles are the same color. Thus

$$P(RR \text{ or } WW) = \frac{3}{10} \cdot \frac{2}{10} + \frac{7}{10} \cdot \frac{6}{10} = \frac{12}{25}$$

is the probability of the same color, the reduced sample space. Therefore

$$p = \frac{21/50}{12/25} = \frac{7}{8} = 0.875$$

4.19. A box contains a fair coin A and a two-headed coin B. A coin is selected at random and tossed twice.

(a) If heads appears both times, find the probability p that the coin is two-headed.

(b) If tails appears both times, find the probability p that the coin is two-headed.

Construct the corresponding stochastic tree diagram as in Fig. 4-11.

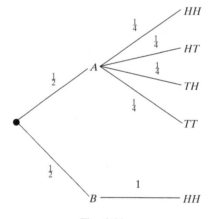

Fig. 4-11

(a) We seek $P(B|HH)$. Heads appears twice only in the top path and in the bottom path. Hence

$$P(HH) = \frac{1}{2} \cdot \frac{1}{4} + \frac{1}{2} \cdot 1 = \frac{5}{8}$$

On the other hand, $P(B \cap HH) = P(B) = \frac{1}{2}$. Thus

$$p = P(B|HH) = \frac{P(B \cap HH)}{P(B)} = \frac{1/2}{5/8} = \frac{4}{5}$$

(b) If tails appears then it could not be the two-headed coin B. Hence $p = 0$.

4.20. Suppose the following two boxes are given:

> Box A contains 3 red and 2 white marbles.
>
> Box B contains 2 red and 5 white marbles.

A box is selected at random; a marble is drawn and put into the other box; then a marble is drawn from the second box. Find the probability p that both marbles drawn are of the same color.

Construct the corresponding stochastic tree diagram as in Fig. 4-12. Note that this is a three-step stochastic process: (1) choosing a box, (2) choosing a marble, (3) choosing a second marble. Note that if box A is selected and a red marble R is drawn and put into box B, then box B will have 3 red marbles and 5 white marbles.

There are 4 paths which lead to 2 marbles of the same color; hence

$$p = \frac{1}{2} \cdot \frac{3}{5} \cdot \frac{3}{8} + \frac{1}{2} \cdot \frac{2}{5} \cdot \frac{3}{4} + \frac{1}{2} \cdot \frac{2}{7} \cdot \frac{2}{3} + \frac{1}{2} \cdot \frac{5}{7} \cdot \frac{1}{2} = \frac{901}{1680} \approx 0.536$$

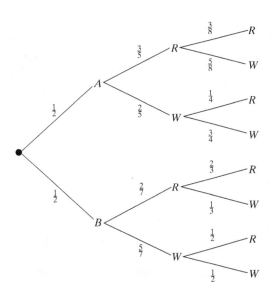

Fig. 4-12

LAW OF TOTAL PROBABILITY, BAYES' RULE

4.21. In a certain city, 40 percent of the people consider themselves Conservatives (C), 35 percent consider themselves to be Liberals (L), and 25 percent consider themselves to be Independents (I). During a particular election, 45 percent of the Conservatives voted, 40 percent of the Liberals voted and 60% of the Independents voted. Suppose a person is randomly selected.

(a) Find the probability that the person voted.
(b) If the person voted, find the probability that the voter is

> (i) Conservative, (ii) Liberal, (iii) Independent.

(a) Let V denote the event that a person voted. We need $P(V)$. By the law of total probability,

$$P(V) = P(C)P(V|C) + P(L)P(V|L) + P(I)P(V|I)$$
$$= (0.40)(0.45) + (0.35)(0.40) + (0.25)(0.60) = 0.47$$

(b) Use Bayes' rule:

(i) $P(C|V) = \dfrac{P(C)P(V|C)}{P(V)} = \dfrac{(0.40)(0.45)}{0.47} = \dfrac{18}{47} \approx 38.3\%$

(ii) $P(L|V) = \dfrac{P(L)P(V|L)}{P(V)} = \dfrac{(0.35)(0.40)}{0.47} = \dfrac{14}{47} \approx 29.8\%$

(iii) $P(I|V) = \dfrac{P(I)P(V|I)}{P(V)} = \dfrac{(0.25)(0.60)}{0.47} = \dfrac{15}{47} \approx 31.9\%$

4.22. In a certain college, 4 percent of the men and 1 percent of the women are taller than 6 feet. Furthermore, 60 percent of the students are women. Suppose a randomly selected student is taller than 6 feet. Find the probability that the student is a woman.

Let A = {students taller than 6 feet}. We seek $P(W|A)$, the probability that a student is a woman, given that the student is taller than 6 feet. By Bayes' formula,

$$P(W|A) = \frac{P(W)P(A|W)}{P(W)P(A|W) + P(M)P(A|M)} = \frac{(0.60)(0.01)}{(0.60)(0.01) + (0.40)(0.04)} = \frac{3}{11}$$

4.23. Three machines A, B, and C produce, respectively, 40 percent, 10 percent, and 50 percent of the items in a factory. The percentage of defective items produced by the machines is, respectively, 2 percent, 3 percent, and 4 percent. An item from the factory is selected at random.

(a) Find the probability that the item is defective.
(b) If the item is defective, find the probability that the item was produced by:
    (i) machine A,
    (ii) machine B,
    (iii) machine C.

(a) Let D denote the event that an item is defective. Then, by the law of total probability,

$$P(D) = P(A)P(D|A) + P(B)P(D|B) + P(C)P(D|C)$$
$$= (0.40)(0.02) + (0.10)(0.03) + (0.50)(0.04) = 0.031 = 3.1\%$$

(b) Use Bayes' formula to obtain

(i) $P(A|D) = \dfrac{P(A)P(D|A)}{P(D)} = \dfrac{(0.40)(0.02)}{0.031} = \dfrac{8}{31} \approx 25.8\%$

(ii) $P(B|D) = \dfrac{P(B)P(B|Y)}{P(D)} = \dfrac{(0.10)(0.03)}{0.031} = \dfrac{3}{31} \approx 9.7\%$

(iii) $P(C|D) = \dfrac{P(C)P(D|C)}{P(D)} = \dfrac{(0.50)(0.04)}{0.031} = \dfrac{20}{31} \approx 64.5\%$

4.24. Suppose a student dormitory in a college consists of:

    (1) 40 percent freshmen of whom 15 percent are New York residents

    (2) 25 percent sophomores of whom 40 percent are New York residents

    (3) 20 percent juniors of whom 25 percent are New York residents

    (4) 15 percent seniors of whom 20 percent are New York residents

A student is randomly selected from the dormitory.

(*a*) Find the probability that the student is a New York resident.

(*b*) If the student is a New York resident, find the probability that the student is a: (i) freshman, (ii) junior.

    Let A, B, C, D denote, respectively, the set of freshmen, sophomores, juniors, and seniors, and let E denote the set of students who are New York residents.

(*a*) We find $P(E)$ by the law of total probability. We have:

$$P(E) = (0.40)(0.15) + (0.25)(0.40) + (0.20)(0.25) + (0.15)(0.20)$$
$$= 0.06 + 0.10 + 0.05 + 0.03 = 0.24 = 24\%$$

(*b*) Use Bayes' formula to obtain:

    (i) $P(A|E) = \dfrac{P(A)P(E|A)}{P(E)} = \dfrac{(0.40)(0.15)}{0.24} = \dfrac{6}{24} = 25\%$

    (ii) $P(C|E) = \dfrac{P(C)P(E|C)}{P(E)} = \dfrac{(0.20)(0.25)}{0.24} = \dfrac{5}{25} = 20\%$

4.25. A box contains 10 coins where 5 coins are two-headed, 3 coins are two-tailed, and 2 are fair coins. A coin is chosen at random and tossed.

(*a*) Find the probability that a head appears.

(*b*) If a head appears, find the probability that the coin is fair.

    Let X, Y, Z denote, respectively, the two-headed coins, the two-tailed coins, and the fair coins. Then $P(X) = 0.5$, $P(Y) = 0.3$, $P(Z) = 0.2$. Note $P(H|X) = 1$, that is, a two-headed coin must yield a head. Similarly, $P(H|Y) = 0$ and $P(H|Z) = 0.5$. Figure 4-13 is a stochastic tree (with the root at the top) describing the given data.

(*a*) By the law of total probability or by adding the probabilities of the three paths in Fig. 4-13 leading to H, we get

$$P(H) = (0.5)(1) + (0.3)(0) + (0.2)(0.5) = 0.6$$

(*b*) By Bayes' rule,

$$P(Z|H) = \frac{P(Z)P(H|Z)}{P(H)} = \frac{(0.2)(0.5)}{0.6} = \frac{1}{6} = 16.7\%$$

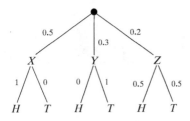

Fig. 4-13

INDEPENDENT EVENTS

4.26. Two men A and B fire at a target. Suppose $P(A) = \frac{1}{3}$ and $P(B) = \frac{1}{5}$ denote their probabilities of hitting the target. (We assume that the events A and B are independent.) Find the probability that:

(a) A does not hit the target.   (c) One of them hits the target.

(b) Both hit the target.   (d) Neither hits the target.

(a) By the complement rule,

$$P(\text{not } A) = P(A^c) = 1 - P(A) = 1 - \frac{1}{3} = \frac{2}{3}$$

(b) Since the events A and B are independent,

$$P(A \text{ and } B) = P(A \cap B) = P(A) \cdot P(B) = \frac{1}{3} \cdot \frac{1}{5} = \frac{1}{15}$$

(c) By the addition rule (Theorem 3.6),

$$P(A \text{ or } B) = P(A \cup B) = P(A) + P(B) - P(A \cap B) = \frac{1}{3} + \frac{1}{5} - \frac{1}{15} = \frac{7}{15}$$

(d) By DeMorgan's law, "neither A nor B" is the complement of $A \cup B$. [See Problem 3.1(b).] Hence

$$P(\text{neither } A \text{ nor } B) = P((A \cup B)^c) = 1 - P(A \cup B) = 1 - \frac{7}{15} = \frac{8}{15}$$

4.27. Box A contains 5 red marbles and 3 blue marbles and Box B contains 3 red and 2 blue. A marble is drawn at random from each box.

(a) Find the probability p that both marbles are red.

(b) Find the probability p that one is red and one is blue.

(a) The probability of choosing a red marble from A is $\frac{5}{8}$ and a red marble from B is $\frac{3}{5}$. Since the events are independent,

$$p = \frac{5}{8} \cdot \frac{3}{5} = \frac{3}{8}$$

(b) There are two (mutually exclusive) events:

   X: a red marble from A and a blue marble from B

   Y: a blue marble from A and a red marble from B

We have

$$P(X) = \frac{5}{8} \cdot \frac{2}{5} = \frac{1}{4} \quad \text{and} \quad P(Y) = \frac{3}{8} \cdot \frac{3}{5} = \frac{9}{40}$$

Accordingly, since X and Y are mutually exclusive,

$$p = P(X) + P(Y) = \frac{1}{4} + \frac{9}{40} = \frac{19}{40}$$

4.28. Let A be the event that a man will live 10 more years, and let B be the event that his wife lives 10 more years. Suppose $P(A) = \frac{1}{4}$ and $P(B) = \frac{1}{3}$. Assuming A and B are independent events, find the probability that, in 10 years

    (a) Both will be alive.   (c) Neither will be alive.

    (b) At least one will be alive.   (d) Only the wife will be alive.

    (a) We seek $P(A \cap B)$. Since A and B are independent events,

$$P(A \cap B) = P(A) \cdot P(B) = \frac{1}{4} \cdot \frac{1}{3} = \frac{1}{12}$$

    (b) We seek $P(A \cup B)$. By the addition rule (Theorem 3.6),

$$P(A \cup B) = P(A) + P(B) - P(A \cap B) = \frac{1}{4} + \frac{1}{3} - \frac{1}{12} = \frac{1}{2}$$

    (c) By DeMorgan's law, "neither A nor B" is the complement of $A \cup B$. [Problem 3.1(b).] Hence

$$P(A^c \cap B^c) = P((A \cup B)^c) = 1 - P(A \cup B) = 1 - \frac{1}{2} = \frac{1}{2}$$

Alternately, we have $P(A^c) = \frac{3}{4}$ and $P(B^c) = \frac{2}{3}$; and, since A^c and B^c are independent,

$$P(A^c \cap B^c) = \frac{3}{4} \cdot \frac{2}{3} = \frac{1}{2}$$

    (d) We seek $P(A^c \cap B)$. Since A^c and B are also independent,

$$P(A^c \cap B) = \frac{3}{4} \cdot \frac{1}{3} = \frac{1}{4}$$

4.29. Consider the following events for a family with children:

$$A = \{\text{children of both sexes}\}, \qquad B = \{\text{at most one boy}\}$$

    (a) Show that A and B are independent events if a family has 3 children.

    (b) Show that A and B are dependent events if a family has only 2 children.

    (a) We have the equiprobable space $S = \{bbb, bbg, bgb, bgg, gbb, gbg, ggb, ggg\}$. Here

$$A = \{bbg, bgb, bgg, gbb, gbg, ggb\} \qquad \text{and so} \qquad P(A) = \frac{6}{8} = \frac{3}{4}$$

$$B = \{bgg, gbg, ggb, ggg\} \qquad \text{and so} \qquad P(B) = \frac{4}{8} = \frac{1}{2}$$

$$A \cap B = \{bgg, gbg, ggb\} \qquad \text{and so} \qquad P(A \cap B) = \frac{3}{8}$$

Since $P(A)P(B) = \frac{3}{4} \cdot \frac{1}{2} = \frac{3}{8} = P(A \cap B)$, A and B are independent.

    (b) We have the equiprobable space $S = \{bb, bg, gb, gg\}$. Here

$$A = \{bg, gb\} \qquad \text{and so} \qquad P(A) = \frac{1}{2}$$

$$B = \{bg, gb, gg\} \qquad \text{and so} \qquad P(B) = \frac{3}{4}$$

$$A \cap B = \{bg, gb\} \qquad \text{and so} \qquad P(A \cap B) = \frac{1}{2}$$

Since $P(A)P(B) \neq P(A \cap B)$, A and B are dependent.

4.30. Three men A, B, C fire at a target. Suppose $P(A) = 1/6$, $P(B) = 1/4$, $P(C) = 1/3$ denote their probabilities of hitting the target. (We assume that the events that A, B, C hit the target are independent.)

(a) Find the probability p that they all hit the target.

(b) Find the probability p that they all miss the target.

(c) Find the probability p that at least one of them hits the target.

(a) We seek $P(A \cap B \cap C)$. Since A, B, C are independent events,

$$P(A \cap B \cap C) = P(A) \cdot P(B) \cdot P(C) = \frac{1}{6} \cdot \frac{1}{4} \cdot \frac{1}{3} = \frac{1}{72} = 1.4\%$$

(b) We seek $P(A^c \cap B^c \cap C^c)$. We have $P(A^c) = 1 - P(A) = 5/6$. Similarly, $P(B^c = 3/4$ and $P(C^c) = 2/3$. Since A, B, C are independent events, so are A^c, B^c, C^c. Hence

$$P(A^c \cap B^c \cap C^c) = P(A^c) \cdot P(B^c) \cdot P(C^c) = \frac{5}{6} \cdot \frac{3}{4} \cdot \frac{2}{3} = \frac{5}{12} = 41.7\%$$

(c) Let D be the event that one or more of them hit the target. Then D is the complement of the event $A^c \cap B^c \cap C^c$, that they all miss the target. Thus

$$P(D) = P((A^c \cap B^c \cap C^c)^c) = 1 - \frac{5}{12} = \frac{7}{12} = 58.3\%$$

4.31. Consider the data in Problem 4.30. (a) Find the probability p that exactly one of them hits the target. (b) If the target is hit only once, find the probability p that it was the first man A.

(a) Let E be the event that exactly one of them hit the target. Then

$$E = (A \cap B^c \cap C^c) \cup (A^c \cap B \cap C^c) \cup (A^c \cap B^c \cap C)$$

That is, if only one man hit the target then it was only A, $A \cap B^c \cap C^c$, or only B, $A^c \cap B \cap C^c$, or only C, $A^c \cap B^c \cap C$. These three events are mutually exclusive. Thus, we obtain (using Problem 4.79)

$$p = P(E) = P(A \cap B^c \cap C^c) + P(A^c \cap B \cap C^c) + P(A^c \cap B^c \cap C)$$

$$= \frac{1}{6} \cdot \frac{3}{4} \cdot \frac{2}{3} + \frac{5}{6} \cdot \frac{1}{4} \cdot \frac{2}{3} + \frac{5}{6} \cdot \frac{3}{4} \cdot \frac{1}{3} = \frac{1}{12} + \frac{5}{36} + \frac{5}{24} = \frac{31}{72} = 43.1\%$$

(b) We seek $P(A|E)$, the probability that A hit the target given that only one man hit the target. Now $A \cap E = A \cap B^c \cap C^c$ is the event that only A hit the target. Also, by (a), $P(A \cap E) = 1/12$ and $P(E) = 31/72$; hence

$$P(A|E) = \frac{P(A \cap E)}{P(E)} = \frac{1/12}{31/72} = \frac{6}{31} = 19.4\%$$

4.32. Let $S = \{a, b, c, d\}$ be an equiprobable space; hence each elementary event has probability 1/4. Consider the events:

$$A = \{a, d\}, \quad B = \{b\,d\}, \quad C = \{c, d\}$$

(a) Show that A, B, C are pairwise independent.

(b) Show that A, B, C are not independent.

(a) Here $P(A) = P(B) = P(C) = 1/2$. Since $A \cap B = \{d\}$,

$$P(A \cap B) = P(\{d\}) = \frac{1}{4} = P(A)P(B)$$

Hence A and B are independent. Similarly, A and C are independent and B and C are independent.

(b) Here $A \cap B \cap C = \{d\}$, and so $P(A \cap B \cap C) = 1/4$. Therefore

$$P(A)P(B)P(C) = \frac{1}{8} \neq P(A \cap B \cap C)$$

Accordingly, A, B, C are not independent.

4.33. Suppose $S = \{1, 2, 3, 4, 5, 6, 7, 8\}$ is an equiprobable space; hence each elementary event has probability 1/8. Consider the events:

$$A = \{1, 2, 3, 4\}, \qquad B = \{2, 3, 4, 5\}, \qquad C = \{4, 6, 7, 8\}$$

(a) Show that $P(A \cap B \cap C) = P(A)P(B)P(C)$.

(b) Show that

   (i) $P(A \cap B) \neq P(A)P(B)$,
   (ii) $P(A \cap C) \neq P(A)P(C)$,
   (iii) $P(B \cap C) \neq P(B)P(C)$.

(a) Here $P(A) = P(B) = P(C) = 4/8 = 1/2$. Since $A \cap B \cap C = \{4\}$,

$$P(A \cap B \cap C) = \frac{1}{8} = P(A)P(B)P(C)$$

(b) (i) $A \cap B = \{3, 4, 5\}$, so $P(A \cap B) = 3/8$. But $P(A)P(B) = 1/4$; hence $P(A \cap B) \neq P(A)P(B)$.
   (ii) $A \cap C = \{4\}$, so $P(A \cap C) = 1/8$. But $P(A)P(C) = 1/4$; hence $P(A \cap C) \neq P(A)P(C)$.
   (iii) $B \cap C = \{4\}$, so $P(B \cap C) = 1/8$. But $P(B)P(C) = 1/4$; hence $P(B \cap C) \neq P(B)P(C)$.

4.34. Prove: Suppose A and B are independent events. Then A^c and B^c are independent events.

We need to show that $P(A^c \cap B^c) = P(A^c) \cdot P(B^c)$. Let $P(A) = x$ and $P(B) = y$. Then $P(A^c) = 1 - x$ and $P(B^c) = 1 - y$. Since A and B are independent, $P(A \cap B) = P(A) \cdot P(B) = xy$. Thus, by the addition rule (Theorem 3.6),

$$P(A \cup B) = P(A) + P(B) - P(A \cap B) = x + y - xy$$

By DeMorgan's law, $(A \cup B)^c = A^c \cap B^c$; hence

$$P(A^c \cap B^c) = P((A \cup B)^c) = 1 - P(A \cup B) = 1 - x - y + xy$$

On the other hand,

$$P(A^c) \cdot P(B^c) = (1 - x)(1 - y) = 1 - x - y + xy$$

Thus, $P(A^c \cap B^c) = P(A^c) \cdot P(B^c)$, and so A^c and B^c are independent.

Similarly, one can show that A and B^c, as well as A^c and B, are independent.

INDEPENDENT REPEATED TRIALS

4.35. A fair coin is tossed three times. Find the probability that there will appear:

(a) three heads, (b) exactly two heads, (c) exactly one head, (d) no heads.

Let H denote a head and T a tail on any toss. The three tosses can be modeled as an equiprobable space in which there are eight possible outcomes:

$$S = \{HHH, HHT, HTH, HTT, THH, THT, TTH, TTT\}$$

However, since the result on any one toss does not depend on the result of any other toss, the three tosses may be modeled as three independent trials in which $P(H) = \frac{1}{2}$ and $P(T) = \frac{1}{2}$ on any one trial. Then

(a) P (three heads) $= P(HHH) = \frac{1}{2} \cdot \frac{1}{2} \cdot \frac{1}{2} = \frac{1}{8}$

(b) P (exactly two heads) $= P(HHT$ or HTH or $THH) = \frac{1}{2} \cdot \frac{1}{2} \cdot \frac{1}{2} + \frac{1}{2} \cdot \frac{1}{2} \cdot \frac{1}{2} + \frac{1}{2} \cdot \frac{1}{2} \cdot \frac{1}{2} = \frac{3}{8}$

(c) As in (b), P (exactly one head) $= P$ (exactly two tails) $= \frac{3}{8}$

(d) As in (a), P (no heads) $= P(TTT) = \frac{1}{8}$

4.36. Suppose only horses a, b, c, d race together yielding the sample space $S = \{a, b, c, d\}$, and suppose the probabilities of winning are as follows:

$$P(a) = 0.2, \quad P(b) = 0.5, \quad P(c) = 0.1, \quad P(d) = 0.2$$

They race three times.

(a) Describe and find the number of elements in the product probability space S_3.

(b) Find the probability that the same horse wins all three races.

(c) Find the probability that a, b, c each wins one race.

(a) By definition, $S_3 = S \times S \times S = \{(x, y, z) : x, y, z \in S\}$ and

$$P((x, y, z) = P(x)P(y)P(z))$$

Thus, in particular, S_3 contains $4^3 = 64$ elements.

(b) Writing xyz for (x, y, z), we seek the probability of the event

$$A = \{aaa, bbb, ccc, ddd\}$$

By definition

$$P(aaa) = (0.2)^3 = 0.008, \qquad P(ccc) = (0.1)^3 = 0.001$$
$$P(bbb) = (0.5)^3 = 0.125, \qquad P(ddd) = (0.2)^3 = 0.008$$

Thus, $P(A) = 0.0008 + 0.125 + 0.001 + 0.008 = 0.142$.

(c) We seek the probability of the event

$$B = \{abc, acb, bac, bca, cab, cba\}$$

Every element in B has the same probability $(0.2)(0.5)(0.1) = 0.01$. Hence $P(B) = 6(0.01) = 0.06$.

4.37. A certain soccer team wins (W) with probability 0.6, loses (L) with probability 0.3, and ties (T) with probability 0.1. The team plays three games over the weekend. (a) Determine the elements of the event A that the team wins at least twice and does not lose; and find $P(A)$. (b) Determine the elements of the event B that the team wins, loses, and ties in some order; and find $P(B)$.

(a) A consists of all ordered triples with at least two W's and no L's. Thus

$$A = \{WWW, WWT, WTW, TWW\}$$

Since these events are mutually exclusive,

$$P(A) = P(WWW) + P(WWT) + P(WTW) + P(TWW)$$
$$= (0.6)(0.6)(0.6) + (0.6)(0.6)(0.1) + (0.6)(0.1)(0.6) + (0.1)(0.6)(0.6)$$
$$= 0.216 + 0.36 + 0.36 + 0.36 = 0.324 = 32.4\%$$

(b) Here $B = \{WLT, WTL, LWT, LTW, TWL, TLW\}$. Each element in B has the probability $(0.6)(0.3)(0.1) = 0.018$. Hence

$$P(B) = 6(0.108) = 0.108 = 10.8\%$$

4.38. A certain type of missile hits its target with probability $p = 0.3$. Find the minimum number n of missiles that should be fired so that there is at least an 80 percent probability of hitting the target.

The probability of missing the target is $q = 1 - p = 0.7$. Hence the probability that n missiles miss the target is $(0.7)^n$. Thus, we seek the smallest n for which

$$1 - (0.7)^n > 0.80 \qquad \text{or equivalent} \qquad (0.7)^n < 0.20$$

Compute:

$$(0.7)^1 = 0.7, \qquad (0.7)^2 = 0.49, \qquad (0.7)^3 = 0.343, \qquad (0.7)^4 = 0.2401, \qquad (0.7)^5 = 0.16807$$

Thus, at least $n = 5$ missiles should be fired.

4.39. The probability that a man hits a target is 1/3. He fires at the target $n = 6$ times. (a) Describe and find the number of elements in the sample space S. (b) Let E be the event that he hits the target exactly $k = 2$ times. List the elements of E and find the number $n(E)$ of elements in E. (c) Find $P(E)$.

(a) S consists of all 6-element sequences consisting of S's (successes) and F's (failures); hence S contains $2^6 = 64$ elements.

(b) E consists of all sequences with two S's and four F's; hence E consists of the following elements:

$$SSFFFF, SFSFFF, SFFSFF, SFFFSF, SFFFFS, FSSFFF, FSFSFF, FSFFSF,$$
$$FSFFFS, FFSSFF, FFSFSF, FFSFFS, FFFSSF, FFFSFS, FFFFSS$$

Observe that the list contains 15 elements. [This is expected since we are distributing $k = 2$ letters S among the $n = 6$ positions in the sequence, and $C(6, 2) = 15$.] Thus $n(E) = 15$.

(c) Here $P(S) = 1/3$, so $P(F) = 1 - P(S) = 2/3$. Thus each of the above sequences occurs with the same probability

$$p = (1/3)^2 (2/3)^4 = \frac{16}{729}$$

Hence $P(E) = 15(16/729) = 80/243 \approx 33\%$.

4.40. Let S be a finite probability space and let T be the probability space of n independent trials in S. Show that T is well defined. That is, show

(i) The probability of each element of T is nonnegative.

(ii) The sum of their probabilities is 1.

Suppose $S = \{a_1, a_2, \ldots, a_r\}$. Then each element of T is of the form

$$a_{i_1} a_{i_2} \cdots a_{i_n} \quad \text{where} \quad i_1, i_2, \ldots, i_n \in \{1, 2, \ldots, r\}$$

Since each $P(a_i) \geq 0$, we have

$$P(a_{i_1} a_{i_2} \cdots a_{i_n}) = P(a_{i_1}) P(a_{i_2}) \cdots P(a_{i_n}) \geq 0$$

for every element of T. Hence (i) holds.

We prove (ii) by induction on n. It is obviously true for $n = 1$. Therefore, we consider $n > 1$ and assume (ii) has been proved for $n - 1$. We have

$$\sum_{i_1,\ldots,i_n=1}^{r} P(a_{i_1}a_{i_2}\cdots a_{i_n}) = \sum_{i_1,\ldots,i_n=1}^{r} P(a_{i_1})P(a_{i_2})\cdots P(a_{i_n}) = \sum_{i_1,\ldots,i_{n-1}=1}^{r} P(a_{i_1})P(a_{i_2})\cdots P(a_{i_{n-1}})\sum_{i_n=1}^{r} P(a_{i_n})$$

$$= \sum_{i_1,\ldots,i_{n-1}=1}^{r} P(a_{i_1})P(a_{i_2})\cdots P(a_{i_{n-1}}) = 1$$

where the last equality follows from the inductive hypothesis. Thus, (ii) also holds.

Supplementary Problems

CONDITIONAL PROBABILITY

4.41. A fair die is tossed. Consider events $A = \{1, 3, 5\}$, $B = \{2, 3, 5\}$, $C = \{1, 2, 3, 4\}$. Find:

(a) $P(A \cap B)$ and $P(A \cup C)$ (c) $P(A \mid C)$ and $P(C \mid A)$

(b) $P(A \mid B)$ and $P(B \mid A)$ (d) $P(B \mid C)$ and $P(C \mid B)$

4.42. A digit is selected at random from the digits 1 through 9. Consider the events $A = \{1, 3, 5, 7, 9\}$, $B = \{2, 3, 5, 7\}$, $C = \{6, 7, 8, 9\}$. Find:

(a) $P(A \cap B)$ and $P(A \cup C)$ (c) $P(A \mid C)$ and $P(C \mid A)$

(b) $P(A \mid B)$ and $P(B \mid A)$ (d) $P(B \mid C)$ and $P(C \mid B)$

4.43. A pair of fair dice is tossed. If the faces appearing are different, find the probability that:

(a) the sum is even, (b) the sum exceeds nine.

4.44. Let A and B be events with $P(A) = 0.6$, $P(B) = 0.3$, and $P(A \cap B) = 0.2$. Find:

(a) $P(A \cup B)$, (b) $P(A \mid B)$, (c) $P(B \mid A)$.

4.45. Referring to Problem 4.44, find: (a) $P(A \cap B^c)$, (b) $P(A \mid B^c)$.

4.46. Let A and B be events with $P(A) = \frac{1}{3}$, $P(B) = \frac{1}{4}$, and $P(A \cup B) = \frac{1}{2}$.

(a) Find $P(A \mid B)$ and $P(B \mid A)$. (b) Are A and B independent?

4.47. A woman is dealt 3 spades from an ordinary deck of 52 cards. (See Fig. 3-4.) If she is given two more cards, find the probability that both of the cards are also spades.

4.48. Two marbles are selected one after the other without replacement from a box containing 3 white marbles and 2 red marbles. Find the probability p that:

(a) The two marbles are white.   (c) The second is white if the first is white.

(b) The two marbles are red.   (d) The second is red if the first is red.

4.49. Two marbles are selected one after the other with replacement from a box containing 3 white marbles and 2 red marbles. Find the probability p that:

(a) The two marbles are white.   (c) The second is white if the first is white.

(b) The two marbles are red.   (d) The second is red if the first is red.

4.50. Two different digits are selected at random from the digits 1 through 5.

    (*a*)   If the sum is odd, what is the probability that 2 is one of the numbers selected?

    (*b*)   If 2 is one of the digits, what is the probability that the sum is odd?

4.51. Three cards are drawn in succession (without replacement) from a 52-card deck. Find the probability that:

    (*a*)   There are three aces.

    (*b*)   If the first is an ace, then the other two are aces.

    (*c*)   If the first two are aces, then the third is an ace.

4.52. A die is weighted to yield the following probability distribution:

Number	1	2	3	4	5	6
Probability	0.2	0.1	0.1	0.3	0.1	0.2

Let $A = \{1, 2, 3\}$, $B = \{2, 3, 5\}$, $C = \{2, 4, 6\}$. Find:

    (*a*) $P(A)$, $P(B)$, $P(C)$ (*d*) $P(A|C)$, $P(C|A)$

    (*b*) $P(A^c)$, $P(B^c)$, $P(C^c)$ (*e*) $P(B|C)$, $P(C|B)$

    (*c*) $P(A|B)$, $P(B|A)$

4.53. In a country club, 65 percent of the members play tennis, 40 percent play golf, and 20 percent play both tennis and golf. A member is chosen at random. Find the probability that the member:

    (*a*)   Plays tennis or golf.   (*c*)   Plays golf if he or she plays tennis.

    (*b*)   Plays neither tennis nor golf.   (*d*)   Plays tennis if he or she plays golf.

4.54. Suppose 60 percent of the freshmen class of a small college are women. Furthermore, suppose 25 percent of the men and 10 percent of the women in the class are studying mathematics. A freshman student is chosen at random. Find the probability that:

    (*a*)   The student is studying mathematics.

    (*b*)   If the student is studying mathematics, then the student is a woman.

4.55. Three students are selected at random one after another from a class with 10 boys and 5 girls. Find the probability that:

    (*a*)   The first two are boys and the third is a girl.

    (*b*)   The first and third are boys and the second is a girl.

    (*c*)   All three are of the same sex.

    (*d*)   Only the first and third are of the same sex.

FINITE STOCHASTIC PROCESSES

4.56. Two boxes are given as follows:

    Box *A* contains 5 red marbles, 3 white marbles, and 8 blue marbles.

    Box *B* contains 3 red marbles and 5 white marbles.

A box is selected at random and a marble is randomly chosen. Find the probability that the marble is:
(*a*) red, (*b*) white, (*c*) blue.

4.57. Refer to Problem 4.56. Find the probability that box A was selected if the marble is:
(a) red, (b) white, (c) blue.

4.58. Consider Box A and Box B in Problem 4.56. A fair die is tossed; if a 3 or 6 appears, a marble is randomly chosen from A, otherwise a marble is chosen from B. Find the probability that the marble is:
(a) red, (b) white, (c) blue.

4.59. Refer to Problem 4.58. Find the probability that box A was selected if the marble is:
(a) red, (b) white, (c) blue.

4.60. A box contains three coins, two of them fair and one two-headed. A coin is randomly selected and tossed twice. If heads appear both times, what is the probability that the coin is two-headed?

4.61. A box contains a fair coin and a two-headed coin. A coin is selected at random and tossed. If heads appears, then the other coin is tossed; if tails appears, then the same coin is tossed a second time. Find the probability that:

(a) Heads appears on the second toss.

(b) If heads appears on the second toss, then it also appeared on the first toss.

4.62. Two boxes are given as follows:

Box A contains x red marbles and y white marbles.

Box B contains z red marbles and t white marbles.

(a) A box is selected at random and a marble is drawn. Find the probability that the marble is red.

(b) A marble is selected from A and put into B, and then a marble is drawn from B. Find the probability that the marble is red.

TOTAL PROBABILITY AND BAYES' FORMULA

4.63. A city is partitioned into districts A, B, C having 20 percent, 40 percent, and 40 percent of the registered voters, respectively. The registered voters listed as Democrats are 50 percent in A, 25 percent in B, and 75 percent in C. A registered voter is chosen randomly in the city.

(a) Find the probability that the voter is a listed Democrat.

(b) If the registered voter is a listed Democrat, find the probability that the voter came from district B.

4.64. Refer to Problem 4.63. Suppose a district is chosen at random, and then a registered voter is randomly chosen from the district.

(a) Find the probability that the voter is a listed Democrat.

(b) If the voter is a listed Democrat, what is the probability that the voter came from district A?

4.65. Women in City College constitute 60 percent of the freshmen, 40 percent of the sophomores, 40 percent of the juniors, and 45 percent of the seniors. The school population is 30 percent freshmen, 25 percent sophomores, 25 percent juniors, and 20 percent seniors. A student from City College is chosen at random.

(a) Find the probability that the student is a woman.

(b) If a student is a woman, what is the probability that she is a sophomore?

4.66. Refer to Problem 4.65. Suppose one of the four classes is chosen at random, and then a student is randomly chosen from the class.

  (a) Find the probability that the student is a woman.

  (b) If the student is a woman, what is the probability that she is a sophomore?

4.67. A company produces lightbulbs at three factories A, B, C.

   Factory A produces 40 percent of the total number of bulbs, of which 2 percent are defective.

   Factory B produces 35 percent of the total number of bulbs, of which 4 percent are defective.

   Factory C produces 25 percent of the total number of bulbs, of which 3 percent are defective.

  A defective bulb is found among the total output. Find the probability that it came from

  (a) factory A, (b) factory B, (c) factory C.

4.68. Refer to Problem 4.67. Suppose a factory is chosen at random, and one of its bulbs is randomly selected. If the bulb is defective, find the probability that it came from (a) factory A, (b) factory B, (c) factory C.

INDEPENDENT EVENTS

4.69. Let A and B be independent events with $P(A) = 0.3$ and $P(B) = 0.4$. Find: (a) $P(A \cap B)$ and $P(A \cup B)$, (b) $P(A|B)$ and $P(B|A)$.

4.70. Box A contains 5 red marbles and 3 blue marbles and Box B contains 2 red and 3 blue. A marble is drawn at random from each box. Find the probability p that (a) Both marbles are red. (b) One is red and one is blue.

4.71. Box A contains 5 red marbles and 3 blue marbles and Box B contains 2 red and 3 blue. Two marbles are drawn at random from each box. Find the probability p that (a) They are all red. (b) They are all the same color.

4.72. Let A and B be independent events with $P(A) = 0.2$ and $P(B) = 0.3$. Find:

  (a) $P(A \cap B)$ and $P(A \cup B)$ (c) $P(A|B)$ and $P(B|A)$

  (b) $P(A \cap B^c)$ and $P(A \cup B^c)$ (d) $P(A|B^c)$ and $P(B^c|A)$

4.73. Let A and B be events with $P(A) = 0.3$, $P(A \cup B) = 0.5$, and $P(B) = p$. Find p if:

  (a) A and B are disjoint, (b) A and B are independent, (c) A is a subset of B.

4.74. The probability that A hits a target is 1/4 and the probability that B hits a target is 1/3. They each fire once at the target. Find the probability that

  (a) They both hit the target.

  (b) The target is hit exactly once.

  (c) If the target is hit only oonce, then A hit the target.

4.75. The probability that A hits a target is 1/4 and the probability that B hits a target is 1/3. They each fire twice. Find the probability that the target is hit: (a) at least once, (b) exactly once.

4.76. The probabilities that three men hit a target are, respectively, 0.3, 0.5, and 0.4. Each fires once at the target. (As usual, assume that the three events that each hits the target are independent.)

  (a) Find the probability that they all: (i) hit the target, (ii) miss the target.

  (b) Find the probability that the target is hit: (i) at least once, (ii) exactly once.

  (c) If only one hits the target, what is the probability that it was the first man?

4.77. Three fair coins are tossed.   Consider the events:

$$A = \{\text{all heads or all tails}\}, \qquad B = \{\text{at least two heads}\}, \qquad C = \{\text{at most two heads}\}$$

Of the pairs (A, B), (A, C), and (B, C), which are independent?

4.78. Suppose A and B are independent events.   Show that A and B^c are independent, and that A^c and B are independent.

4.79. Suppose A, B, C are independent events.   Show that:

(a) A^c, B, C are independent; (b) A^c, B^c, C are independent; (c) A^c, B^c, C^c are independent.

4.80. Suppose A, B, C are independent events.   Show that A and $B \cup C$ are independent.

INDEPENDENT REPEATED TRIALS

4.81. Whenever horses a, b, c race together, their respective probabilities of winning are $0.3, 0.5, 0.2$.   They race three times.

(a)   Find the probability that the same horse wins all three races.

(b)   Find the probability that a, b, c each wins one race.

4.82. A team wins (W) with probability 0.5, loses (L) with probability 0.3, and ties (T) with probability 0.2.   The team plays twice.   (a) Determine the sample space S and the probability of each elementary event.   (b) Find the probability that the team wins at least once.

4.83. A certain type of missile hits its target with probability $p = \frac{1}{3}$.   (a) If 3 missiles are fired, find the probability that the target is hit at least once.   (b) Find the minimum number n of missiles that should be fired so that there is at least a 90 percent probability of hitting the target.

4.84. In any game, the probability that the Hornets (H) will defeat the Rockets (R) is 0.6.   Find the probability that the Hornets will win a best-out-of-three series.

4.85. The batting average of a baseball player is .300.   He comes to bat 4 times.   Find the probability that he will get: (a) exactly two hits, (b) at least one hit.

4.86. Consider a countably infinite probability space $S = \{a_1, a_2, \ldots\}$.   Let

$$T = S^n = \{(s_1, s_2, \ldots, s_n) : s_i \in S\} \quad \text{and} \quad P(s_1, s_2, \ldots, s_n) = P(s_1)P(s_2) \cdots P(s_n)$$

Show that T is also a countably infinite probability space.   (This generalizes the definition of independent trials to a countably infinite space.)

Answers to Supplementary Problems

4.41   (a) 2/6, 5/6; (b) 2/3, 2/3; (c) 1/2, 2/3; (d) 1/2, 2/3.

4.42.   (a) 3/9, 7/9; (b) 3/4, 3/5; (c) 1/2, 2/5; (d) 1/4, 1/4.

4.43.   (a) 12/30; (b) 4/30.

4.44.   (a) 0.7; (b) 2/3; (c) 1/3.

4.45.   (*a*) 0.4; (*b*) 4/7.

4.46.   (*a*) 1/3; 1/4; (*b*) No.

4.47. $C(10, 2)/C(49, 2)$.

4.48.   (*a*) 3/10; (*b*) 1/10; (*c*) 1/2; (*d*) 1/4.

4.49.   (*a*) 9/25; (*b*) 4/25; (*c*) 3/5; (*d*) 2/5.

4.50.   (*a*) 1/3; (*b*) 3/4.

4.51.   (*a*) $1/(13 \cdot 17 \cdot 25) = 0.014\%$; (*b*) $1/1275 = 0.08\%$; (*c*) $1/50 = 2\%$.

4.52.   (*a*) 0.4, 0.3, 0.6; (*b*) 0.6, 0.7, 0.4; (*c*) 2/3, 1/2; (*d*) 1/6, 1/4; (*e*) 1/6, 1/3.

4.53.   (*a*) 85%; (*b*) 15%; (*c*) $20/65 \approx 30.1\%$; (*d*) $1/2 = 50\%$.

4.54.   (*a*) 16%; (*b*) $6/16 = 37.5\%$.

4.55.   (*a*) $15/91 \approx 16.5\%$; (*b*) $15/91 \approx 16.5\%$; (*c*) $5/21 \approx 23.8\%$.

4.56.   (*a*) 11/32; (*b*) 13/32; (*c*) 8/32.

4.57.   (*a*) 5/11; (*b*) 3/13; (*c*) 1.

4.58.   (*a*) $17/48 \approx 35.4\%$; (*b*) $23/48 \approx 47.9\%$; (*c*) $8/48 \approx 16.7\%$.

4.59.   (*a*) $5/17 \approx 29.4\%$; (*b*) $3/23 \approx 13.0\%$; (*c*) 1.

4.60.   2/3.

4.61.   (*a*) 5/8; (*b*) 4/5.

4.62.   (*a*) $\frac{1}{2}\left(\frac{x}{x+y} + \frac{z}{z+t}\right)$; (*b*) $\frac{xz + x + yz}{(x+y)(z+t+1)}$.

4.63.   (*a*) 50%; (*b*) 20%.

4.64.   (*a*) 50%; (*b*) 1/3.

4.65.   (*a*) 47%; (*b*) $10/47 \approx 21.3\%$.

4.66.   (*a*) 46.25%; (*b*) 21.6%.

4.67.   (*a*) $80/295 \approx 27.1\%$; (*b*) $140/295 \approx 47.5\%$; (*c*) $75/295 \approx 25.574\%$.

4.68.   (*a*) 2/9; (*b*) 4/9; (*c*) 3/9.

4.69.   (*a*) 0.12, 0.58; (*b*) 0.3, 0.4.

4.70.   (*a*) 1/4; (*b*) 21/40.

4.71.   (*a*) 1/28; (*b*) $1/28 + 9/280 = 19/280$.

4.72.   (*a*) 0.06, 0.44; (*b*) 0.14, 0.76; (*c*) 0.25, 0.30; (*d*) 0.20, 0.80.

4.73.   (*a*) 0.2; (*b*) 2/7; (*c*) 0.5.

4.74.   (*a*) 1/12; (*b*) 5/12; (*c*) 2/5.

4.75.   (*a*) $1 - \frac{1}{4} = \frac{3}{4}$; (*b*) $\frac{1}{6} + \frac{1}{4} = \frac{5}{12}$.

4.76.   (*a*) 6%, 21%; (*b*) 79%, 44%; (*c*) $9/44 \approx 20.45\%$.

4.77.   Only A and B are independent.

4.81.   (*a*) $P(aaa$ or bbb or $ccc) = 0.26$; (*b*) $6(0.03) = 0.18$.

4.82.   (*a*) $S = \{WW, WL, WT, LW, LL, LT, TW, TL, TT\}$; 0.25, 0.15, 0.10, 0.15, 0.09, 0.06, 0.10, 0.06, 0.04;
            (*b*) $1 - 0.25 = 0.75$.

4.83.   (*a*) $1 - (2/3)^3 = 19/27$; (*b*) $(2/3)^n < 10\%$ so $n > 6$.

4.84. $P(HH$ or HRH or $RHH) = 64.8\%$.

4.85.   (*a*) $6(0.44) = 26.5\%$; (*b*) $1 - P(MMMM) \approx 76\%$.

Random Variables

5.1 INTRODUCTION

Random variables play an important role in probability. This chapter formally defines a random variable and presents its basic properties. The next chapter treats special types of random variables.

A random variable is a special kind of a function, so we recall some notation and definitions about functions. Let S and T be sets. Suppose to each $s \in S$ there is assigned a unique element of T; the collection f of such assignments is called a *function* from S into T, and it is written

$$f: S \to T$$

We write $f(s)$ for the element of T that f assigns to $s \in S$, and $f(s)$ is called the *image* of s under f or the *value* of f at s. The *image* $f(A)$ of any subset A of S, and the *pre-image* $f^{-1}(B)$ of any subset B of T are defined as follows:

$$f(A) = \{f(s): s \in S\} \quad \text{and} \quad f^{-1}(B) = \{s: f(s) \in B\}$$

In words, $f(A)$ consists of the images of the points in A, and $f^{-1}(B)$ consists of those points in S whose image belongs to B. In particular, the set $f(S)$ of all the image points of elements in S is called the *image set* (or *image* or *range*) of the function f.

5.2 RANDOM VARIABLES

Let S be a sample space of an experiment. As noted previously, the outcome of the experiment, or the points in S, need not be numbers. For example, in tossing a coin, the outcomes are H (heads) or T (tails), and in tossing a pair of dice, the outcome are pairs of integers. However, we frequently wish to assign a specific number to each outcome of the experiment. For example, in the tossing of a pair of dice, we may want to assign the sum of the two integers to the outcome. Such an assignment of numerical values to the points in S is called a *random variable*. Specifically, we have the following definition.

Definition: A *random variable* X on a sample space S is a function from S into the set $\mathbf{R}$ of real numbers such that the pre-image of any interval of $\mathbf{R}$ is an event in S.

We emphasize that if S is a discrete sample space in which every subset of S is an event, then clearly every real-valued function on S is a random variable. On the other hand, if S is uncountable, then it can be shown that certain real-valued functions on S are not random variables.

The notation R_X will be used to denote the image of a random variable X, that is, R_X is the set of those numbers assigned by X to a sample space S. We will refer to R_X as the *range space* of X. This chapter will mainly investigate *discrete* random variables, where the range space R_X is finite or countable. *Continuous* random variables are those where the range space R_X is a continuum of numbers such as an interval or a union of intervals. Such random variables, which may require some calculus for their investigation, will be treated near the end of the chapter.

EXAMPLE 5.1 A pair of fair dice is tossed. (See Example 3.2.) The sample space S consists of the 36 ordered pairs (a, b) where a and b can be any integers between 1 and 6, that is,

$$S = \{(1, 1), (1, 2), \ldots, (6, 6)\}$$

Let X assign to each point (a, b) the maximum of its numbers, that is, $X(a, b) = \max(a, b)$. For example,

$$X(1, 1) = 1, \qquad X(3, 4) = 4, \qquad X(5, 2) = 5, \qquad X(6, 6) = 6$$

Then X is a random variable where any number between 1 and 6 could occur, and no other number can occur. Thus, the range space R_X of X is as follows:

$$R_X = \{1, 2, 3, 4, 5, 6\}$$

Now let Y assign to each point (a, b) the sum of its numbers, that is, $Y(a, b) = a + b$. For example,

$$Y(1, 1) = 2, \qquad Y(3, 4) = 7, \qquad Y(6, 3) = 9, \qquad Y(6, 6) = 12$$

Then, Y is a random variable where any number between 2 and 12 could occur, and no other number can occur. Thus, the range space R_Y of Y is as follows:

$$R_Y = \{2, 3, 4, 5, 6, 7, 8, 9, 10, 11, 12\}$$

EXAMPLE 5.2

(a) A box contains 12 items of which 3 are defective. A sample of 3 items is selected from the box. The sample space S consists of the $C(12, 3) = 220$ different samples of size 3. Let X denote the number of defective items in the sample; then X is a random variable with range space $R_X = \{0, 1, 2, 3\}$.

(b) A coin is tossed until a head occurs. The sample space follows:

$$S = \{H, TH, TTH, TTTH, TTTTH, \ldots\}$$

Let X denote the number of times the coin is tossed. Then, X is a random variable with range space

$$R_X = \{1, 2, 3, 4, \ldots, \infty\}$$

(We include the number ∞ for the case that only tails occurs.) Here X is an infinite but discrete random variable.

(c) A point is chosen in a circle C of radius r. Let X denote the distance of the point from the center. Then, X is a random variable whose value can be any number between 0 and r, inclusive. Thus, the range space R_X of X is a closed interval:

$$R_X = [0, r] = \{x : 0 \leq x \leq r\}$$

Here, X is a continuous random variable.

Sums and Products of Random Variable

Let X and Y be a random variable on the same sample space S. Then $X + Y$, $X + k$, kX, and XY (where k is a real number) are the functions on S defined as follows (where s is any point in S):

$$(X + Y)(s) = X(s) + Y(s), \qquad (kX)(s) = kX(s),$$
$$(X + k)(s) = X(s) + k, \qquad (XY)(s) = X(s)Y(s)$$

More generally, for any polynomial, exponential, or continuous function $h(t)$, we define $h(X)$ to be the function on S defined by

$$[h(X)](s) = h[X(s)]$$

One can show that these are also random variables on S. (This is trivially true for the case that every subset of S is an event.)

We use the short notation $P(X = a)$ and $P(a \leq X \leq b)$, respectively, for the probability that "X maps into a" and "X maps into the interval $[a, b]$". That is,

$$P(X = a) \qquad \text{is short for} \qquad P(\{s \in S : X(s) = a\}$$
$$P(a \leq X \leq b) \qquad \text{is short for} \qquad P(\{s \in S : a \leq X(s) \leq b\}$$

Analogous meanings are given to

$$P(X \leq a), \qquad P(X = a, Y = b), \qquad P(a \leq X \leq b), c \leq Y \leq d)$$

and so on.

5.3 PROBABILITY DISTRIBUTION OF A FINITE RANDOM VARIABLE

Let X be a *finite random variable* on a sample spacea S, that is, X assigns only a finite number of values to S. Say,

$$R_X = \{x_1, x_2, \ldots, x_n\}$$

(We assume that $x_1 < x_2 < \cdots < x_n$.) Then, X induces a function f which assigns probabilities to the points in R_X as follows:

$$f(x_k) = P(X = x_k) = P(\{s \in S : X(s) = x_k\}$$

The set of ordered pairs $[x_i, f(x_i)]$ is usually given in the form of a table as follows:

x	x_1	x_2	x_3	$\cdots$	x_n
$f(x)$	$f(x_1)$	$f(x_2)$	$f(x_3)$	$\cdots$	$f(x_n)$

This function f is called the *probability distribution* or, simply, *distribution*, of the random variable X; it satisfies the following two conditions:

(i) $f(x_k) \geq 0$ and (ii) $\sum_k f(x_k) = 1$

Accordingly, the range space R_X with the above assignment of probabilities is a probability space.

Remark: It is convenient sometimes to extend a probability distribution f to all real numbers by defining $f(x) = 0$ when x does not belong to R_X. A graph of such a function $f(x)$ is called a *probability graph*.

Notation: Sometimes a probability distribution will be given using the pairs $[x_i, p_i]$ or $[x_i, P(x_i)]$ or $[x, P(X = x)]$ rather than the functional notation $[x, f(x)]$.

Equiprobable Spaces

Now suppose X is a random variable on a finite equiprobable space S. Then, X is a finite random variable, and the following theorem tells us how to obtain the distribution of X:

Theorem 5.1: Let S be a finite equiprobable space, and let X be a random variable on S with range space $R_X = \{x_1, x_2, \ldots, x_t\}$. Then

$$p_k = P(x_k) = \frac{\text{number of points in } S \text{ whose image is } x_k}{\text{number of points in } S}$$

The proof appears in Problem 5.41. It essentially follows from the fact that S is an equiprobable space, and hence

$$p_k = P(\{s : X(s) = x_k\}) = \frac{n(\{s : X(s) = x_k\})}{n(S)}$$

We apply this theorem in the next examples.

EXAMPLE 5.3 Let S be the sample space when a pair of fair dice is tossed. Then S is a finite equiprobable space consisting of the 36 ordered pairs (a, b) where a and b are any integers between 1 and 6:

$$S = \{(1, 1), (1, 2), (1, 3), \ldots, (6, 6)\}$$

Let X and Y be the random variables on S in Example 5.1, that is, X denotes the maximum of the numbers, $X(a, b) = \max(a, b)$, and Y denotes the sum of the numbers, $Y(a, b) = a + b$.

(a) Find the distribution f of X.   (b) Find the distribution g of Y.

Also, exhibit their probability graphs.

Here S is an equiprobable space with 36 points so we can use Theorem 5.1 and simply count the number of points with the given numerical value.

(a) *Random Variable X.* We compute the distribution f of X as follows:

(1) Only one toss $(1, 1)$ has the maximum value 1; hence $f(1) = \frac{1}{36}$.
(2) Three tosses, $(1, 2), (2, 2), (2, 1)$, have the maximum value 2; hence $f(2) = \frac{3}{36}$.
(3) Five tosses, $(1, 3), (2, 3), (3, 3), (3, 2), (3, 1)$, have the maximum value 3; hence $f(3) = \frac{5}{36}$.

Similarly, $f(4) = \dfrac{7}{36}$, $f(5) = \dfrac{9}{36}$, $f(6) = \dfrac{11}{36}$ Thus, the distribution f of X is as follows:

x	1	2	3	4	5	6
$f(x)$	$\frac{1}{36}$	$\frac{3}{36}$	$\frac{5}{36}$	$\frac{7}{36}$	$\frac{9}{36}$	$\frac{11}{36}$

The probability graph of X is pictured in Fig. 5-1(a).

(b) *Random Variable Y.* The distribution g of the random variable Y is as follows:

y	2	3	4	5	6	7	8	9	10	11	12
$g(y)$	$\frac{1}{36}$	$\frac{2}{36}$	$\frac{3}{36}$	$\frac{4}{36}$	$\frac{5}{36}$	$\frac{6}{36}$	$\frac{5}{36}$	$\frac{4}{36}$	$\frac{3}{36}$	$\frac{2}{36}$	$\frac{1}{36}$

We obtain, for example, $g(6) = \frac{5}{36}$ from the fact that exactly five of the tosses have sum 6:

$$(1, 5), (2, 4), (3, 3), (4, 2), (5, 1)$$

The other entries are obtained similarly. The probability graph of Y is pictured in Fig. 5-1(b).

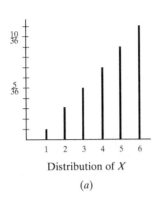

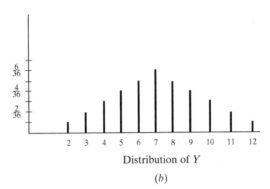

Distribution of X Distribution of Y

(a)　　　　　　　　　　　　　　　　　　　　　　(b)

Fig. 5-1

EXAMPLE 5.4　Suppose a fair coin is tossed three times yielding the following sample space:

$$S = \{HHH, HHT, HTH, HTT, THH, THT, TTH, TTT\}$$

Let X be the random variable which assigns to each point in S the number of heads.　Then clearly X can only be 0, 1, 2, or 3.　That is, the following is its range space:

$$R_X = \{0, 1, 2, 3\}$$

Observe that:

  (i)　There is only one point TTT where $X = 0$.

  (ii)　There are three points, HTT, THT, TTH, where $X = 1$.

 (iii)　There are three points, HHT, HTH, THH, where $X = 2$.

 (iv)　There is only one point HHH where $X = 3$.

Since the coin is fair, S is an 8-element equiprobable space.　Hence Theorem 5.1 tells us that the distribution f of X is as follows:

x	0	1	2	3
$f(x)$	$\frac{1}{8}$	$\frac{3}{8}$	$\frac{3}{8}$	$\frac{1}{8}$

EXAMPLE 5.5　Suppose a coin is tossed three times, but now suppose the coin is weighted so that $P(H) = \frac{2}{3}$ and $P(T) = \frac{1}{3}$.　The sample space is again

$$S = \{HHH, HHT, HTH, HTT, THH, THT, TTH, TTT\}$$

Let X be the random variable which assigns to each point in S its number of heads.　Find the distribution f of X.
　　Now S is not an equiprobable space.　Specifically, the probabilities of the points in S are as follows:

$$P(HHH) = \frac{8}{27}, \qquad P(HHT) = \frac{4}{27}, \qquad P(HTH) = \frac{4}{27}, \qquad P(HTT) = \frac{2}{27}$$

$$P(THH) = \frac{4}{27}, \qquad P(THT) = \frac{2}{27}, \qquad P(TTH) = \frac{2}{27}, \qquad P(TTT) = \frac{1}{27}$$

Since S is not an equiprobable space, we cannot use Theorem 5.1 to find the distribution f of X.　We find f directly by using its definition.　Namely,

$$f(0) = P(TTT) = \frac{1}{27}, \qquad f(1) = P(\{HTT, THT, TTH\}) = \frac{2}{27} + \frac{2}{27} + \frac{2}{27} = \frac{6}{27}$$

$$f(2) = P(\{HHT, HTH, THH\}) = \frac{4}{27} + \frac{4}{27} + \frac{4}{27} = \frac{12}{27}, \qquad f(3) = P(HHH) = \frac{8}{27}$$

Thus, the distribution f of X is as follows:

x	0	1	2	3
$f(x)$	$\frac{1}{27}$	$\frac{6}{27}$	$\frac{12}{27}$	$\frac{8}{27}$

The probability graph of f is shown in Fig. 5-2(a). An alternate picture of f is by a histogram which appears in Fig. 5-2(b). One may view the histogram as making the random variable continuous where $X = 1$ means X lies between 0.5 and 1.5.

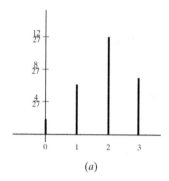
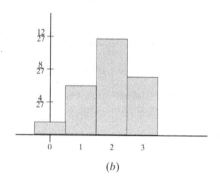

(a)   (b)

Fig. 5-2

5.4 EXPECTATION OF A FINITE RANDOM VARIABLE

Let X be a finite random variable, and suppose the following is its distribution:

x	x_1	x_2	x_3	$\cdots$	x_n
$f(x)$	$f(x_1)$	$f(x_2)$	$f(x_3)$	$\cdots$	$f(x_n)$

Then the *mean*, or *expectation* (or *expected value*) of X, denoted by $E(X)$, or simply E, is defined by

$$E = E(X) = x_1 f(x_1) + x_2 f(x_2) + \cdots + x_n f(x_n) = \sum x_i f(x_i)$$

Equivalently, when the notation $[x_i, p_i]$ is used instead of $[x, f(x)]$,

$$E = E(X) = x_1 p_1 + x_2 p_2 + \cdots + x_n p_n = \sum x_i p_i$$

Roughly speaking, if the x_i are numerical outcomes of an experiment, then E is the expected value of the experiment. We may also view E as the *weighted average* of the outcomes where each outcome is weighted by its probability. (For notation convenience we omit the limits of the index in the above summations.)

EXAMPLE 5.6 A pair of fair dice are tossed. Let X and Y be the random variables in Example 5.1, that is, X denotes the maximum of the numbers, $X(a, b) = \max(a, b)$ and Y denotes the sum of the numbers, $Y(a, b) = a + b$. Using the distribution of X, which appears in Example 5.3, the expectation of X is computed as follows:

$$E(X) = 1\left(\frac{1}{36}\right) + 2\left(\frac{3}{36}\right) + 3\left(\frac{5}{36}\right) + 4\left(\frac{7}{36}\right) + 5\left(\frac{9}{36}\right) + 6\left(\frac{11}{36}\right) = \frac{161}{36} \approx 4.47$$

Using the distribution of Y, which also appears in Example 5.3, the expectation of Y is computed as follows:

$$E(Y) = 2\left(\frac{1}{36}\right) + 3\left(\frac{2}{36}\right) + 4\left(\frac{3}{36}\right) + \cdots + 12\left(\frac{1}{36}\right) = \frac{252}{36} = 7$$

EXAMPLE 5.7 Let X and Y be random variables with the following respective distributions:

x_i	2	3	6	10
p_i	0.2	0.2	0.5	0.1

y_i	−8	−2	0	3	7
p_i	0.2	0.3	0.1	0.3	0.1

Then

$$E(X) = \sum x_i p_i = 2(0.2) + 3(0.2) + 6(0.5) + 10(0.1)$$
$$= 0.4 + 0.6 + 3.0 + 1.0 = 5$$

$$E(Y) = \sum y_i p_i = -8(0.2) - 2(0.3) + 0(0.1) + 3(0.3) + 7(0.1)$$
$$= -1.6 - 0.6 + 0 + 0.9 + 0.7 = -0.6$$

Remark: The above Example 5.7 shows that the expectation of a random variable may be negative. It also shows that we can talk about the distribution and expectation of a random variable X without any reference to the original sample space S.

EXAMPLE 5.8 Suppose a fair coin is tossed 6 times. One can show (Section 6.2) that the number x_i of heads occurs with probability p_i as follows:

x_i	0	1	2	3	4	5	6
p_i	$\frac{1}{64}$	$\frac{6}{64}$	$\frac{15}{64}$	$\frac{30}{64}$	$\frac{15}{64}$	$\frac{6}{64}$	$\frac{1}{64}$

Then the expected number E of heads is as follows:

$$E = 0\left(\frac{1}{64}\right) + 1\left(\frac{6}{64}\right) + 2\left(\frac{15}{64}\right) + 3\left(\frac{20}{64}\right) + 4\left(\frac{15}{64}\right) + 5\left(\frac{6}{64}\right) + 6\left(\frac{1}{64}\right) = 3$$

This agrees with our intuition that, when a fair coin is repeatedly tossed, about half of the tosses should be heads.

The following theorems (proved in Problems 5.44 and 5.45) relate the notion of expectation to operations on random variables defined in Section 5.2.

Theorem 5.2: Let X be a random variable and let k be a real number. Then

(i) $E(kX) = kE(X)$ and   (ii) $E(X + k) = E(X) + k$

Thus, for any real numbers a and b,

$$E(aX + b) = E(aX) + b = aE(X) + b$$

Theorem 5.3: Let X and Y be random variables on the same sample space S. Then

$$E(X + Y) = E(X) + E(Y)$$

A simple induction argument yields the following:

Corollary 5.4: Let $X_1, X_2, \ldots, X_n$ be random variables on the same sample space S. Then

$$E(X_1 + X_2 + \cdots + X_n) = E(X_1) + E(X_2) + \cdots + E(X_n)$$

Expectation and Games of Chance

Frequently, a game of chance consists of n outcomes $a_1, a_2, \ldots, a_n$ occurring with respective probabilities $p_1, p_2, \ldots, p_n$. Suppose the payoff to a player is w_i for the outcome a_i, where a positive w_i denotes a win for the player and a negative w_i denotes a loss. Then the expected value E of the game for the player is the quantity

$$E = w_1 p_1 + w_2 p_2 + \cdots + w_n p_n$$

The assignment of w_i to a_i may be viewed as a random variable X, and the expectation $E(X)$ of X is the expected value of the game. The game is *fair* if $E = 0$, *favorable* to the player if E is positive, and *unfavorable* to the player if E is negative.

EXAMPLE 5.9 A fair die is tossed. If 2, 3, or 5 occurs, the player wins that number of dollars, but if 1, 4, or 6 occurs, the player loses that number of dollars. The possible payoffs for the player and their respective probabilities follow:

x	2	3	5	-1	-4	-6
$f(x)$	$\frac{1}{6}$	$\frac{1}{6}$	$\frac{1}{6}$	$\frac{1}{6}$	$\frac{1}{6}$	$\frac{1}{6}$

The negative numbers -1, -4, -6 refer to the fact that the player loses when 1, 4, or 6 occurs. Then the expected value E of the game is as follows:

$$E = 2\left(\frac{1}{6}\right) + 3\left(\frac{1}{6}\right) + 5\left(\frac{1}{6}\right) - 1\left(\frac{1}{6}\right) - 4\left(\frac{1}{6}\right) - 6\left(\frac{1}{6}\right) = -\frac{1}{6}$$

Thus, the game is unfavorable to the player since the expected value E is negative.

Mean and Expectation

Suppose X is a random variable with n distinct values $x_1, x_2, \ldots, x_n$ and suppose each x_i occurs with the same probability p_i. Then each $p_i = \frac{1}{n}$. Accordingly

$$E(X) = x_1\left(\frac{1}{n}\right) + x_2\left(\frac{1}{n}\right) + \cdots + x_n\left(\frac{1}{n}\right) = \frac{x_1 + x_2 + \cdots + x_n}{n}$$

This is precisely the *average* or *mean* value of the numbers $x_1, x_2, \ldots, x_n$. (See Appendix A.) For this reason $E(X)$ is called the *mean* of the random variable X. Furthermore, since the Greek letter μ (read "mu") is used for the mean value of a population, we also use μ for the expectation of X. That is,

$$\boxed{\mu = \mu_X = E(X)}$$

The mean μ is an important parameter for a probability distribution, and in the next section, Section 5.4, we introduce another important parameter, denoted by the Greek letter σ (read "sigma"), called the *standard deviation* of X.

5.5 VARIANCE AND STANDARD DEVIATION

The mean of a random variable X measures, in a certain sense, the "average" value of X. The concepts in this section, variance and standard deviation, measure the "spread" or "dispersion" of X.

Let X be a random variable with mean $\mu = E(X)$ and the following probability distribution:

x	x_1	x_2	x_3	$\cdots$	x_n
$f(x)$	$f(x_1)$	$f(x_2)$	$f(x_3)$	$\cdots$	$f(x_n)$

The *variance* of X, denoted by var(X), is defined by

$$\text{var}(X) = (x_1 - \mu)^2 f(x_1) + (x_2 - \mu)^2 f(x_2) + \cdots + (x_n - \mu)^2 f(x_n)$$

$$= \sum (x_i - \mu)^2 f(x_i) = E((X - \mu)^2)$$

The *standard deviation* of X, denoted by σ_X or simply σ, is the nonnegative square root of var(X), that is

$$\sigma_X = \sqrt{\text{var}(X)}$$

Accordingly, var$(X) = \sigma_X^2$. Both var(X) and σ_X^2 or simply σ^2 are used to denote the variance of a random variable X.

The next theorem gives us an alternate and sometimes more useful formula for calculating the variance of a random variable X.

Theorem 5.5: $\text{var}(X) = x_1^2 f(x_1) + x_2^2 f(x_2) + \cdots + x_n^2 f(x_n) - \mu^2 = \sum x_i^2 f(x_i) - \mu^2 = E(X^2) - \mu^2$

Proof: Using $\sum x_i f(x_i) = \mu$ and $\sum f(x_i) = 1$, we obtain

$$\sum (x_i - \mu)^2 f(x_i) = \sum (x_i^2 - 2\mu x_i + \mu^2) f(x_i)$$

$$= \sum x_i^2 f(x_i) - 2\mu \sum x_i f(x_i) + \mu^2 \sum f(x_i)$$

$$= \sum x_i^2 f(x_i) - 2\mu^2 + \mu^2 = \sum x_i^2 f(x_i) - \mu^2$$

This proves the theorem.

Remark: Both the variance var$(X) = \sigma^2$ and the standard deviation σ measure the weighted spread of the values x_i about the mean μ; however, one advantage of the standard deviation σ is that it has the same units as μ.

EXAMPLE 5.10   Let X denote the number of times heads occurs when a fair coin is tossed 6 times. The distribution of X appears in Example 5.8 where its mean $\mu = 3$ is computed. The variance of X is computed using its definition as follows:

$$\text{var}(X) = (0 - 3)^2 \frac{1}{64} + (1 - 3)^2 \frac{6}{64} + (2 - 3)^2 \frac{15}{64} + \cdots + (6 - 3)^2 \frac{1}{64} = 1.5$$

Alternately, by Theorem 5.5,

$$\text{var}(X) = 0^2 \frac{1}{64} + 1^2 \frac{6}{64} + 2^2 \frac{15}{64} + 3^2 \frac{20}{64} + 4^2 \frac{15}{64} + 5^2 \frac{6}{64} + 6^2 \frac{1}{64} = 1.5$$

Thus, the standard deviation is $\sigma = \sqrt{1.5} \approx 1.225$ (heads).

EXAMPLE 5.11   A pair of fair dice are tossed.   Let X and Y be the random variables in Example 5.1, that is, X denotes the maximum of the numbers, $X(a, b) = \max(a, b)$, and Y denotes the sum of the numbers, $Y(a, b) = a + b$.   The distributions of X and Y appear in Example 5.3 and their expectations were computed in Example 5.6 yielding:

$$\mu_X = E(X) = 4.47 \quad \text{and} \quad \mu_Y = E(Y) = 7$$

Find the variance and standard deviation of: (a) X, (b) Y.

(a)   First we compute $E(X^2)$ as follows:

$$E(X^2) = \sum x_i^2 f(x_i) = 1^2\left(\frac{1}{36}\right) + 2^2\left(\frac{3}{36}\right) + 3^2\left(\frac{5}{36}\right) + 4^2\left(\frac{7}{36}\right) + 5^2\left(\frac{9}{36}\right) + 6^2\left(\frac{11}{36}\right) = \frac{791}{36} = 21.97$$

Hence

$$\text{var}(X) = E(X^2) - \mu_X^2 = 21.97 - 20.25 = 1.99 \quad \text{and} \quad \sigma_X = \sqrt{1.99} = 1.4$$

(b)   First we compute $E(Y^2)$ as follows:

$$E(Y^2) = \sum y_i^2 f(y_i) = 2^2\left(\frac{1}{36}\right) + 3^2\left(\frac{2}{36}\right) + 4^2\left(\frac{3}{36}\right) + \cdots + 12^2\left(\frac{1}{36}\right) = \frac{1974}{36} = 54.8$$

Hence

$$\text{var}(Y) = E(Y^2) - \mu_Y^2 = 54.8 - 49 = 5.8 \quad \text{and} \quad \sigma_Y = \sqrt{5.8} = 2.4$$

We establish (Problem 5.46) an important property of the variance and standard deviation.

Theorem 5.6:   Let X be a random variable and let a and b be constants.   Then

$$\text{var}(aX + b) = a^2 \, \text{var}(X) \quad \text{and} \quad \sigma_{aX+b} = |a| \sigma_X$$

There are two special cases of Theorem 5.6 which occur frequently; the first where $a = 1$ and the second where $b = 0$.   Specifically, for any constant k

(i) $\text{var}(X + k) = \text{var}(X)$ and hence $\sigma_{X+k} = \sigma_X$.

(ii) $\text{var}(kX) = k^2 \, \text{var}(X)$ and hence $\sigma_{kX} = |k| \sigma_X$.

Remark: There are physical interpretations of the mean and variance.   Suppose the x axis is a thin wire and at each point x_i there is a unit with mass p_i.   Then, if a fulcrum or pivot is placed at the point μ [Fig. 5-3(a)], the system will be balanced.   Hence μ is called the *center of mass* of the system.   On the other hand, if the system were rotating about the center of mass μ [Fig. 5-3(b)], then the variance σ^2 measures the system's resistance to stopping, called the *moment of inertia* of the system.

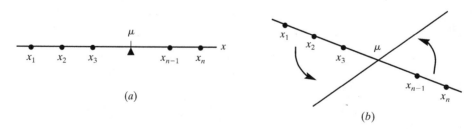

Fig. 5-3

Standardized Random Variable

Let X be a random variable with mean μ and standard deviation $\sigma > 0$. The *standardized random variable Z* is defined by

$$Z = \frac{X - \mu}{\sigma}$$

Important properties of Z are contained in the next theorem (proved in Problem 5.48).

Theorem 5.7: The standardized random variable Z has mean $\mu_Z = 0$ and standard deviation $\sigma_Z = 1$.

EXAMPLE 5.12 Suppose a random variable X has the following distribution:

x	2	4	6	8
$f(x)$	0.1	0.2	0.3	0.4

(a)   Compute the mean μ and standard deviation σ of X.

(b)   Find the probability distribution of the standardized random variable. $Z = (X - \mu)/\sigma$, and show that $\mu_Z = 0$ and $\sigma_Z = 1$, as predicted by Theorem 5.6.

(a)   We have:

$$\mu = E(X) = \sum x_i f(x_i) = 2(0.1) + 4(0.2) + 6(0.3) + 8(0.4) = 6$$

$$E(X^2) = \sum x_i^2 f(x_i) = 2^2(0.1) + 4^2(0.2) + 6^2(0.3) + 8^2(0.4) = 40$$

Now using Theorem 5.5, we obtain

$$\sigma^2 = \text{var}(X) = E(X^2) - \mu^2 = 40 - 6^2 = 4 \quad \text{and} \quad \sigma = 2$$

(b)   Using $z = (x - \mu)/\sigma = (x - 6)/2$ and $f(z) = f(x)$, we obtain the following distribution for Z:

z	−2	−1	0	1
$f(z)$	0.1	0.2	0.3	0.4

Then

$$\mu_Z = E(Z) = \sum z_i f(z_i) = -2(0.1) - 1(0.2) + 0(0.3) + 1(0.4) = 0$$

$$E(Z^2) = \sum z_i^2 f(z_i) = (-2)^2(0.1) + (-1)^2(0.2) + 0^2(0.3) + 1^2(0.4) = 1$$

Using Theorem 5.5, we obtain

$$\sigma_Z^2 = \text{var}(Z) = E(Z^2) - \mu^2 = 1 - 0^2 = 1 \quad \text{and} \quad \sigma_Z = 1$$

(The results $\mu_Z = 0$ and $\sigma_Z = 1$ were predicted by Theorem 5.7.)

5.6   JOINT DISTRIBUTION OF RANDOM VARIABLES

Let X and Y be random variables on the same sample space S with respective range spaces

$$R_X = \{x_1, x_2, \ldots, x_n\} \quad \text{and} \quad R_Y = \{y_1, y_2, \ldots, y_m\}$$

The *joint distribution* or *joint probability function* of X and Y is the function h on the product space $R_X \times R_Y$ defined by

$$h(x_i, y_j) \equiv P(X = x_i, Y = y_j) \equiv P(\{s \in S : X(s) = x_i, Y(s) = y_j\})$$

The function h is usually given in the form of a table as in Fig. 5-4. The function h has the properties

$$\text{(i)} \quad h(x_i, y_j) \geq 0 \qquad \text{(ii)} \quad \sum_i \sum_j h(x_i, y_j) = 1$$

Thus, h defines a probability space on the product space $R_X \times R_Y$.

X \ Y	y_1	y_2	$\cdots$	y_j	$\cdots$	y_m	Sum
x_1	$h(x_1 y_1)$	$h(x_1, y_2)$	$\cdots$	$h(x_1, y_j)$	$\cdots$	$h(x_1, y_m)$	$f(x_1)$
x_2	$h(x_2, y_1)$	$h(x_2, y_2)$	$\cdots$	$h(x_2, y_j)$	$\cdots$	$h(x_2, y_m)$	$f(x_2)$
$\cdots$							$\cdots$
x_i	$h(x_i, y_1)$	$h(x_i, y_2)$	$\cdots$	$h(x_i, y_j)$	$\cdots$	$h(x_i, y_m)$	$f(x_i)$
$\cdots$							$\cdots$
x_m	$h(x_n, y_1)$	$h(x_n, y_2)$	$\cdots$	$h(x_n, y_j)$	$\cdots$	$h(x_n, y_m)$	$f(x_n)$
Sum	$g(y_1)$	$g(y_2)$	$\cdots$	$g(y_j)$	$\cdots$	$g(y_m)$	

Fig. 5-4

The functions f and g on the right side and the bottom side, respectively, of the joint distribution table in Fig. 5-4 are defined by

$$f(x_i) = \sum_j h(x_i, y_j) \quad \text{and} \quad g(y_j) = \sum_i h(x_i, y_j)$$

That is, $f(x_i)$ is the sum of the entries in the ith row and $g(y_j)$ is the sum of the entries in the jth column. They are called the *marginal distributions*, and are, in fact, the (individual) distributions of X and Y, respectively (Problem 5.42).

Covariance and Correlation

Let X and Y be random variables with the joint distribution $h(x, y)$, and respective means μ_X and μ_Y. The *covariance* of X and Y, denoted by $\text{cov}(X, Y)$, is defined by

$$\text{cov}(X, Y) = \sum_{i,j} (x_i - \mu_X)(y_j - \mu_Y)h(x_i, y_j) = E[(X - \mu_X)(Y - \mu_Y)]$$

Equivalently (Problem 5.47),

$$\text{cov}(X, Y) = \sum_{i,j} x_i y_j h(x_i, y_j) - \mu_X \mu_Y = E(XY) - \mu_X \mu_Y$$

The *correlation* of X and Y, denoted by $\rho(X, Y)$, is defined by

$$\rho(X, Y) = \frac{\text{cov}(X, Y)}{\mu_X \mu_Y}$$

The correlation ρ is dimensionless and has the following properties:

(i) $\rho(X, Y) = \rho(Y, X)$,   (iii) $\rho(X, X) = 1$, $\rho(X, -X) = -1$,

(ii) $-1 \leq \rho \leq 1$,   (iv) $\rho(aX + b, cY + d) = \rho(X, Y)$ if $a, c \neq 0$.

We note (Example 5.14) that a pair of random variables with identical individual distributions can have distinct covariances and correlations. Thus, $cov(X, Y)$ and $\rho(X, Y)$ are measurements of the way that X and Y are interrelated.

EXAMPLE 5.13 Let S be the sample space when a pair of fair dice is tossed, and let X and Y be the random variables on S in Example 5.1. That is, to each point (a, b) in S, X assigns the maximum of the numbers and Y assigns the sum of the numbers:

$$X(a, b) = \max(a, b) \quad \text{and} \quad Y(a, b) = a + b$$

The joint distribution of X and Y appears in Fig. 5-5. The entry $h(3, 5) = \frac{2}{36}$ comes from the fact that $(3, 2)$ and $(2, 3)$ are the only points in S whose maximum number is 3 and whose sum is 5, that is,

$$h(3, 5) \equiv P(X = 3, Y = 5) = P\{(3, 2), (2, 3)\} = \frac{2}{36}$$

The other entries are obtained in a similar manner.

X \ Y	2	3	4	5	6	7	8	9	10	11	12	Sum
1	$\frac{1}{36}$	0	0	0	0	0	0	0	0	0	0	$\frac{1}{36}$
2	0	$\frac{2}{36}$	$\frac{1}{36}$	0	0	0	0	0	0	0	0	$\frac{3}{36}$
3	0	0	$\frac{2}{36}$	$\frac{2}{36}$	$\frac{1}{36}$	0	0	0	0	0	0	$\frac{5}{36}$
4	0	0	0	$\frac{2}{36}$	$\frac{2}{36}$	$\frac{2}{36}$	$\frac{1}{36}$	0	0	0	0	$\frac{7}{36}$
5	0	0	0	0	$\frac{2}{36}$	$\frac{2}{36}$	$\frac{2}{36}$	$\frac{2}{36}$	$\frac{1}{36}$	0	0	$\frac{9}{36}$
6	0	0	0	0	0	$\frac{2}{36}$	$\frac{2}{36}$	$\frac{2}{36}$	$\frac{2}{36}$	$\frac{2}{36}$	$\frac{1}{36}$	$\frac{11}{36}$
Sum	$\frac{1}{36}$	$\frac{2}{36}$	$\frac{3}{36}$	$\frac{4}{36}$	$\frac{5}{36}$	$\frac{6}{36}$	$\frac{5}{36}$	$\frac{4}{36}$	$\frac{3}{36}$	$\frac{2}{36}$	$\frac{1}{36}$	

Fig. 5-5

Observe that the right side sum column does give the distribution f of X and the bottom sum row does give the distribution g of Y which appear in Example 5.3.

We compute the covariance and correlation of X and Y. First we compute $E(XY)$ as follows:

$$E(XY) = \sum x_i y_j h(x_i, y_j)$$

$$= 1(2)\left(\frac{1}{36}\right) + 2(3)\left(\frac{2}{36}\right) + 2(4)\left(\frac{1}{36}\right) + \cdots + 6(12)\left(\frac{1}{36}\right) = \frac{1232}{36} \approx 34.2$$

By Example 5.6, $\mu_X = 4.47$ and $\mu_Y = 7$, and by Example 5.11, $\sigma_X = 1.4$ and $\sigma_Y = 2.4$; hence

$$cov(X, Y) = E(XY) - \mu_X \mu_Y = 34.2 - (4.47)(7) = 2.9$$

and

$$\rho(X, Y) = \frac{cov(X, Y)}{\mu_X \mu_Y} = \frac{2.9}{(1.4)(2.4)} = 0.86$$

EXAMPLE 5.14 Let X and Y be random variables with joint distribution as shown in Fig. 5-6(a), and let X' and Y' be random variables with joint distribution as shown in Fig. 5-6(b). The marginal entries in Fig. 5-6 tell us that X and X' have the same distribution, and that Y and Y' have the same distribution, as follows:

x	1	3
$f(x)$	$\frac{1}{2}$	$\frac{1}{2}$

y	4	10
$g(y)$	$\frac{1}{2}$	$\frac{1}{2}$

Distribution of X and X' Distribution of Y and Y'

Thus

$$\mu_X = \mu_{X'} = 1\left(\frac{1}{2}\right) + 3\left(\frac{1}{2}\right) = 2 \quad \text{and} \quad \mu_Y = \mu_{Y'} = 4\left(\frac{1}{2}\right) + 10\left(\frac{1}{2}\right) = 7$$

We show that $\operatorname{cov}(X, Y) \neq \operatorname{cov}(X', Y')$ and hence $\rho(X, Y) \neq \rho(X', Y')$. First we compute $E(XY)$ and $E(X'Y')$ as follows:

$$E(XY) = 1(4)\left(\frac{1}{4}\right) + 1(10)\left(\frac{1}{4}\right) + 3(4)\left(\frac{1}{4}\right) + 3(10)\left(\frac{1}{4}\right) = 14$$

$$E(X'Y') = 1(4)(0) + 1(10)\left(\frac{1}{2}\right) + 3(4)\left(\frac{1}{4}\right) + 3(10)(0) = 11$$

Since $\mu_X = \mu_{X'} = 2$ and $\mu_Y = \mu_{Y'} = 7$, we obtain

$$\operatorname{cov}(X, Y) = E(XY) - \mu_X \mu_Y = 0 \quad \text{and} \quad \operatorname{cov}(X', Y') = E(XY) - \mu_X \mu_Y = -3$$

X \ Y	4	10	Sum
1	1/4	1/4	1/2
3	1/4	1/4	1/2
Sum	1/2	1/2	

X' \ Y'	4	10	Sum
1	0	1/2	1/2
3	1/2	0	1/2
Sum	1/2	1/2	

(a)   (b)

Fig. 5-6

Remark: The notion of a joint distribution h is extended to any finite number of random variables $X, Y, \ldots, Z$ in the obvious way, that is, h is a function on the product set $R_X \times R_Y \times \cdots \times R_Z$ defined by

$$h(x_i, y_j, \ldots, z_k) \equiv P(X = x_i, Y = y_j, \ldots, Z = z_k)$$

As in the case of two random variables, h defines a probability space on the product set $R_X \times R_Y \times \cdots \times R_Z$.

5.7 INDEPENDENT RANDOM VARIABLES

Let $X, Y, \ldots, Z$ be random variables on the same sample space S. Then $X, Y, \ldots, Z$ are said to be *independent* if, for any values $x_i, y_j, \ldots, z_k$, we have

$$P(X = x_i, Y = y_j, \ldots, Z = z_k) = P(X = x_i)P(Y = y_j) \cdots P(Z = z_k)$$

In particular, X and Y are independent if

$$P(X = x_i, Y = y_j) = P(X = x_i)P(Y = y_j)$$

Now suppose X and Y have respective distribution f and g and joint distribution h. Then the above equation may be written as

$$h(x_i, y_j) = f(x_i)g(y_j)$$

Thus, random variables X and Y are independent if each entry $h(x_i, y_j)$ in the joint distribution table is the product of its marginal entries $f(x_i)$ and $g(y_j)$.

EXAMPLE 5.15 Let X and Y be random variables with the joint distribution in Fig. 5-7. Then X and Y are independent random variables since each entry in the joint distribution can be obtained by multiplying its marginal entries. For example,

$$P(1, 2) = P(X = 1)P(Y = 2) = (0.30)(0.20) = 0.06$$
$$P(1, 3) = P(X = 1)P(Y = 3) = (0.30)(0.50) = 0.15$$
$$P(1, 4) = P(X = 1)P(Y = 4) = (0.30)(0.30) = 0.09$$

And so on.

X \ Y	2	3	4	Sum
1	0.06	0.15	0.09	0.30
2	0.14	0.35	0.21	0.70
Sum	0.20	0.50	0.30	

Fig. 5-7

EXAMPLE 5.16 A fair coin is tossed twice giving the equiprobable space $S = \{HH, HT, TH, TT\}$.

(a) Let X and Y be random variables on S defined as follows:

    (i) $X = 1$ if the first toss is H and $X = 0$ otherwise.

    (ii) $Y = 1$ if both tosses are H and $Y = 0$ otherwise.

The joint distribution of X and Y appears in Fig. 5-8(a). Note that X and Y are not independent random variables. For example, $P(0, 0)$ is not equal to the product of the marginal entries. Namely,

$$P(0, 0) = \tfrac{1}{2} \neq (\tfrac{1}{2})(\tfrac{3}{4}) = P(X = 0)P(Y = 0)$$

(b) Now let X and Y be random variables on S defined as follows:

    (i) $X = 1$ if the first toss is H and $X = 0$ otherwise.

    (ii) $Y = 1$ if the second toss is H and $Y = 0$ otherwise.

The joint distribution of X and Y appears in Fig. 5-8(b). Note that X and Y are now independent. Specifically, each of the four entries is $\tfrac{1}{4}$, and each entry is the product of its marginal entries:

$$P(i, j) = \tfrac{1}{4} = (\tfrac{1}{2})(\tfrac{1}{2}) = P(X = i)P(Y = j)$$

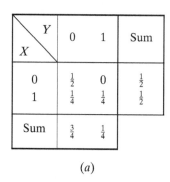

(a)   (b)

Fig. 5-8

The following theorems (proved in Problems 5.49 and 5.50) give important properties of independent random variables which do not hold in general.

Theorem 5.8: Let X and Y be independent random variables. Then:

      (i) $E(XY) = E(X)E(Y)$,

      (ii) $\text{var}(X + Y) = \text{var}(X) + \text{var}(Y)$,

      (iii) $\text{cov}(X, Y) = 0$.

Part (ii) in the above theorem generalizes as follows:

Theorem 5.9: Let $X_1, X_2, \ldots, X_n$ be independent random variables. Then

$$\text{var}(X_1 + \cdots + X_n) = \text{var}(X_1) + \cdots + \text{var}(X_n)$$

5.8 FUNCTIONS OF A RANDOM VARIABLE

Let X and Y be random variables on the same sample space S. Then Y is said to be a function of X if Y can be represented $Y = \Phi(X)$ for some real-valued function Φ of a real variable, that is, if $Y(s) = \Phi[X(s)]$ for every $s \in S$. For example, kX, X^2, $X + k$, and $(X + k)^2$ are all functions of X with $\Phi(x) = kx$, x^2, $x + k$, and $(x + k)^2$, respectively. We have the following fundamental result (proved in Problem 5.43):

Theorem 5.10: Let X and Y be random variables on the same sample space S with $Y = \Phi(X)$.

Then

$$E(Y) = \sum_{i=1}^{n} \Phi(x_i) f(x_i)$$

where f is the distribution function of X.

Similarly, a random variable Z is said to be a function of X and Y if Z can be represented $Z = \Phi(X, Y)$ where Φ is a real-valued function of two real variables, that is, if

$$Z(s) = \Phi[X(s), Y(s)]$$

for every $s \in S$. For example, $X + Y$ is a function of X and Y with $\Phi(x, y) = x + y$.

Corresponding to the above theorem, we have the following analogous result:

Theorem 5.11:　Let X, Y, Z be random variables on the same sample space S with $Z = \Phi(X, Y)$. Then

$$E(Z) = \sum_{i,j} \Phi(x_i, y_j) h(x_i, y_j)$$

where h is the joint distribution of X and Y.

We note that the above two theorems have been used implicitly in the preceding discussion and theorems. The proof of Theorem 5.11 will be given as a supplementary problem; it generalizes to a function of n random variables in the obvious way.

EXAMPLE 5.17　Let X and Y be the *dependent* (nonindependent) random variables in Example 5.16(a), and let $Z = X + Y$. We show that

$$E(Z) = E(X) + E(Y) \qquad \text{but} \qquad \text{var}(Z) \neq \text{var}(X) + \text{var}(Y)$$

[Thus, Theorem 5.8 need not hold for dependent (nonindependent) random variables.]

The joint distribution of X and Y appears in Fig. 5-8(a). The right marginal distribution is the distribution of X; hence

$$\mu_X = E(X) = 0\left(\frac{1}{2}\right) + 1\left(\frac{1}{2}\right) = \frac{1}{2} \quad \text{and} \quad E(X^2) = 0^2\left(\frac{1}{2}\right) + 1^2\left(\frac{1}{2}\right) = \frac{1}{2}$$

$$\text{Var}(X) = E(X^2) - \mu_X^2 = \frac{1}{2} - \frac{1}{4} = \frac{1}{4}$$

The bottom marginal distribution is the distribution of Y; hence

$$\mu_Y = E(Y) = 0\left(\frac{3}{4}\right) + 1\left(\frac{1}{4}\right) = \frac{1}{4} \quad \text{and} \quad E(Y^2) = 0^2\left(\frac{3}{4}\right) + 1^2\left(\frac{1}{4}\right) = \frac{1}{4}$$

$$\text{var}(Y) = E(Y^2) - \mu_Y^2 = \frac{1}{4} - \frac{1}{16} = \frac{3}{16}$$

The random variable $Z = X + Y$ assumes the values 0, 1, 2 with respective probabilities $\frac{1}{2}, \frac{1}{4}, \frac{1}{4}$. Thus

$$\mu_Z = E(Z) = 0\left(\frac{1}{2}\right) + 1\left(\frac{1}{4}\right) + 2\left(\frac{1}{4}\right) = \frac{3}{4} \quad \text{and} \quad E(Z^2) = 0^2\left(\frac{1}{2}\right) + 1^2\left(\frac{1}{4}\right) + 2^2\left(\frac{1}{4}\right) = \frac{5}{4}$$

$$\text{var}(Z) = E(Z^2) - \mu_Z^2 = \frac{5}{4} - \frac{9}{16} = \frac{11}{16}$$

Therefore

$$E(X) + E(Y) = \frac{1}{2} + \frac{1}{4} = \frac{3}{4} = E(Z) \qquad \text{but} \qquad \text{var}(X) + \text{var}(Y) = \frac{1}{4} + \frac{3}{16} = \frac{7}{16} \neq \frac{11}{16} = \text{var}(Z)$$

EXAMPLE 5.18　Let X and Y be the independent random variables in Example 5.16(b), and let $Z = X + Y$. We show that

$$E(Z) = E(X) + E(Y) \quad \text{and} \quad \text{var}(Z) = \text{var}(X) + \text{var}(Y)$$

The equation for $E(Z)$ is always true, and the equation for $\text{var}(Z)$ is expected since X and Y are independent.

The joint distribution of X and Y appears in Fig. 5-8(b). The right and bottom marginal distributions are the distributions of X and Y, respectively, and they are identical. Thus

$$\mu_X = \mu_Y = 0\left(\frac{1}{2}\right) + 1\left(\frac{1}{2}\right) = \frac{1}{2} \quad \text{and} \quad E(X^2) = E(Y^2) = 0^2\left(\frac{1}{2}\right) + 1^2\left(\frac{1}{2}\right) = \frac{1}{2}$$

$$\text{var}(X) = \text{var}(Y) = E(X^2) - \mu_X^2 = \frac{1}{2} - \frac{1}{4} = \frac{1}{4}$$

The random variable $Z = X + Y$ assumes the values 0, 1, 2 but now with respective probabilities 1/4, 1/2, 1/4. Thus:

$$\mu_Z = E(Z) = 0\left(\frac{1}{4}\right) + 1\left(\frac{1}{2}\right) + 2\left(\frac{1}{4}\right) = 1 \quad \text{and} \quad E(Z^2) = 0^2\left(\frac{1}{4}\right) + 1^2\left(\frac{1}{2}\right) + 2^2\left(\frac{1}{4}\right) = \frac{3}{2}$$

$$\text{var}(Z) = E(Z^2) - \mu_Z^2 = \frac{3}{2} - 1 = \frac{1}{2}$$

Therefore

$$E(X) + E(Y) = \frac{1}{2} + \frac{1}{2} = 1 = E(Z)$$

and

$$\text{var}(X) + \text{var}(Y) = \frac{1}{4} + \frac{1}{4} = \frac{1}{2} = \text{var}(Z)$$

5.9 DISCRETE RANDOM VARIABLES IN GENERAL

Now suppose X is a random variable on a sample space S with a countable infinite range space, say $R_S = \{x_1, x_2, \ldots\}$. As in the finite case, X induces a function f on R_X, called the *distribution* of X, defined by

$$f(x_i) \equiv P(X = x_i)$$

The distribution is frequently presented by a table as follows:

x	x_1	x_2	x_3	$\cdots$
$f(x)$	$f(x_1)$	$f(x_2)$	$f(x_3)$	$\cdots$

The distribution f has the following two properties:

$$\text{(i)} \quad f(x_i) \geq 0, \qquad \text{(ii)} \quad \sum_{i=1}^{\infty} f(x_i) = 1$$

Thus, R_X with the above assignment of probabilities is a probability space.

The *expectation* $E(X)$ and *variance* var(X) of the above random variable X are defined by the following series when the relevant series converge absolutely:

$$E(X) = x_1 f(x_1) + x_2 f(x_2) + x_3 f(x_3) + \cdots = \sum_{i=1}^{\infty} x_i f(x_i)$$

$$\text{var}(X) = (x_1 - \mu)^2 f(x_1) + (x_2 - \mu)^2 f(x_2) + \cdots = \sum_{i=1}^{\infty} (x_i - \mu)^2 f(x_i)$$

It can be shown that var(X) exists if and only if $\mu = E(X)$ and $E(X^2)$ both exist and in this case the following formula holds just as in the finite case:

$$\text{var}(X) = E(X^2) - \mu^2$$

When var(X) does exist, the *standard deviation* σ_X is defined just as in the finite case:

$$\sigma_X = \sqrt{\text{var}(X)}$$

The notions of joint distribution, independent random variables, and functions of random variables are the same as in the finite case. Moreover, suppose X and Y are defined on the same sample space S and $\operatorname{var}(X)$ and $\operatorname{var}(Y)$ both exist.   Then the *covariance* of X and Y, written $\operatorname{cov}(X, Y)$, is defined by the following series which can also be shown to converge absolutely:

$$\operatorname{cov}(X, Y) = \sum_{i,j} (x_i - \mu_X)(y_j - \mu_Y)h(x_i, y_j)$$

In addition, the relation

$$\operatorname{cov}(X, Y) = \sum_{i,j} x_i y_j h(x_i, y_j) - \mu_X \mu_Y = E(XY) - \mu_X \mu_Y$$

holds just as in the finite case.

Remark: To avoid technical difficulties, we will establish many of the theorems in this chapter only for finite random variables.

5.10 CONTINUOUS RANDOM VARIABLES

Suppose that X is a random variable on a sample space S whose range space R_X is a continuum of numbers such as an interval.   Recall from the definition of a random variable that the set $\{a \leq X \leq b\}$ is an event in S and therefore the probability $P(a \leq X \leq b)$ is well defined.   We assume there is a piecewise continuous function $f: \mathbf{R} \rightarrow \mathbf{R}$ such that $P(a \leq X \leq b)$ is equal to the area under the graph of f between $x = a$ and $x = b$, as shown in Fig. 5-9.   In the language of calculus

$$P(a \leq X \leq b) = \int_a^b f(x)\,dx$$

In this case X is said to be a *continuous random variable*.   The function f is called the *distribution* or the *continuous probability function* (or *density function*) of X; it satisfies the conditions:

$$\text{(i)} \quad f(x) \geq 0, \qquad \text{and} \qquad \text{(ii)} \quad \int_{-\infty}^{\infty} f(x)\,dx = 1$$

That is, f is nonnegative and the total area under its graph is 1.

The *expectation* $E(X)$ for a continuous random variable X is defined by the following integral when it exists:

$$E(X) = \int_{-\infty}^{\infty} x f(x)\,dx$$

Functions of random variables are defined just as in the discrete case.   Furthermore, if $Y = \Phi(X)$ then it can be shown that

$$E(Y) = \int_{-\infty}^{\infty} \Phi(x) f(x)\,dx$$

when it exists.

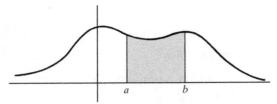

$P(a \leq X \leq b)$ = area of shaded region.

Fig. 5-9

The *variance* var(X) is defined by the following integral when it exists:

$$\text{var}(X) = E((X - \mu)^2) = \int_{-\infty}^{\infty} (x - \mu)^2 f(x)\, dx$$

Just as in the discrete case, it can be shown that var(X) exists if and only if $\mu = E(X)$ and $E(X^2)$ both exist and then

$$\text{var}(X) = E(X^2) - \mu^2 = \int_{-\infty}^{\infty} x^2 f(x)\, dx - \mu^2$$

When var(X) does exist, the *standard deviation* σ_X is defined as in the discrete case by

$$\sigma_X = \sqrt{\text{var}(X)}$$

EXAMPLE 5.19 Let X be a random variable with the following distribution function f:

$$f(x) = \begin{cases} \frac{1}{2}x & \text{if } 0 \le x \le 2 \\ 0 & \text{elsewhere} \end{cases}$$

The graph of f appears in Fig. 5-10. Then

$$P(1 \le X \le 1.5) = \text{area of shaded region in diagram} = \frac{5}{16}$$

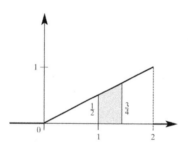

Fig. 5-10

Using calculus, we are able to compute the expectation, variance, and standard deviation of X as follows:

$$E(X) = \int_{-\infty}^{\infty} xf(x)\, dx = \int_{0}^{2} \frac{1}{2}x^2\, dx = \left[\frac{x^3}{6}\right]_{0}^{2} = \frac{4}{3}$$

$$E(X^2) = \int_{-\infty}^{\infty} x^2 f(x)\, dx = \int_{0}^{2} \frac{1}{2}x^3\, dx = \left[\frac{x^4}{8}\right]_{0}^{2} = 2$$

$$\text{var}(X) = E(X^2) - \mu^2 = 2 - \frac{16}{9} = \frac{2}{9} \quad \text{and} \quad \sigma_X = \sqrt{\frac{2}{9}} = \frac{1}{3}\sqrt{2}$$

Independent Continuous Random Variables

A finite number of continuous random variables $X, Y, \ldots, Z$ are said to be *independent* if, for any intervals $[a, a']$, $[b, b']$, ..., $[c, c']$, we have

$$P(a \le X \le a', b \le Y \le b', \ldots, c \le Z \le c') = P(a \le X \le a')P(b \le Y \le b') \cdots P(c \le Z \le c')$$

Observe that intervals play the same role in the continuous case as points did in the discrete case.

5.11 CUMULATIVE DISTRIBUTION FUNCTION

Let X be a random variable (discrete or continuous). The *cumulative distribution function* F of X is the function $F: \mathbf{R} \to \mathbf{R}$ defined by

$$F(a) = P(X \leq a)$$

Suppose X is a discrete random variable with distribution f. Then F is the "step function" defined by

$$F(x) = \sum_{x_i \leq x} f(x_i)$$

On the other hand, suppose X is a continuous random variable with distribution f. Then

$$F(x) = \int_{-\infty}^{x} f(t)\, dt$$

In either case, F has the following two properties:

(i) F is *monotonically increasing*, that is,

$$F(a) \leq F(b) \quad \text{whenever} \quad a \leq b$$

(ii)   The limit of F to the left is 0 and to the right is 1:

$$\lim_{x \to -\infty} F(x) = 0 \quad \text{and} \quad \lim_{x \to \infty} F(x) = 1$$

EXAMPLE 5.20   Let X be a discrete random variable with the following distribution function f:

x	-2	1	2	4
$f(x)$	1/4	1/8	1/2	1/8

The graph of the cumulative distribution function F of X appears in Fig. 5-11. Observe that F is a "step function" with a step at x_i with height $f(x_i)$.

EXAMPLE 5.21   Let X be a continuous random variable with the following distribution function f:

$$f(x) = \begin{cases} \frac{1}{2}x & \text{if } 0 \leq x \leq 2 \\ 0 & \text{elsewhere} \end{cases}$$

The cumulative distribution function F of X follows:

$$F(x) = \begin{cases} 0 & \text{if } x < 0 \\ \frac{1}{4}x^2 & \text{if } 0 \leq x \leq 2 \\ 1 & \text{if } x > 2 \end{cases}$$

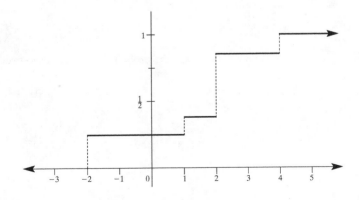

Fig. 5-11. Graph of F.

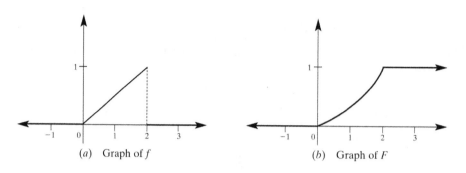

(a) Graph of f (b) Graph of F

Fig. 5-12

Here we use the fact that, for $0 \le x \le 2$,

$$F(x) = \int_0^x \frac{1}{2} t \, dt = \frac{1}{4} x^2$$

The graph of f appears in Fig. 5-12(a) and the graph of F appears in Fig. 5-12(b).

5.12 CHEBYSHEV'S INEQUALITY AND THE LAW OF LARGE NUMBERS

The standard deviation σ of a random variable X measures the spread of the values of X about the mean μ of X.   Accordingly, for smaller values of σ, we would expect that X will be closer to its mean μ.   This intuitive expectation is made more precise by the following inequality, named after the Russian mathematician P. L. Chebyshev (1921–1994):

Theorem 5.12   (Chebyshev's Inequality): Let X be a random variable with mean μ and standard deviation σ.   Then, for any positive number k, the probability that a value of X lies in the interval $[\mu - k\sigma, \mu + k\sigma]$ is at least $1 - 1/k^2$.   That is,

$$P(\mu - k\sigma \le X \le \mu + k\sigma) \ge 1 - \frac{1}{k^2}$$

A proof of this important theorem is given in Problem 5.51.   We illustrate the use of this theorem in the next example.

EXAMPLE 5.22   Suppose X is a random variable with mean $\mu = 100$ and standard deviation $\sigma = 5$.

(a)   Find the conclusion that one can derive from Chebyshev's inequality for $k = 2$ and $k = 3$.
      Setting $k = 2$, we get

$$\mu - k\sigma = 100 - 2(5) = 90, \qquad \mu + k\sigma = 100 + 2(5) = 110, \qquad 1 - \frac{1}{k^2} = 1 - \frac{1}{2^2} = \frac{3}{4}$$

Thus, from Chebyshev's inequality, we can conclude that the probability that X lies between 90 and 110 is at least 3/4, that is

$$P(90 \le X \le 110) \ge \frac{3}{4}$$

Similarly, setting $k = 3$, we get

$$\mu - k\sigma = 100 - 3(5) = 85, \qquad \mu + k\sigma = 100 + 3(5) = 115, \qquad 1 - \frac{1}{k^2} = \frac{8}{9}$$

Thus $$P(85 \le X \le 115) \ge \frac{8}{9}$$

(b) Estimate the probability that X lies between $100 - 20 = 80$ and $100 + 20 = 120$.

Here $k\sigma = 20$. Since $\sigma = 5$, we get $5k = 20$ and so $k = 4$. Thus, by Chebyshev's inequality,

$$P(80 \le X \le 120) \ge 1 - \frac{1}{k^2} = 1 - \frac{1}{4^2} = \frac{15}{16} \approx 0.94$$

(c) Find an interval $[a, b]$ about the mean $\mu = 100$ for which the probability that X lies in the interval is at least 99 percent.

Here we set $1 - 1/k^2 = 0.99$ and solve for k. This yields

$$1 - 0.99 = \frac{1}{k^2} \quad \text{or} \quad 0.01 = \frac{1}{k^2} \quad \text{or} \quad k^2 = \frac{1}{0.1} = 100 \quad \text{or} \quad k = 10$$

Thus, the desired interval is

$$[a, b] = [\mu - k\sigma, \mu + k\sigma] = [100 - 10(5), 100 + 10(5)] = [50, 150]$$

Sample Mean and the Law of Large Numbers

The intuitive idea of probability is the so-called law of averages, that is, if an event A occurs with probability p, then the "average number of occurrences of A" approaches p as the number n of (independent) trials increases. This concept is made precise by the law of large numbers which is stated below. First, however, we need to define the notion of the sample mean.

Let X be the random variable corresponding to some experiment. The notion of n independent trials of the experiment was defined above. We may view the numerical value of each particular trial to be a random variable with the same mean as X. Specifically, we let X_k denote the outcome of the kth trial where $k = 1, 2, \ldots, n$. The average value of all n outcomes is also a random variable, denoted by $\bar{X}_n$ and called the *sample mean*. That is,

$$\bar{X}_n = \frac{X_1 + X_2 + \cdots + X_n}{n}$$

The law of large numbers says that as n increases, the probability that the value of sample mean $\bar{X}_n$ is close to μ approaches 1.

EXAMPLE 5.23 Suppose a fair die is tossed 8 times with the following outcomes:

$$x_1 = 2, \; x_2 = 5, \; x_3 = 4, \; x_4 = 1, \; x_5 = 4, \; x_6 = 6, \; x_7 = 3, \; x_8 = 2$$

Then the corresponding value of the sample mean $\bar{X}_8$ follows:

$$\bar{x}_8 = \frac{2 + 5 + 4 + 1 + 4 + 6 + 3 + 2}{8} = \frac{27}{8} = 3.375$$

For a fair die, the mean $\mu = 3.5$. The law of large numbers tells us that as n gets larger, the probability that the sample mean $\bar{X}_n$ will get close to 3.5 becomes larger and, in fact, approaches one.

A technical statement of the law of large numbers follows.

Theorem 5.13 (Law of Large Numbers): For any positive number α, no matter how small,

$$P(\mu - \alpha \le \bar{X}_n \le \mu + \alpha) \to 1 \quad \text{as} \quad n \to \infty$$

That is, the probability that the sample mean $\bar{X}_n$ has a value in the interval $[\mu - \alpha, \mu + \alpha]$ approaches 1 as n approaches infinity.

The following remarks are in order.

Remark 1:   We prove Chebyshev's inequality only for the discrete case.   The continuous case follows from an analogous proof which uses integrals instead of summations.

Remark 2:   We prove the law of large numbers only in the case that the variance var(X_i) of the X_i exists, that is, does not diverge.   We note that the theorem is true whenever the expectation $E(X_i)$ exists.

Remark 3:   The above law of large numbers is proved in Problem 5.52 using Chebyshev's inequality.   A stronger version of the theorem, called the strong law of large numbers, is given in more advanced treatments of probability theory.

Solved Problems

RANDOM VARIABLES AND EXPECTED VALUE

5.1.   Suppose a random variable X takes on the values -3, -1, 2, and 5 with respective probabilities

$$\frac{2k-3}{10}, \qquad \frac{k-2}{10}, \qquad \frac{k-1}{10}, \qquad \frac{k+1}{10}$$

(*a*) Determine the distribution of X.   (*b*) Find the expected value $E(X)$ of X.

(*a*)   Set the sum of the probabilities equal to 1, and solve for k obtaining $k = 3$.   Then put $k = 3$ into the above probabilities yielding 0.3, 0.4, 0.2, 0.1.   Thus, the distribution of X is as follows:

x	-3	-1	2	5
$P(X = x)$	0.3	0.1	0.2	0.4

(*b*)   The expected value $E(X)$ is obtained by multiplying each value of X by its probability and taking the sum.   Thus

$$E(X) = (-3)(0.3) + (-1)(0.1) + 2(0.2) + 5(0.4) = 1.4$$

5.2.   A fair coin is tossed 4 times.   Let X denote the number of heads occurring.   Find:

(*a*) distribution f of X,   (*b*) $E(X)$,   (*c*) probability graph and histogram of X.

The sample space S is an equiprobable space consisting of $2^4 = 16$ sequences made up of H's and T's.

(*a*)   Since X is the number of heads, and each sequence consists of four elements, X takes on the values of 0, 1, 2, 3, 4, that is, $R_X = \{0, 1, 2, 3, 4\}$.

  (i)   One point $TTTT$ has 0 heads; hence $f(1) = 1/16$.
  (ii)   Four points, $HTTT, THTT, TTHT, TTTH$, have 1 head; hence $f(1) = 4/16$.
  (iii)   Six points, $HHTT, HTHT, HTTH, THHT, THTH, TTHH$, have 2 heads; hence $f(2) = 6/16$.
  (iv)   Four points, $HHHT, HHTH, HTHH, THHH$, have 1 head; hence $f(1) = 4/16$.
  (v)   One point, $HHHH$, has 4 heads; hence $f(4) = 1/16$.

The distribution f of X follows:

x	0	1	2	3	4
$f(x)$	$\frac{1}{16}$	$\frac{4}{16}$	$\frac{6}{16}$	$\frac{4}{16}$	$\frac{1}{16}$

(b) The expected value $E(X)$ is obtained by multiplying each value of X by its probability and taking the sum. Hence

$$E(X) = 0\left(\frac{1}{16}\right) + 1\left(\frac{4}{16}\right) + 2\left(\frac{6}{16}\right) + 3\left(\frac{4}{16}\right) + 4\left(\frac{1}{16}\right) = 2$$

This agrees with our intuition that, when a fair coin is repeatedly tossed, about half of the tosses should be heads.

(c) The probability bar chart of X appears in Fig. 5-13(a), and the probability histogram appears in Fig. 5-13(b). One may view the histogram as making the random variable continuous where $X = 1$ means X lies between 0.5 and 1.5.

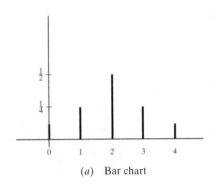

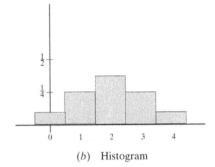

(a) Bar chart　　　　　　　　　　　　　　(b) Histogram

Fig. 5-13

5.3. A fair coin is tossed until a head or five tails occurs. Find the expected number E of tosses of the coin.

The sample space S consists of the six points

$$H,\ TH,\ TTH,\ TTTH,\ TTTTH,\ TTTTT$$

with respective probabilities (independent trials)

$$\frac{1}{2},\ \left(\frac{1}{2}\right)^2 = \frac{1}{4},\ \left(\frac{1}{2}\right)^3 = \frac{1}{8},\ \left(\frac{1}{2}\right)^4 = \frac{1}{16},\ \left(\frac{1}{2}\right)^5 = \frac{1}{16},\ \left(\frac{1}{2}\right)^5 = \frac{1}{16}$$

The random variable X of interest is the number of tosses in each outcome. Thus

$$\begin{array}{lll} X(H) = 1 & X(TTH) = 3 & X(TTTTH) = 5 \\ X(TH) = 2 & X(TTTH) = 4 & X(TTTTT) = 5 \end{array}$$

These X values are assigned the following probabilities:

$$P(1) \equiv P(H) = \frac{1}{2}, \quad P(2) \equiv P(TH) = \frac{1}{4}, \quad P(3) \equiv P(TTH) = \frac{1}{8}$$

$$P(4) \equiv P(TTTH) = \frac{1}{16}, \quad P(5) \equiv P(\{TTTTH, TTTTT\}) = \frac{1}{32} + \frac{1}{32} = \frac{1}{16}$$

Accordingly,　　　　　$E = E(X) = 1\left(\frac{1}{2}\right) + 2\left(\frac{1}{4}\right) + 3\left(\frac{1}{8}\right) + 4\left(\frac{1}{16}\right) + 5\left(\frac{1}{16}\right) \approx 1.9$

5.4. A random sample with replacement of size $n = 2$ is chosen from the set $\{1, 2, 3\}$ yielding the 9-element equiprobable space

$$S = \{(1,1),\ (1,2),\ (1,3),\ (2,1),\ (2,2),\ (2,3),\ (3,1),\ (3,2),\ (3,3)\}$$

Let X denote the sum of the two numbers. (a) Find the distribution f of X. (b) Find the expected value $E(X)$.

(a)   The random variable X assumes the values 2, 3, 4, 5, 6, that is, $R_X = \{2, 3, 4, 5, 6\}$. We compute the distribution f of X:

   (i)   One point $(1, 1)$ has sum 2; hence $f(2) = 1/9$.

  (ii)   Two points, $(1, 2)$, $(2, 1)$, have sum 3; hence $f(3) = 2/9$.

 (iii)   Three points, $(1, 3)$, $(2, 2)$, $(1, 3)$, have sum 4; hence $f(4) = 3/9$.

 (iv)   Two points, $(2, 3)$, $(3, 2)$, have sum 5; hence $f(5) = 2/9$.

  (v)   One point $(3, 3)$ has sum 6; hence $f(6) = 1/9$.

Thus, the distribution f of X is as follows:

x	2	3	4	5	6
$f(x)$	1/9	2/9	3/9	2/9	1/9

(b)   The expected value $E(X)$ is obtained by multiplying each value of x by its probability and taking the sum. Thus

$$E(X) = 2\left(\frac{1}{9}\right) + 3\left(\frac{2}{9}\right) + 4\left(\frac{3}{9}\right) + 5\left(\frac{2}{9}\right) + 6\left(\frac{1}{9}\right) = 4$$

5.5. Let Y denote the minimum of the two numbers in each element of the probability space S in Problem 5.4. (a) Find the distribution g of Y. (b) Find the expected value $E(Y)$.

(a)   The random variable Y only assumes the values 1, 2, 3, that is, $R_Y = \{1, 2, 3\}$. We compute the distribution g of Y:

   (i)   Five points, $(1, 1)$, $(1, 2)$, $(1, 3)$, $(2, 1)$, $(3, 1)$, have minimum 1; hence $g(1) = 5/9$.

  (ii)   Three points, $(2, 2)$, $(2, 3)$, $(3, 2)$, have minimum 2; hence $g(2) = 3/9$.

 (iii)   One point $(3, 3)$ has minimum 3; hence $g(3) = 1$.

Thus the distribution g of Y is as follows:

y	1	2	3
$g(y)$	5/9	3/9	1/9

(b)   Multiply each value of y by its probability and take the sum obtaining:

$$E(Y) = 1\left(\frac{5}{9}\right) + 2\left(\frac{3}{9}\right) + 3\left(\frac{1}{9}\right) = \frac{12}{9} \approx 1.33$$

5.6. Five cards are numbered 1 to 5. Two cards are drawn at random (without replacement) to yield the following equiprobable space S with $C(5, 2) = 10$ elements:

$$S = [\{1, 2\}, \{1, 3\}, \{1, 4\}, \{1, 5\}, \{2, 3\}, \{2, 4\}, \{2, 5\}, \{3, 4\}, \{3, 5\}, \{4, 5\}]$$

Let X denote the sum of the numbers drawn. (a) Find the distribution f of X. (b) Find $E(X)$.

(a) The random variable X assumes the values 3, 4, 5, 6, 7, 8, 9, that is, $R_X = \{3, 4, 5, 6, 7, 8, 9\}$. The distribution f of X is obtained as follows:

    (i)   One point, $\{1, 2\}$, has sum 3; hence $f(3) = 0.1$.

    (ii) One point, $\{1, 3\}$, has sum 4; hence $f(4) = 0.1$.

    (iii) Two points, $\{1, 4\}$, $\{2, 3\}$, have sum 5; hence $f(5) = 0.2$.

And so on. This yields the following distribution f of X:

x	3	4	5	6	7	8	9
$f(x)$	0.1	0.1	0.2	0.2	0.2	0.1	0.1

(b) The expected value $E(X)$ is obtained by multiplying each value x of X by its probability $f(x)$ and taking the sum. Thus

$$E(X) = 3(0.1) + 4(0.1) + 5(0.2) + 6(0.2) + 7(0.2) + 8(0.1) + 9(0.1) = 6$$

5.7. Let Y denote the minimum of the two numbers in each element of the probability space S in Problem 5.6. (a) Find the distribution g of Y. (b) Find $E(Y)$.

(a) The random variable Y only assumes the values 1, 2, 3, 4, that is, $R_Y = \{1, 2, 3, 4\}$. The distribution g of Y is obtained as follows:

    (i)   Four points, $\{1, 2\}$, $\{1, 3\}$, $\{1, 4\}$, $\{1, 5\}$, have minimum 1; hence $g(1) = 0.4$.

    (ii) Three points, $\{2, 3\}$, $\{2, 4\}$, $\{2, 5\}$, have minimum 2; hence $g(2) = 0.3$.

    (iii) Two points, $\{3, 4\}$, $\{3, 5\}$, have minimum 31; hence $g(3) = 0.2$.

    (iv) One point, $\{4, 5\}$, has minimum 4; hence $g(4) = 0.1$.

Thus, the distribution g of Y is as follows:

y	1	2	3	4
$g(y)$	0.4	0.3	0.2	0.1

(b) Multiply each value of y by its probability $g(y)$ and take the sum, obtaining

$$E(Y) = 1(0.4) + 2(0.3) + 3(0.2) + 4(0.1) = 2.0$$

5.8. A player tosses two fair coins yielding the equiprobable space

$$S = \{HH, HT, TH, TT\}$$

The player wins \$2 if 2 heads occur and \$1 if 1 head occurs. On the other hand, the player loses \$3 is no heads occur. Find the expected value E of the game. Is the game fair? (The game is *fair*, *favorable*, or *unfavorable* to the player accordingly as $E = 0$, $E > 0$, or $E < 0$.)

Let X denote the player's gain to yield

$$X(HH) = \$2, \ X(HT) = X(TH) = \$1, \ X(TT) = -\$3$$

Thus, the distribution of X is as follows:

x	2	1	-3
$P(X = x)$	1/4	2/4	1/4

The expectation of X follows:

$$E = E(X) = 2\left(\frac{1}{4}\right) + 1\left(\frac{2}{4}\right) - 3\left(\frac{1}{4}\right) = \frac{1}{4} = \$0.25$$

Since $E(X) > 0$, the game is favorable to the player.

5.9.   A player tosses two fair coins. The player wins \$3 if 2 heads occur and \$1 if 1 head occurs. For the game to be fair, how much should the player lose if no heads occur?

Let Y denote the player's gain; then the distribution of Y is as follows where k denotes the unknown payoff to the player:

y	3	1	k
$P(Y = y)$	1/4	2/4	1/4

Thus

$$E = E(Y) = 3\left(\frac{1}{4}\right) + 1\left(\frac{2}{4}\right) + k\left(\frac{1}{4}\right) = \frac{5 + k}{4}$$

For a fair game, $E(Y)$ should be zero. This yields $k = -5$. Thus, the player should lose \$5 if no heads occur.

5.10.   A box contains 8 lightbulbs of which 3 are defective. A bulb is selected from the box and tested. If it is defective, another bulb is selected and tested, until a nondefective bulb is chosen. Find the expected number E of bulbs chosen.

Writing D for defective and N for nondefective, the sample space S has the four elements

$$N, \quad DN, \quad DDN, \quad DDDN$$

with respective probabilities

$$\frac{5}{8}, \qquad \frac{3}{8} \cdot \frac{5}{7} = \frac{15}{56}, \qquad \frac{3}{8} \cdot \frac{2}{7} \cdot \frac{5}{6} = \frac{5}{56}, \qquad \frac{3}{8} \cdot \frac{2}{7} \cdot \frac{1}{6} \cdot \frac{5}{5} = \frac{1}{56}$$

The number X of bulbs chosen has the values

$$X(N) = 1, \qquad X(DN) = 2, \qquad X(DDN) = 3, \qquad X(DDDN) = 4$$

with the above respective probabilities. Hence

$$E(X) = 1\left(\frac{5}{8}\right) + 2\left(\frac{15}{56}\right) + 3\left(\frac{5}{56}\right) + 4\left(\frac{1}{56}\right) = \frac{3}{2} = 1.5$$

5.11.   A coin is weighted so that $P(H) = \frac{3}{4}$ and $P(T) = \frac{1}{4}$. The coin is tossed 3 times yielding the following 8-element probability space:

$$S = \{HHH, HHT, HTH, HTT, THH, THT, TTH, TTT\}$$

Let X denote the number of heads that appears. (a) Find the distribution f of X. (b) Find $E(X)$.

(a)   The points in S have the following respective probabilities:

$$P(HHH) = \frac{3}{4} \cdot \frac{3}{4} \cdot \frac{3}{4} = \frac{27}{64}, \qquad P(THH) = \frac{1}{4} \cdot \frac{3}{4} \cdot \frac{3}{4} = \frac{9}{64}$$

$$P(HHT) = \frac{3}{4} \cdot \frac{3}{4} \cdot \frac{1}{4} = \frac{9}{64}, \qquad P(THT) = \frac{1}{4} \cdot \frac{3}{4} \cdot \frac{1}{4} = \frac{3}{64}$$

$$P(HTH) = \frac{3}{4} \cdot \frac{1}{4} \cdot \frac{3}{4} = \frac{9}{64}, \qquad P(TTH) = \frac{1}{4} \cdot \frac{1}{4} \cdot \frac{3}{4} = \frac{3}{64}$$

$$P(HTT) = \frac{3}{4} \cdot \frac{1}{4} \cdot \frac{1}{4} = \frac{3}{64}, \qquad P(TTT) = \frac{1}{4} \cdot \frac{1}{4} \cdot \frac{1}{4} = \frac{1}{64}$$

Since X denotes the number of heads,

$$X(HHH) = 3 \qquad X(HHT) = X(HTH) = X(THH) = 2$$
$$X(HHT) = X(THT) = X(TTH) = 1 \qquad X(TTT) = 0$$

Thus, $R_X = \{0, 1, 2, 3\}$ is the range space of X.   Also,

$$f(0) = P(HHH) = \frac{1}{16}, \qquad\qquad\qquad f(2) = P(HHT) + P(HTH) + P(THH) = \frac{27}{64}$$

$$f(1) = P(HTT) + P(THT) + P(TTH) = \frac{9}{64}, \qquad f(3) = P(HHH) = \frac{27}{64}$$

The distribution of X follows:

x	0	1	2	3
$f(x)$	$\frac{1}{64}$	$\frac{9}{64}$	$\frac{27}{64}$	$\frac{27}{64}$

(b)   Multiply each value of x by its probability $f(x)$ and take the sum to obtain

$$E(X) = 0\left(\frac{1}{64}\right) + 1\left(\frac{9}{64}\right) + 2\left(\frac{27}{64}\right) + 3\left(\frac{27}{64}\right) = \frac{144}{64} = 2.25$$

5.12.   Concentric circles of radius 1 and 3 in are drawn on a circular target of radius 5 in as pictured in Fig. 5-14.   A person fires at the target and, as indicated by Fig. 5-14, receives 10, 5, or 3 points according to whether the target is hit inside the smaller circle, inside the middle annular region, or inside the outer annular region, respectively.   Suppose the person hits the target with probability $\frac{1}{2}$, and then is just as likely to hit one point of the target as the other.   Find the expected number E of points scored each time the person fires.

The probability of scoring 10, 5, 3, or 0 points follows:

$$f(10) = \frac{1}{2} \cdot \frac{\text{area of 10 points}}{\text{area of target}} = \frac{1}{2} \cdot \frac{\pi(1)^2}{\pi(5)^2} = \frac{1}{50}$$

$$f(5) = \frac{1}{2} \cdot \frac{\text{area of 5 points}}{\text{area of target}} = \frac{1}{2} \cdot \frac{\pi(3)^2 - \pi(1)^2}{\pi(5)^2} = \frac{8}{50}$$

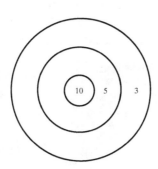

Fig. 5-14

$$f(3) = \frac{1}{2} \cdot \frac{\text{area of 3 points}}{\text{area of target}} = \frac{1}{2} \cdot \frac{\pi(5)^2 - \pi(3)^2}{\pi(5)^2} = \frac{16}{50}$$

$$f(0) = \frac{1}{2}$$

Thus, $E = 10\left(\frac{1}{50}\right) + 8\left(\frac{8}{50}\right) + 3\left(\frac{16}{50}\right) + 0\left(\frac{1}{2}\right) = \frac{98}{50} = 1.96$.

5.13. A coin is weighted so that $P(H) = p$ and hence $P(T) = q = 1 - p$. The coin is tossed until a head appears.

Let E denote the expected number of tosses. Prove $E = 1/p$.
That sample space is

$$S = \{H, TH, TTH, \ldots, T^n H, \ldots, T^\infty\}$$

where T^n denotes n tosses of H and T^∞ denotes the case that heads never appears. (This would happen if the coin were two-headed.) Let X denote the number of tosses. Accordingly, X assumes the values $1, 2, 3, \ldots, \infty$ with corresponding probabilities

$$p, qp, q^2 p, \ldots, q^n p, \ldots, 0$$

Thus
$$E = \sum nq^{n-1}p = p\left(\sum nq^{n-1}\right)$$

where the sum is from 1 to ∞. Let

$$y = \sum q^n = \frac{1}{1-q}$$

The derivative with respect to q yields

$$\frac{dy}{dq} = \sum nq^{n-1} = \frac{1}{(1-q)^2}$$

Substituting this value of $\sum nq^{n-1}$ in the formula for E yields

$$E = \frac{p}{(1-q)^2} = \frac{p}{p^2} = \frac{1}{p}$$

Note that calculus is used to evaluate the infinite series.

Remark: This is an example of an infinite discrete sample space.

5.14. A linear array *EMPLOYEE* has n elements. Suppose *NAME* appears randomly in the array, and there is a linear search to find the location K of *NAME*, that is, to find K such that *EMPLOYEE*[K] = *NAME*. Let $f(n)$ denote the number of comparisons in the linear search.

(a) Find the expected value of $f(n)$.

(b) Find the maximum value (worst case) of $f(n)$.

(a) Let X denote the number of comparisons. Since *NAME* can appear in any position in the array with the same probability of $1/n$, we have $X = 1, 2, 3, \ldots, n$, each with probability $1/n$. Hence

$$f(n) = E(X) = 1 \cdot \frac{1}{n} + 2 \cdot \frac{1}{n} + 3 \cdot \frac{1}{n} + \cdots + n \cdot \frac{1}{n}$$

$$= (1 + 2 + \cdots + n) \cdot \frac{1}{n} = \frac{n(n+1)}{2} \cdot \frac{1}{n} = \frac{n+1}{2}$$

(b) If *NAME* appears at the end of the array, then $f(n) = n$, which is the worst possible case.

MEAN, VARIANCE, AND STANDARD DEVIATION

5.15. Find the mean $\mu = E(X)$, variance $\sigma^2 = \text{var}(X)$, and standard deviation $\sigma = \sigma_X$ of each distribution:

(a)

x	2	3	11
$f(x)$	1/3	1/2	1/6

(b)

x	1	3	4	5
$f(x)$	0.4	0.1	0.2	0.3

Use the formulas,

$$\mu = E(X) = x_1 f(x_1) + x_2 f(x_2) + \cdots + x_m f(x_m) = \Sigma x_i f(x_i)$$
$$E(X^2) = x_1^2 f(x_1) + x_2^2 f(x_2) + \cdots + x_m^2 f(x_m) = \Sigma x_i^2 f(x_i)$$

Then use the formulas

$$\sigma^2 = \text{var}(X) = E(X^2) - \mu^2 \quad \text{and} \quad \sigma = \sigma_X = \sqrt{\text{var}(X)}$$

to obtain $\sigma^2 = \text{var}(X)$ and σ.

(a) Use the above formulas to first obtain:

$$\mu = \Sigma x_i f(x_i) = 2\left(\frac{1}{3}\right) + 3\left(\frac{1}{2}\right) + 11\left(\frac{1}{6}\right) = 4$$

$$E(X^2) = \Sigma x_i^2 f(x_i) = 2^2\left(\frac{1}{3}\right) + 3^2\left(\frac{1}{2}\right) + 11^2\left(\frac{1}{6}\right) = 26$$

Then

$$\sigma^2 = \text{var}(X) = E(X^2) - \mu^2 = 26 - 16 = 10$$
$$\sigma = \sqrt{\text{var}(X)} = \sqrt{10} = 3.2$$

(b) Use the above formulas to first obtain:

$$\mu = \Sigma x_i f(x_i) = 1(0.4) + 3(0.1) + 4(0.2) + 5(0.3) = 3$$
$$E(X^2) = \Sigma x_i^2 f(x_i) = 1(0.4) + 9(0.1) + 16(0.2) + 25(0.3) = 12$$

Then

$$\sigma^2 = \text{var}(X) = E(X^2) - \mu^2 = 12 - 9 = 3$$
$$\sigma = \sqrt{\text{var}(X)} = \sqrt{3} = 1.7$$

5.16. Find the mean $\mu = E(X)$, variance $\sigma^2 = \text{var}(X)$, and standard deviation $\sigma = \sigma_X$ of each distribution:

(a)

x_i	−5	−4	1	2
p_i	1/4	1/8	1/2	1/8

(b)

x_i	1	3	5	7
p_i	0.3	0.1	0.4	0.2

Here the distribution is presented using x_i and p_i instead of x and $f(x)$. The following are the analogous formulas:

$$\mu = E(X) = x_1 p_1 + x_2 p_2 + \cdots + x_m p_m = \Sigma x_i p_i$$
$$E(X^2) = x_1^2 p_1 + x_2^2 p_2 + \cdots + x_m^2 p_m = \Sigma x_i^2 p_i$$

Then, as before,

$$\sigma^2 = \text{var}(X) = E(X^2) - \mu^2 \quad \text{and} \quad \sigma = \sigma_X = \sqrt{\text{var}(X)}$$

(a)

$$\mu = E(X) = \Sigma x_i p_i = -5\left(\frac{1}{4}\right) - 4\left(\frac{1}{8}\right) + 1\left(\frac{1}{2}\right) + 2\left(\frac{1}{8}\right) = -1$$

$$E(X^2) = \Sigma x_i^2 p_i = 25\left(\frac{1}{4}\right) + 16\left(\frac{1}{8}\right) + 1\left(\frac{1}{2}\right) + 4\left(\frac{1}{8}\right) = 9.25$$

Then
$$\sigma^2 = \text{var}(X) = E(X^2) - \mu^2 = 9.25 - (-1)^2 = 8.25$$
$$\sigma = \sqrt{\text{var}(X)} = \sqrt{8.25} = 2.9$$

(b)
$$\mu = E(X) = \Sigma x_i p_i = 1(0.3) + 3(0.1) + 5(0.4) + 7(0.2) = 4.0$$
$$E(X^2) = \Sigma x_i^2 p_i = 1^2(0.3) + 3^2(0.1) + 5^2(0.4) + 7^2(0.2) = 21.0$$

Then
$$\sigma^2 = \text{var}(X) = E(X^2) - \mu^2 = 21 - (4)^2 = 5$$
$$\sigma = \sqrt{\text{var}(X)} = \sqrt{5} = 2.24$$

5.17. A fair die is tossed yielding the equiprobable space

$$S = \{1, 2, 3, 4, 5, 6\}$$

Let X denote twice the number appearing. Find the distribution f, mean μ_X, variance σ_X^2, and standard deviation σ_X of X.

Here $X(1) = 2$, $X(2) = 4$, $X(3) = 6$, $X(4) = 8$, $X(5) = 10$, $X(6) = 12$. Also, each number has probability 1/6. Thus, the following is the distribution f of X:

x	2	4	6	8	10	12
$f(x)$	1/6	1/6	1/6	1/6	1/6	1/6

Accordingly

$$\mu_X = E(X) = \Sigma x_i f(x_i)$$
$$= 2\left(\frac{1}{6}\right) + 4\left(\frac{1}{6}\right) + 6\left(\frac{1}{6}\right) + 8\left(\frac{1}{6}\right) + 10\left(\frac{1}{6}\right) + 12\left(\frac{1}{6}\right) = \frac{42}{6} = 7$$
$$E(X^2) = \Sigma x_i^2 f(x_i)$$
$$= 4\left(\frac{1}{6}\right) + 16\left(\frac{1}{6}\right) + 36\left(\frac{1}{6}\right) + 64\left(\frac{1}{6}\right) + 100\left(\frac{1}{6}\right) + 144\left(\frac{1}{6}\right) = \frac{354}{6} = 60.7$$

Then
$$\sigma_X^2 = \text{var}(X) = E(X^2) - \mu_X^2 = 60.7 - (7)^2 = 11.7$$
$$\sigma_X = \sqrt{\text{var}(X)} = \sqrt{11.7} = 3.4$$

5.18. A fair die is tossed yielding the equiprobable space

$$S = \{1, 2, 3, 4, 5, 6\}$$

Let Y be 1 or 3 accordingly as an odd or even number appears. Find the distribution g, expectation μ_Y, variance σ_Y^2, and standard deviation σ_Y of Y.

Here $Y(1) = 1$, $Y(2) = 3$, $Y(3) = 1$, $Y(4) = 3$, $Y(5) = 1$, $Y(6) = 3$. Then $R_Y = \{1, 3\}$ is the range space of Y. Therefore

$$g(1) = P(Y = 1) = P(\{1, 3, 5\}) = \frac{3}{6} = \frac{1}{2} \quad \text{and} \quad g(3) = P(Y = 3) = P(\{2, 4, 6\}) = \frac{3}{6} = \frac{1}{2}$$

Thus, the distribution g of Y is as follows:

y	1	3
$g(y)$	1/2	1/2

Accordingly,

$$\mu_Y = E(Y) = \sum y_i g(y_i) = 1\left(\frac{1}{2}\right) + 3\left(\frac{1}{2}\right) = 2$$

$$E(Y^2) = \sum y_i^2 g(y_i) = 1\left(\frac{1}{2}\right) + 9\left(\frac{1}{2}\right) = 5$$

Then

$$\sigma_Y^2 = \text{var}(Y) = E(Y^2) - \mu_Y^2 = 5 - (1)^2 = 1$$
$$\sigma_Y = \sqrt{\text{var}(Y)} = \sqrt{1} = 1$$

5.19. Let X and Y be the random variables in Problems 5.17 and 5.18 which are defined on the same sample space S. Recall that $Z = X + Y$ is the random variable on S defined by

$$Z(s) = X(s) + Y(s)$$

Find the distribution, expectation, variance, and standard deviation of $Z = X + Y$. Also, verify that $E(Z) = E(X + Y) = E(X) + E(Y)$.

The sample space is still $S = \{1, 2, 3, 4, 5, 6\}$ and each sample point still has probability 1/6. Use $Z(s) = X(s) + Y(s)$ and the values of X and Y from Problems 5.17 and 5.18 to obtain

$Z(1) = X(1) + Y(1) = 2 + 1 = 3$	$Z(4) = X(4) + Y(4) = 8 + 3 = 11$
$Z(2) = X(2) + Y(2) = 4 + 3 = 7$	$Z(5) = X(5) + Y(5) = 10 + 1 = 11$
$Z(3) = X(3) + Y(3) = 6 + 1 = 7$	$Z(6) = X(6) + Y(6) = 12 + 3 = 15$

The range space of Z is $R_Z = \{3, 7, 11, 15\}$. Also, 3 and 15 are each assumed at only one sample point and hence have a probability 1/6; whereas 7 and 11 are each assumed at two sample points and hence have a probability 2/6. Thus, the distribution of $Z = X + Y$ is as follows:

z_i	3	7	11	15
$P(z_i)$	1/6	2/6	2/6	1/6

Therefore

$$\mu_Z = E(Z) = \sum z_i P(z_i) = 3\left(\frac{1}{6}\right) + 7\left(\frac{2}{6}\right) + 11\left(\frac{2}{6}\right) + 15\left(\frac{1}{6}\right) = \frac{54}{6} = 9$$

$$E(Z^2) = \sum z_i^2 P(z_i) = 9\left(\frac{1}{6}\right) + 49\left(\frac{2}{6}\right) + 121\left(\frac{2}{6}\right) + 225\left(\frac{1}{6}\right) = \frac{574}{6} = 95.7$$

Then

$$\sigma_Z^2 = \text{var}(Z) = E(Z^2) - \mu_Z^2 = 95.7 - (9)^2 = 14.7$$
$$\sigma_Z = \sqrt{\text{var}(Z)} = \sqrt{14.7} = 3.8$$

Moreover, $E(Z) = E(X + Y) = 9 = 7 + 2 = E(X) + E(Y)$. On the other hand,

$$\text{var}(X) + \text{var}(Y) = 11.7 + 1 = 12.7 \neq \text{var}(Z)$$

5.20. Let X and Y be the random variable in Problems 5.17 and 5.18 which are defined on the same sample space S. Recall that $W = XY$ is the random variable on S defined by

$$W(s) = X(s) \cdot Y(s)$$

Find the distribution, expectation, variance, and standard deviation of $W = XY$.

The sample space is still $S = \{1, 2, 3, 4, 5, 6\}$ and each sample point still has a probability 1/6. Use $W(s) = (XY)(s) = X(s)Y(s)$ and the values of X and Y from Problems 5.17 and 5.18 to obtain

$W(1) = X(1)Y(1) = 2(1) = 2,$	$W(4) = X(4)Y(4) = 8(3) = 24$
$W(2) = X(2)Y(2) = 4(3) = 12,$	$W(5) = X(5)Y(5) = 10(1) = 10$
$W(3) = X(3)Y(3) = 6(1) = 6,$	$W(6) = X(6)Y(6) = 12(3) = 36$

Each value of $W = XY$ is assumed at just one sample point; hence the distribution of W is as follows:

w_i	2	6	10	12	24	36
$P(w_i)$	1/6	1/6	1/6	1/6	1/6	1/6

Therefore

$$\mu_w = E(W) = \sum w_i P(w_i) = \frac{2}{6} + \frac{6}{6} + \frac{10}{6} + \frac{12}{6} + \frac{24}{6} + \frac{36}{6} = \frac{90}{6} = 15$$

$$E(W^2) = \sum w_i^2 P(w_i) = \frac{4}{6} + \frac{36}{6} + \frac{100}{6} + \frac{144}{6} + \frac{576}{6} + \frac{1296}{6} = \frac{2156}{6} = 359.3$$

Then
$$\sigma_W^2 = \text{var}(W) = E(W^2) - \mu_W^2 = 359.3 - (15)^2 = 134.3$$
$$\sigma_W = \sqrt{\text{var}(W)} = \sqrt{134.3} = 11.6$$

[Note: $E(W) = E(XY) = 15 \neq E(X)E(Y) = 7(2) = 14$.]

5.21. Let X be a random variable with distribution

x	1	2	3
$P(x)$	0.3	0.5	0.2

Find the mean μ_X, variance σ_X^2, and standard deviation σ_X of X.

The formulas for μ_X and $E(X^2)$ yield

$$\mu_X = E(X) = \sum x_i P(x_i) = 1(0.3) + 2(0.5) + 3(0.2) = 1.9$$
$$E(X^2) = \sum x_i^2 P(x_i) = 1^2(0.3) + 2^2(0.5) + 3^2(0.2) = 4.1$$

Then
$$\sigma_X^2 = \text{var}(X) = E(X^2) - \mu^2 = 4.1 - (1.9)^2 = 0.49$$
$$\sigma_X = \sqrt{\text{var}(X)} = \sqrt{0.49} = 0.7$$

5.22. Consider the random variables X in the preceding Problem 5.21. Find the distribution, mean μ_Y, variance σ_Y^2, and standard deviation σ_Y of the random variable $Y = \Phi(X)$ where (a) $\Phi(x) = x^3$, (b) $\Phi(x) = 2^x$, (c) $\Phi(x) = x^2 + 3x + 4$.

The distribution of any arbitrary random variable $Y = \Phi(X)$ where $P(y) = P(x)$ is as follows:

y	$\Phi(1)$	$\Phi(2)$	$\Phi(3)$
$P(y)$	0.3	0.5	0.2

(a) Using $1^3 = 1$, $2^3 = 8$, $3^3 = 27$, the distribution of $Y = X^3$ is as follows:

y	1	8	27
$P(y)$	0.3	0.5	0.2

Therefore

$$\mu_Y = E(Y) = \sum \Phi(x_i)P(x_i) = \sum y_i P(y_i) = 1(0.3) + 8(0.5) + 27(0.2) = 9.7$$
$$E(Y^2) = \sum y_i^2 P(y_i) = 1^2(0.3) + 8^2(0.5) + 27^2(0.2) = 178.1$$

Then
$$\sigma_Y^2 = \text{var}(Y) = E(Y^2) - \mu^2 = 178.1 - (9.7)^2 = 84.0$$
$$\sigma_Y = \sqrt{\text{var}(Y)} = \sqrt{84.0} = 9.17$$

(b) Using $2^1 = 2$, $2^2 = 4$, $2^3 = 8$, the distribution of $Y = 2^X$ is as follows:

y	2	4	8
$P(y)$	0.3	0.5	0.2

Therefore
$$\mu_Y = E(Y) = \sum y_i P(y_i) = 2(0.3) + 4(0.5) + 8(0.2) = 4.2$$
$$E(Y^2) = \sum y_i^2 P(y_i) = 2^2(0.3) + 4^2(0.5) + 8^2(0.2) = 41.2$$

Then
$$\sigma_Y^2 = \text{var}(Y) = E(Y^2) - \mu^2 = 41.2 - (4.2)^2 = 23.6$$
$$\sigma_Y = \sqrt{\text{var}(Y)} = \sqrt{23.6} = 4.86$$

(c) Substitute $x = 1, 2, 3$ in $\Phi(x) = x^2 + 3x + 4$ to obtain $\Phi(1) = 8$, $\Phi(2) = 14$, $\Phi(3) = 22$. Then the distribution of $Y = X^2 + 3X + 4$ is as follows:

y	8	14	22
$P(y)$	0.3	0.5	0.2

Therefore
$$\mu_Y = E(Y) = \sum y_i P(y_i) = 8(0.3) + 14(0.5) + 22(0.2) = 13.9$$
$$E(Y^2) = \sum y_i^2 P(y_i) = 8^2(0.3) + 14^2(0.5) + 22^2(0.2) = 214$$

Then
$$\sigma_Y^2 = \text{var}(Y) = E(Y^2) - \mu^2 = 214 - (13.9)^2 = 20.8$$
$$\sigma_Y = \sqrt{\text{var}(Y)} = \sqrt{20.8} = 4.56$$

5.23. Let X be a random variable with distribution

x	1	3	5	7
$P(X = x)$	0.4	0.3	0.2	0.1

(a) Find the mean μ_X, variance σ_X^2, and standard deviation σ_X of X.

(b) Find the distribution of the standardized random variables $Z = (X - \mu)/\sigma$ of X, and show that $\mu_Z = 0$ and $\sigma_Z = 1$ (as predicted by Theorem 5.6).

(a) The formulas for μ_X and $E(X^2)$ yield
$$\mu_X = E(X) = \sum x_i P(x_i) = 1(0.4) + 3(0.3) + 5(0.2) + 7(0.1) = 3$$
$$E(X^2) = \sum x_i^2 P(x_i) = 1^2(0.4) + 3^2(0.3) + 5^2(0.2) + 7^2(0.1) = 13$$

Then
$$\sigma_X^2 = E(X^2) - \mu^2 = 13 - (3)^2 = 4 \qquad \text{and} \qquad \sigma_X = 2$$

(b) Using $Z = (X - \mu)/\sigma = (x - 3)/2$ and $P(z) = P(x)$, we obtain the following distribution of Z:

z	−1	0	1	2
$P(z)$	0.4	0.3	0.2	0.1

Therefore
$$\mu_Z = \sum z_i P(z_i) = -1(0.4) + 0(0.3) + 1(0.2) + 2(0.1) = 0$$
$$E(Z^2) = \sum z_i^2 P(z_i) = (-1)^2(0.4) + 0^2(0.3) + 1^2(0.2) + 2^2(0.1) = 1$$

Then
$$\sigma_Z^2 = E(Z^2) - \mu_Z^2 = 1 - (0)^2 = 1 \qquad \text{and} \qquad \sigma_Z = 1$$

5.24. Let X denote the number of heads when a fair coin is tossed 4 times. By Problem 5.2, the mean $\mu_X = 2$ and its distribution is as follows:

x	0	1	2	3	4
$P(x)$	1/16	4/16	6/16	4/16	1/16

(a) Find the standard deviation σ_X of X.

(b) Find the distribution of the standardized random variable $Z = (X - \mu)/\sigma$ of X, and show that $\mu_Z = 0$ and $\sigma_Z = 1$ (as predicted by Theorem 5.6).

(a) First compute $E(X^2)$ as follows:

$$E(X^2) = \Sigma x_i^2 p_i = 0^2\left(\frac{1}{16}\right) + 1^2\left(\frac{4}{16}\right) + 2^2\left(\frac{6}{16}\right) + 3^2\left(\frac{4}{16}\right) + 4^2\left(\frac{1}{16}\right) = \frac{80}{16} = 5$$

Using $\mu_X = 2$, we obtain

$$\sigma_X^2 = E(X^2) - \mu^2 = 5 - (2)^2 = 1 \qquad \text{and} \qquad \sigma_X = 1$$

(b) Using $Z = (X - \mu)/\sigma = (x - 2)/1$ and $P(z) = P(x)$, we obtain the following distribution of Z:

z	-2	-1	0	1	2
$P(z)$	1/16	4/16	6/16	4/16	1/16

Therefore

$$\mu_Z = \Sigma z_i P(z_i) = -2\left(\frac{1}{16}\right) - 1\left(\frac{4}{16}\right) + 0\left(\frac{6}{16}\right) + 1\left(\frac{4}{16}\right) + 2\left(\frac{1}{16}\right) = \frac{0}{16} = 0$$

$$E(Z^2) = \Sigma z_i^2 P(z_i) = 4\left(\frac{1}{16}\right) + 1\left(\frac{4}{16}\right) + 0\left(\frac{6}{16}\right) + 1\left(\frac{4}{16}\right) + 4\left(\frac{1}{16}\right) = \frac{16}{16} = 1$$

Then
$$\sigma_Z^2 = E(Z^2) - \mu_Z^2 = 1 - (0)^2 = 1 \quad \text{and} \quad \sigma_Z = 1$$

JOINT DISTRIBUTION AND INDEPENDENT RANDOM VARIABLES

5.25. Let X and Y be random variables with joint distribution as shown in Fig. 5-15.

(a) Find the distributions of X and Y.

(b) Find $\text{cov}(X, Y)$, the covariance of X and Y.

(c) Find $\rho(X, Y)$, the correlation of X and Y.

(d) Are X and Y independent random variables?

X \ Y	-3	2	4	Sum
1	0.1	0.2	0.2	0.5
3	0.3	0.1	0.1	0.5
Sum	0.4	0.3	0.3	

Fig. 5-15

(a) The marginal distribution on the right of the joint distribution is the distribution of X, and the marginal distribution on the bottom is the distribution of Y. Thus, the distributions of X and Y are as follows:

x	1	3
$f(x)$	0.5	0.5

Distribution of X

y	-3	2	4
$g(y)$	0.4	0.3	0.3

Distribution of Y

(b) First compute μ_X and μ_Y as follows:

$$\mu_X = \sum x_i f(x_i) = 1(0.5) + 3(0.5) = 2$$
$$\mu_Y = \sum y_i g(y_i) = -3(0.4) + 2(0.3) + 4(0.3) = 0.6$$

Next compute $E(XY)$ as follows:

$$E(XY) = \sum x_i y_j f(x_i) g(y_j)$$
$$= 1(-3)(0.1) + 1(2)(0.2) + 1(4)(0.2) + 3(-3)(0.3) + 3(2)(0.1) + 3(4)(0.1) = 0$$

Then $\qquad \text{cov}(X, Y) = E(XY) - \mu_X \mu_Y = 0 - 2(0.6) = 1.2$

(c) First compute σ_X as follows:

$$E(X^2) = \sum x_i^2 f(x_i) = 1^2(0.5) + 3^2(0.5) = 5$$
$$\sigma_X^2 = E(X^2) - \mu_X^2 = 5 - 2^2 = 1 \qquad \text{and} \qquad \sigma_X = 1$$

Next compute σ_Y as follows:

$$E(Y^2) = \sum y_i^2 g(y_i) = (-3)^2(0.4) + 2^2(0.3) + 4^2(0.3) = 9.6$$
$$\sigma_Y^2 = E(Y^2) - \mu_Y^2 = 9.6 - (0.6)^2 = 9.24 \qquad \text{and} \qquad \sigma_Y = 3.0$$

Then $\qquad\qquad\qquad \rho(X, Y) = \dfrac{\text{cov}(X, Y)}{\sigma_X \sigma_Y} = \dfrac{-1.2}{1(3.0)} = -0.4$

(d) X and Y are not independent since the entry $h(1, -3) = 0.1$ is not equal to $f(1)g(-3) = (0.5)(0.4) = 0.2$, the product of its marginal entries, that is,

$$P(X = 1, Y = -3) \neq P(X = 1)P(Y = -3)$$

5.26. Let X and Y be independent random variables with the following distributions:

x	1	2
$f(x)$	0.6	0.4

Distribution of X

y	5	10	15
$g(y)$	0.2	0.5	0.3

Distribution of Y

Find the joint distribution h of X and Y.

Since X and Y are independent, the joint distribution h can be obtained from the marginal distributions f and g. Specifically, first construct the joint distribution table with only the marginal distributions as shown in Fig. 5-16(a). Then multiply the marginal entries to obtain the interior entries, that is, set $h(x_i, y_j) = f(x_i)g(y_j)$. This yields the joint distribution of X and Y appearing in Fig. 5-16(b).

Y \\ X	5	10	15	Sum
1				0.6
2				0.4
Sum	0.2	0.5	0.3	

Y \\ X	5	10	15	Sum
1	0.12	0.30	0.18	0.6
2	0.08	0.20	0.12	0.4
Sum	0.2	0.5	0.3	

(a)   (b)

Fig. 5-16

5.27. A fair coin is tossed 3 times yielding the following 8-element equiprobable space:

$$S = \{HHH, HHT, HTH, HTT, THH, THT, TTH, TTT\}$$

(Thus, each point in S occurs with probability 1/8.) Let X equal 0 or 1 accordingly as a head or a tail occurs on the first toss, and let Y equal the total number of heads that occurs.

(a) Find the distribution f of X and the distribution g of Y.

(b) Find the joint distribution h of X and Y.

(c) Determine whether or not X and Y are independent.

(d) Find $\text{cov}(X, Y)$, the covariance of X and Y.

(a) We have $X(HHH) = 0$, $X(HHT) = 0$, $X(HTH) = 0$, $X(HTT) = 0$
        $X(THH) = 1$, $X(THT) = 1$, $X(TTH) = 1$, $X(TTT) = 1$

Also $Y(HHH) = 3$, $Y(HHT) = 2$, $Y(HTH) = 2$, $Y(HTT) = 1$
        $Y(THH) = 2$, $Y(THT) = 1$, $Y(TTH) = 1$, $Y(TTT) = 0$

Thus, the distributions of X and Y are as follows:

x	0	1
$f(x)$	$\frac{1}{2}$	$\frac{1}{2}$

Distribution of X

y	0	1	2	3
$g(y)$	$\frac{1}{8}$	$\frac{3}{8}$	$\frac{3}{8}$	$\frac{1}{8}$

Distribution of Y

(b) The joint distribution h of X and Y appears in Fig. 5-17. The entry $h(0, 2)$ is obtained using

$$h(0, 2) \equiv P(X = 0, Y = 2) = P(\{HTH, HHT\}) = \tfrac{2}{8}$$

The other entries are obtained similarly.

Y \\ X	0	1	2	3	Sum
0	0	$\frac{1}{8}$	$\frac{2}{8}$	$\frac{1}{8}$	$\frac{1}{2}$
1	$\frac{1}{8}$	$\frac{2}{8}$	$\frac{1}{8}$	0	$\frac{1}{2}$
Sum	$\frac{1}{8}$	$\frac{3}{8}$	$\frac{3}{8}$	$\frac{1}{8}$	

Fig. 5-17

(c) X and Y are not independent. For example, the entry $h(0,0) = 0$ is not equal to $f(0)g(0) = \frac{1}{2} \cdot \frac{1}{8}$, the product of the marginal entries. That is,

$$P(X = 0, Y = 0) \neq P(X = 0)P(Y = 0)$$

(d)   First compute μ_X, μ_Y, and $E(XY)$ as follows:

$$\mu_X = \sum x_i f(x_i) = 0\left(\frac{1}{2}\right) + 1\left(\frac{1}{2}\right) = \frac{1}{2}$$

$$\mu_Y = \sum y_i g(y_i) = 0\left(\frac{1}{8}\right) + 1\left(\frac{3}{8}\right) + 2\left(\frac{3}{8}\right) + 3\left(\frac{1}{8}\right) = \frac{3}{2}$$

$$E(XY) = \sum x_i y_j f(x_i)g(y_j) = 1(1)\left(\frac{2}{8}\right) + 1(2)\left(\frac{1}{8}\right) + \text{terms with a factor } 0 = \frac{1}{2}$$

Then $$\text{cov}(X, Y) = E(XY) - \mu_X \mu_Y = \frac{1}{2} - \frac{1}{2}\left(\frac{3}{2}\right) = -\frac{1}{4}$$

5.28.   Let X and Y be the random variables in Problem 5.27. Recall that $Z = X + Y$ is the random variable on the same sample space S defined by

$$Z(s) = X(s) + Y(s)$$

(a)   Find the distribution Z.

(b)   Show that $E(Z) = E(X + Y) = E(X) + E(Y)$.

(c)   Find var(X), var(Y), and var(Z), and compare var(Z) with var(X) + var(Y).

(a)   Here X assumes the values 0 and 1, and Y assumes the values 0, 1, 2, 3; hence Z can only assume the values 0, 1, 2, 3, 4. Hence the range space of Z is $R_Z = \{0, 1, 2, 3, 4\}$. To find $P(Z = z) = P(X + Y = z)$, we add up the corresponding probabilities from the joint distribution of X and Y. For instance,

$$P(Z = 3) = P(X + Y = 3) = P(0, 3) + P(1, 2) = \tfrac{1}{8} + \tfrac{1}{8} = \tfrac{1}{4}$$

Similarly, we obtain the following distribution of Z:

z	0	1	2	3	4
$P(z)$	0	$\frac{1}{4}$	$\frac{1}{2}$	$\frac{1}{4}$	0

[Since $P(Z = 0) = 0$ and $P(Z = 4) = 0$, we may delete the first and last entries in the distribution of Z.]

(b)   From the distribution of Z we obtain

$$\mu_Z = E(Z) = 1\left(\frac{1}{4}\right) + 2\left(\frac{1}{2}\right) + 3\left(\frac{1}{4}\right) = 2$$

From Problem 5.27, $E(X) = \mu_X = \frac{1}{2}$ and $E(Y) = \mu_Y = \frac{3}{2}$. Hence $E(Z) = E(X) + E(Y)$ (which is expected from Theorem 5.3).

(c)   From the distributions of X, Y, Z we obtain

$$E(X^2) = 0^2\left(\frac{1}{2}\right) + 1^2\left(\frac{1}{2}\right) = \frac{1}{2}$$

$$E(Y^2) = 0^2\left(\frac{1}{8}\right) + 1^2\left(\frac{3}{8}\right) + 2^2\left(\frac{3}{8}\right) + 3^2\left(\frac{1}{8}\right) = \frac{24}{8} = 3$$

$$E(Z^2) = 1^2\left(\frac{1}{4}\right) + 2^2\left(\frac{1}{2}\right) + 3^2\left(\frac{1}{4}\right) = \frac{9}{2}$$

Therefore: $\qquad$ $var(X) = E(X^2) - \mu_X^2 = \dfrac{1}{2} - \left(\dfrac{1}{2}\right)^2 = \dfrac{1}{4}$

$$var(Y) = E(Y^2) - \mu_Y^2 = 3 - \left(\dfrac{3}{2}\right)^2 = \dfrac{3}{4}$$

$$var(Z) = E(Z^2) - \mu_Z^2 = \dfrac{9}{2} - 2^2 = \dfrac{1}{2}$$

Thus, $var(Z) \neq var(X) + var(Y)$. (This may be expected since X and Y are not independent random variables.)

5.29. A sample with replacement of size $n = 2$ is randomly selected from the numbers 1 to 5. This then yields the equiprobable space S consisting of all 25 ordered pairs (a, b) of numbers from 1 to 5. That is,

$$S = \{(1, 1), (1, 2), \ldots, (1, 5), (2, 1), \ldots, (5, 5)\}$$

Let $X = 0$ if the first number is even and $X = 1$ otherwise; let $Y = 1$ if the second number is odd and $Y = 0$ otherwise.

(*a*) Find the distributions of X and Y.

(*b*) Find the joint distribution of X and Y.

(*c*) Determine if X and Y are independent.

(*a*) There are 10 sample points in which the first entry is even, that is, where

$$a = 2 \ \text{ or } \ 4 \qquad \text{and} \qquad b = 1, 2, 3, 4, 5$$

Thus, $P(X = 0) = 10/25 = 0.4$, and so $P(X = 1) = 0.6$. There are 15 sample points in which the second entry is odd, that is, where

$$a = 1, 2, 3, 4, 5 \qquad \text{and} \qquad b = 1, 3, 5$$

Thus, $P(Y = 1) = 15/25 = 0.6$, and so $P(Y = 0) = 0.4$. Therefore, the distributions of X and Y are as follows:

x	0	1
$P(x)$	0.4	0.6

y	0	1
$P(y)$	0.4	0.6

(Note that X and Y are identically distributed.)

(*b*) For the joint distribution of X and Y, we have

$$P(0, 0) = P(a \text{ even}, b \text{ even}) = P\{(2, 2), (2, 4), (4, 2), (4, 4)\} = 4/25 = 0.16$$

$$P(0, 1) = P(a \text{ even}, b \text{ odd}) = P\{(2, 1), (2, 3), (2, 5), (4, 1), (4, 3), (4, 5)\} = 6/25 = 0.24$$

Similarly $P(1, 0) = 6/25 = 0.24$ and $P(1, 1) = 9/25 = 0.36$. Thus, Fig. 5-18 gives the joint distribution of X and Y.

Y \ X	0	1	Sum
0	0.16	0.24	0.4
1	0.24	0.36	0.6
Sum	0.4	0.6	

Fig. 5-18

(c) The product of the marginal entries do give the four interior entries; for example,

$$P(0,0) = 0.16 = (0.4)(0.4) = P(X = 0)P(Y = 0)$$

Thus, X and Y are independent random variables, even though they are identically distributed.

5.30. Let X and Y be the random variables in Problem 5.29, and let $Z = X + Y$.

(a) Find the distribution Z.

(b) Show that $E(Z) = E(X + Y) = E(X) + E(Y)$.

(c) Find var(X), var(Y), and var(Z), and compare var(Z) with var$(X) + $ var(Y).

(a) Here X assumes the values 0 and 1 and Y assumes the values 0 and 1; hence $Z = X + Y$ can only assume the values 0, 1, 2. Hence the range space of Z is $R_Z = \{0, 1, 2\}$. To find $P(Z = z) = P(X + Y = z)$, we add up the corresponding probabilities from the joint distribution of X and Y. Thus

$$P(Z = 0) = P(X = 0, Y = 0) = 0.16$$
$$P(Z = 1) = P(X = 0, Y = 1) + P(X = 1, Y = 0) = 0.24 + 0.24 = 0.48$$
$$P(Z = 2) = P(X = 1, Y = 1) = 0.36$$

Thus, the distribution of Z is as follows:

z	0	1	2
$P(z)$	0.16	0.48	0.36

(b) From the distributions of X, Y, Z we obtain

$$\mu_X = E(X) = 0(0.4) + 1(0.6) = 0.6 \qquad \mu_Y = E(Y) = 0(0.4) + 1(0.6) = 0.6$$
$$\mu_Z = E(Z) = 0(0.16) + 1(0.48) + 2(0.36) = 1.2$$

Hence $E(Z) = E(X) + E(Y)$ (which is expected from Theorem 5.3).

(c) From the distributions of X, Y, Z we obtain

$$E(X^2) = 0^2(0.4) + 1^2(0.6) = 0.6, \quad E(Y^2) = 0^2(0.4) + 1^2(0.6) = 0.6$$
$$E(Z^2) = 0^2(0.16) + 1^2(0.48) + 2^2(0.36) = 1.92$$

Accordingly,
$$\text{var}(X) = E(X^2) - \mu_X^2 = 0.6 - (0.6)^2 = 0.24$$
$$\text{var}(Y) = E(Y^2) - \mu_Y^2 = 0.6 - (0.6)^2 = 0.24$$
$$\text{var}(Z) = E(Z^2) - \mu_Z^2 = 1.92 - (1.2)^2 = 0.48$$

Thus var$(Z) = $ var$(X) + $ var(Y). (This is expected since X and Y are independent random variables.)

5.31. Let X be the random variable with the following distribution, and let $Y = X^2$:

x	-2	-1	1	2
$f(x)$	$\frac{1}{4}$	$\frac{1}{4}$	$\frac{1}{4}$	$\frac{1}{4}$

(a) Find the distribution g of Y.

(b) Find the joint distribution h of X and Y.

(c) Find cov(X, Y) and $\rho(X, Y)$.

(d) Determine whether or not X and Y are independent.

(a) Since $Y = X^2$, the random variable Y only has the values 4 and 1, and each occurs with probability $\frac{1}{4} + \frac{1}{4} = \frac{1}{2}$. Thus, the distribution of Y is as follows:

y	1	4
$g(y)$	$\frac{1}{2}$	$\frac{1}{2}$

(b) The joint distribution h of X and Y appears in Fig. 5-19. Note that if $X = -2$, then $Y = 4$. Therefore

$$h(-2, 1) = 0 \quad \text{and} \quad h(-2, 4) = f(-2) = \frac{1}{4}$$

The other entries are obtained in a similar manner.

X \ Y	1	4	Sum
-2	0	1/4	1/4
-1	1/4	0	1/4
1	1/4	0	1/4
2	0	1/4	1/4
Sum	1/2	1/2	

Fig. 5-19

(c) First compute μ_X, μ_Y, and $E(XY)$ as follows:

$$\mu_X = \Sigma x_i f(x_i) = -2\left(\frac{1}{4}\right) - 1\left(\frac{1}{4}\right) + 1\left(\frac{1}{4}\right) + 2\left(\frac{1}{4}\right) = 0$$

$$\mu_Y = \Sigma y_i g(y_i) = 1\left(\frac{1}{2}\right) + 4\left(\frac{1}{2}\right) = \frac{5}{2}$$

$$E(XY) = \Sigma x_i y_j f(x_i) g(y_j) = -8\left(\frac{1}{4}\right) - 1\left(\frac{1}{4}\right) + 1\left(\frac{1}{4}\right) + 8\left(\frac{1}{4}\right) = 0$$

Then $\qquad \text{cov}(X, Y) = E(XY) - \mu_X \mu_Y = 0 - 0 \cdot \frac{5}{2} = 0 \quad \text{and so} \quad \rho(X, Y) = 0$

(d) X and Y are not independent. For example, the entry $h(-2, 1) = 0$ is not equal to $f(-2)g(1) = \frac{1}{2} \cdot \frac{1}{4}$, the product of the marginal entries. That is,

$$P(X = -2, Y = 1) \quad \neq \quad P(X = -2)P(Y = 1)$$

Remark: Although X and Y are not independent and, in particular, Y is a function of X, this example shows that it is still possible for the covariance and correlation to be 0, which is always true when X and Y are independent.

5.32. Let X_1, X_2, X_3 be independent random variables that are identically distributed with mean $\mu = 100$ and standard deviation $\sigma = 4$. Let $Y = (X_1 + X_2 + X_3)/3$. Find:
(a) the mean μ_Y of Y and (b) the standard deviation σ_Y of Y.

(a) Theorem 5.2 and Corollary 5.4 yield

$$\mu_Y = E(Y) = E\left(\frac{X_1 + X_2 + X_3}{3}\right) = \frac{E(X_1 + X_2 + X_3)}{3}$$

$$= \frac{E(X_1) + E(X_2) + E(X_3)}{3} = \frac{100 + 100 + 100}{3} = 100$$

Note $\mu_Y = \mu$.

(b) We use Theorem 5.6 and, since X_1, X_2, X_3 are independent, we can also use Theorem 5.9 to obtain

$$\sigma_Y^2 = \text{var}(Y) = \text{var}\left(\frac{X_1 + X_2 + X_3}{3}\right) = \frac{\text{var}(X_1 + X_2 + X_3)}{9}$$

$$= \frac{\text{var}(X_1) + \text{var}(X_2) + \text{var}(X_3)}{9} = \frac{2^2 + 2^2 + 2^2}{9} = \frac{12}{9} = \frac{4}{3}$$

Thus

$$\sigma_Y = \sqrt{\frac{4}{3}} = \frac{2}{\sqrt{3}} = \frac{\sigma}{\sqrt{3}}$$

Remark: Suppose Y were the sum of n independent, identically distributed random variables with mean μ and standard deviation σ. Then one can similarly show that

$$\mu_Y = \mu \quad \text{and} \quad \sigma_Y = \frac{\sigma}{\sqrt{n}}$$

That is, the above result is true in general.

CHEBYSHEV'S INEQUALITY

5.33. Suppose a random variable X has mean $\mu = 25$ and standard deviation $\sigma = 2$. Use Chebyshev's inequality to estimate: (a) $P(X \le 35)$ and (b) $P(X \ge 20)$.

(a) By Chebyshev's inequality (Section 5.12),

$$P(\mu - k\sigma \le X \le \mu + k\sigma) \ge 1 - \frac{1}{k^2}$$

Substitute $\mu = 25$, $\sigma = 2$ in $\mu + k\sigma$ and solve the equation $25 + 2k = 35$ for k, getting $k = 5$. Then

$$1 - \frac{1}{k^2} = 1 - \frac{1}{25} = \frac{24}{25} = 0.96$$

Since $\mu - k\sigma = 25 - 10 = 15$, Chebyshev's inequality gives

$$P(15 \le X \le 35) \ge 0.96$$

The event corresponding to $X \le 35$ contains as a subset the event corresponding to $15 \le X \le 35$. Therefore,

$$P(X \le 35) \ge P(15 \le X \le 35) \ge 0.96$$

Hence, the probability that X is less than or equal to 35 is at least 96 percent.

(b) Substitute $\mu = 25$, $\sigma = 2$ in $\mu - k\sigma$ and solve the equation $25 - 2k = 20$ for k, getting $k = 2.5$. Then

$$1 - \frac{1}{k^2} = 1 - \frac{1}{6.25} = 0.84$$

Since $\mu + 2\sigma = 25 + 5 = 30$, Chebyshev's inequality gives

$$P(20 \leq X \leq 30) \; \geq \; 0.84$$

The event corresponding to $X \geq 20$ contains as a subset the event corresponding to $20 \leq X \leq 30$. Therefore,

$$P(X \geq 20) \; \geq \; P(20 \leq X \leq 30) \; \geq \; 0.84$$

which says that the probability that X is greater than or equal to 20 is at least 84 percent.

Remark: This problem illustrates that Chebyshev's inequality can be used to estimate $P(X \leq b)$ when $b \geq \mu$, and to estimate $P(X \geq a)$ when $a \leq \mu$.

5.34. Let X be a random variable with mean $\mu = 40$ and standard deviation $\sigma = 5$. Use Chebyshev's inequality to find a value b for which $P(40 - b \leq X \leq 40 + b) \; \geq \; 0.95$.

First solve $1 - \dfrac{1}{k^2} = 0.95$ for k as follows:

$$0.05 = \frac{1}{k^2} \qquad k^2 = \frac{1}{0.05} = 20 \qquad k = \sqrt{20} = 2\sqrt{5}$$

Then, by Chebyshev's inequality, $b = k\sigma = 10\sqrt{5} \approx 23.4$. Hence, $P(16.6 \leq X \leq 63.6) \; \geq \; 0.95$.

5.35. Let X be a random variable with mean $\mu = 80$ and unknown standard deviation σ. Use Chebyshev's inequality to find a value of σ for which $P(75 \leq X \leq 85) \geq 0.9$.

First solve $1 - \dfrac{1}{k^2} = 0.9$ for k as follows:

$$0.1 = \frac{1}{k^2} \quad \text{or} \quad k^2 = \frac{1}{0.1} = 10 \quad \text{or} \quad k = \sqrt{10}$$

Now, since 75 is 5 units to the left of $\mu = 80$ and 85 is 5 units to the right of μ, we can solve either $\mu - k\sigma = 75$ or $\mu + k\sigma = 85$ for σ. From the latter equation, we get

$$80 + \sqrt{10}\,\sigma = 85 \quad \text{or} \quad \sigma = \frac{5}{\sqrt{10}} \approx 1.58$$

MISCELLANEOUS PROBLEMS

5.36. Let X be a continuous random variable with the following distribution:

$$f(x) = \begin{cases} \frac{1}{6}x + k & \text{if } 0 \leq x \leq k \\ 0 & \text{elsewhere} \end{cases}$$

(a) Evaluate k. (b) Find $P(1 \leq X \leq 2)$.

(a) The graph of f is drawn in Fig. 5-20(a). Since f is a continuous probability function, the shaded region A must have area 1. Note that A forms a trapezoid with parallel bases of length k and $k + \frac{1}{2}$ and altitude 3. Setting the area of A equal to 1 yields

$$\frac{1}{2}\left(k + k + \frac{1}{2}\right) = 1 \qquad \text{or} \qquad k = \frac{1}{12}$$

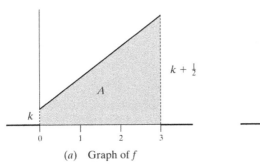

(a)   Graph of f (b) $P(1 \leq X \leq 2)$ = area of B

Fig. 5-20

(b) $P(1 \leq X \leq 2)$ is equal to the area of B which is under the graph of f between $x = 1$ and $x = 2$, as shown in Fig. 5-20(b).   Note:

$$f(1) = \frac{1}{6} + \frac{1}{12} = \frac{3}{12} \quad \text{and} \quad f(2) = \frac{1}{8} + \frac{1}{12} = \frac{5}{12}$$

Hence $P(1 \leq X \leq 2) \quad = \quad \text{area of } B \quad = \quad \frac{1}{2}\left(\frac{3}{12} + \frac{5}{12}\right)(1) = \frac{1}{3}$

5.37.   Let X be the continuous random variable whose distribution function f forms an isosceles triangle above the unit interval $\mathbf{I} = [0, 1]$ and 0 elsewhere (as pictured in Fig. 5-21).

(a)   Find k, the height of the triangle.   (b)   Find the formula which defines f.

(c)   Find the mean $\mu = E(X)$ of X.

(a)   The shaded region A in Fig. 5-21 must have area 1.   Hence

$$\tfrac{1}{2}(1)k = 1 \quad \text{or} \quad k = 2$$

(b)   Note that f is linear between $x = 0$ and $x = 1/2$ with slope $m = (2/(1/2)) = 4$, and f is linear between $x = 1/2$ and $x = 1$ with slope $m = -4$.   Hence

$$f(x) = \begin{cases} 4x & \text{if } 0 \leq x \leq 1/2 \\ -4x + 4 & \text{if } 1/2 \leq x \leq 1 \\ 0 & \text{elsewhere} \end{cases}$$

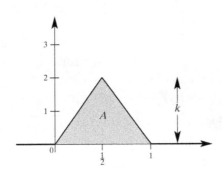

Fig. 5-21

(c) Recall that we may view probability as weight or mass and the mean as the center of gravity. Since the triangle is symmetric, it is intuitively clear that

$$\mu = \tfrac{1}{2}$$

the midpoint of the base of the triangle between 0 and 1. We verify this mathematically using calculus:

$$\mu = E(X) = \int_{-\infty}^{\infty} x f(x)\, dx = \int_{0}^{1/2} x(4x)\, dx + \int_{1/2}^{1} x(-4x + 4)\, dx$$

$$= \int_{0}^{1/2} 4x^2\, dx + \int_{1/2}^{1} (-4x^2 + 4x)\, dx$$

$$= \Big[[4x^3]/3 \Big]_{0}^{1/2} + \Big[[-4x^3]/3 + 2x^2 \Big]_{1/2}^{1}$$

$$= (1/6) + [(-4/3) + 2 + (1/6) - (1/2)] = 1/2$$

5.38. Let h be the joint distribution of random variables X and Y.

(a) Show that the distribution f of the sum $Z = X + Y$ can be obtained by summing the probabilities along the diagonal lines $x + y = z_k$, that is,

$$f(z) = \sum_{z_k = x_i + y_j} h(x_i, y_j) = \sum_{x_i} h(x_i, z_k - x_i)$$

(b) Let X and Y be random variables whose joint distribution h appears in Fig. 5-22 (where the marginal entries have been omitted). Apply (a) to obtain the distribution f of the sum $Z = X + Y$.

X \ Y	−2	−1	0	1	2	3
0	0.05	0.05	0.10	0	0.05	0.05
1	0.10	0.05	0.05	0.10	0	0.05
2	0.03	0.12	0.07	0.06	0.03	0.04

Fig. 5-22

(a) The events $\{X = x_i, Y = y_j; x_i + y_j = z_k\}$ are disjoint. Therefore

$$f(z) = P(Z = z_k) = \sum_{z_k = x_i + y_j} P(X = x_i, Y = y_j) = \sum_{z_k = x_i + y_j} h(x_i, y_j) = \sum_{x_i} h(x_i, z_k - x_i)$$

(b) Note first that $Z = X + Y$ takes on all integer values between $z = -2$ (obtained when $X = 0$ and $Y = -2$) and $z = 5$ (obtained when $X = 2$ and $Y = 3$). Adding along the diagonal lines in Fig. 5-22 (from lower left to upper right) yields

$f(-2) = 0.05,$ $f(2) = 0.07 + 0.10 + 0.05 = 0.22$

$f(-1) = 0.10 + 0.05 = 0.15,$ $f(3) = 0.06 + 0 + 0.05 = 0.11$

$f(0) = 0.03 + 0.05 + 0.10 = 0.18,$ $f(4) = 0.03 + 0.05 = 0.08$

$f(1) = 0.12 + 0.05 + 0 = 0.17,$ $f(5) = 0.04$

Thus, the distribution f of $Z = X + Y$ is as follows:

z	−2	−1	0	1	2	3	4	5
$f(z)$	0.05	0.15	0.18	0.17	0.22	0.11	0.08	0.04

5.39. Let X be a discrete random variable with distribution function f. The rth *moment* M_r of X is defined by

$$M_r = E(X^r) = \sum x_i^r f(x_i)$$

Find the first five moments of X if X has the following distribution:

x	-2	1	3
$f(x)$	0.3	0.5	0.2

(Note that M_1 is the mean of X and M_2 is used in computing the variance and standard deviation of X.)

Use the formula for M_r to obtain

$$M_1 = \Sigma x_i f(x_i) = -2(0.3) + 1(0.5) + 3(0.2) = 0.5$$
$$M_2 = \Sigma x_i^2 f(x_i) = -2^2(0.3) + 1^2(0.5) + 3^2(0.2) = 3.5$$
$$M_3 = \Sigma x_i^3 f(x_i) = -2^3(0.3) + 1^3(0.5) + 3^3(0.2) = 3.5$$
$$M_4 = \Sigma x_i^4 f(x_i) = -2^4(0.3) + 1^4(0.5) + 3^4(0.2) = 21.5$$
$$M_5 = \Sigma x_i^5 f(x_i) = -2^5(0.3) + 1(0.5) + 3^5(0.2) = 49.5$$

5.40. Find the distribution function f of the continuous random variable X whose cumulative distribution function F follows:

(a) $F(x) = \begin{cases} 0 & x < 0 \\ x^3 & 0 \le x \le 1 \\ 1 & x > 1 \end{cases}$ (b) $F(x) = \begin{cases} 0 & x < 0 \\ \sin x & 0 \le x \le \pi/2 \\ 1 & x > \pi/2 \end{cases}$

Recall that $F(x) = \int_{-\infty}^{x} f(t)\, dt$. Calculus tells us that $f(x) = F'(x)$, the derivative of $F(x)$. Thus:

(a) $f(x) = \begin{cases} 0 & x < 0 \\ 3x^2 & 0 \le x \le 1 \\ 0 & x > 1 \end{cases}$

(b) $f(x) = \begin{cases} 0 & x < 0 \\ \cos x & 0 \le x \le \pi/2 \\ 0 & x > \pi/2 \end{cases}$

PROOF OF THEOREMS

Remark: In all proofs, X is a random variable with distribution f, Y is a random variable with distribution g, and h is their joint distribution.

5.41. Prove Theorem 5.1. Let S be an equiprobable space, and let X be a random variable on S with range space $R_X = \{x_1, x_2, \ldots, x_t\}$. Then

$$p_k = P(x_k) = \frac{\text{number of points in } S \text{ whose image is } x_k}{\text{number of points in } S}$$

Let S have n points and let $s_1, s_2, \ldots, s_r$ be the points in S with image x_i. We wish to show that $p_i = f(x_i) = r/n$. By definition,

$$p_i = f(x_i) = \text{sum of the probabilities of the points in } S \text{ whose image is } x_i$$
$$= P(s_1) + P(s_2) + \cdots + P(s_r)$$

Since S is an equiprobable space, each of the n points in S has probability $1/n$. Hence

$$p_i = f(x_i) = \overbrace{\frac{1}{n} + \frac{1}{n} + \cdots + \frac{1}{n}}^{r\text{ times}} = \frac{r}{n}$$

5.42. Show that the marginal distributions, $f(x_i) = \Sigma_j h(x_i, y_j)$ and $g(y_j) = \Sigma_i h(x_i, y_j)$, are the (individual) distributions of X and Y.

Let $A_i \equiv \{X = x_i\}$ and $B_j \equiv \{Y = y_j\}$, that is, let $A_i = X^{-1}(x_i)$ and $B_j = Y^{-1}(y_j)$. Thus, the B_j are disjoint and $S = \cup_j B_j$. Hence

$$A_i = A_i \cap S = A_i \cap (\cup_j b_j) = \cup_j (A_i \cap B_j)$$

where the $A_i \cap B_j$ are also disjoint. Accordingly

$$f(x_i) \;=\; P(X = x_i) \;=\; P(A_i) \;=\; \sum_j P(A_i \cap B_j)$$
$$=\; \sum_j P(X = x_i, Y = y_j) \;=\; \sum_j h(x_i, y_j)$$

The proof for g is similar.

5.43. Prove Theorem 5.10. Let X and Y be random variables on S with $Y = \Phi(X)$. Then $E(Y) = \Sigma_i \Phi(x_i) f(x_i)$ where f is the distribution of X.
(Proof is given for the case where X is discrete and finite.)

Suppose X takes on the values $x_i, \ldots, x_n$ and $\Phi(x_i)$ takes on the values $y_1, \ldots, y_n$ as i runs from 1 to n. Then clearly the possible values of $Y = \Phi(X)$ are $y_1, \ldots, y_n$ and the distribution g of Y is given by

$$g(y_j) = \sum_{\{i:\Phi(x_i)=y_j\}} f(x_i)$$

Therefore

$$E(Y) = \sum_{j=1}^{m} y_j g(y_j) = \sum_{j=1}^{m} y_j \sum_{\{i:0(x_i)=y_j\}} f(x_i)$$
$$= \sum_{i=1}^{n} f(x_i) \sum_{\{j:\Phi(x_i)=y_j\}} y_j = \sum_{i=1}^{n} f(x_i)\Phi(x_i)$$

which proves the theorem.

5.44. Prove Theorem 5.2. Let X be a random variable and let K be a real number. Then

(i) $E(kx) = kE(X)$.   (ii) $E(X + k) = E(X) + k$.

(Proof is given for the general discrete case and the assumption that $E(X)$ and $E(Y)$ both exist.)

(i)   Now $kX = \Phi(X)$ where $\Phi(x) = kx$. Therefore, by Theorem 5.10 (Problem 5.43),

$$E(kX) = \Sigma_i kx_i f(x_i) = k\,\Sigma_i x_i f(x_i) = kE(X)$$

(ii)   Here $X + k = \Phi(X)$ where $\Phi(x) = x + k$. Therefore, using $\Sigma_i f(x_i) = 1$,

$$E(X + k) = \Sigma_i (x_i + k) f(x_i) = \Sigma_i x_i f(x_i) + k\,\Sigma_i f(x_i) = E(X) + k$$

5.45. Prove Theorem 5.3. Let X and Y be random variables on S. Then

$$E(X + Y) = E(X) + E(Y)$$

(Proof is given for the general discrete case and the assumption that $E(X)$ and $E(Y)$ both exist.)

Now $X + Y = \Phi(X, Y)$ where $\Phi(x, y) = x + y$. Therefore, by Theorem 5.10 (Problem 5.43),

$$E(X + Y) = \sum_i \sum_j (x_i + y_j) h(x_i, y_j)$$

$$= \sum_i \sum_j x_i h(x_i, y_j) + \sum_i \sum_j y_j h(x_i, y_j)$$

By Problem 5.42, $f(x_i) = \Sigma_j h(x_i, y_j)$ and $g(y_j) = \Sigma_i h(x_i, y_j)$. Thus

$$E(X + Y) = \sum_i x_i f(x_i) + \sum_j y_j g(y_j) = E(X) + E(Y)$$

5.46. Prove Theorem 5.6. $\operatorname{var}(aX + b) = a_2 \operatorname{var}(X)$.

We prove separately that: (i) $\operatorname{var}(X + k) = \operatorname{var}(X)$, and (ii) $\operatorname{var}(kX) = k^2 \operatorname{var}(X)$, from which the theorem follows. By Theorem 5.2, $\mu_{X+k} = \mu_X + k$ and $\mu_{kX} = k\mu_X$. Also, $\Sigma x_i f(x_i) = \mu_X$ and $\Sigma f(x_i) = 1$. Therefore

$$\operatorname{var}(X + k) = \Sigma(x_i + k)^2 f(x_i) - \mu_{X+k}^2$$

$$= \Sigma x_i^2 f(x_i) + 2k \Sigma x_i f(x_i) + k^2 \Sigma f(x_i) - (\mu_X + k)^2$$

$$= \Sigma x_i^2 f(x_i) + 2k\mu_X + k^2 - (\mu_X^2 + 2k\mu_X + k^2)$$

$$= \Sigma x_i f(x_i) - \mu_X^2 = \operatorname{var}(X)$$

and

$$\operatorname{var}(kX) = \Sigma(kx_i)^2 f(x_i) - \mu_{kX}^2 = k^2 \Sigma x_i^2 f(x_i) - (k\mu_X)^2$$

$$= k^2 \Sigma x_i^2 f(x_i) - k^2 \mu_X^2 = k^2(\Sigma x_i^2 f(x_i) - \mu_X^2) = k^2 \operatorname{var}(X)$$

5.47. Show that

$$\operatorname{cov}(X, Y) = \sum_{i,j} (x_i - \mu_X)(y_j - \mu_Y) h(x_i, y_j) = \sum_{i,j} x_i y_j h(x_i, y_j) - \mu_X \mu_Y$$

(Note that the last term is $E(XY) - \mu_X \mu_Y$.)

(Proof is given for the case that X and Y are discrete and finite.)
We have

$$\sum_{i,j} y_j h(x_i, y_j) = \sum_j y_j g(y_j) = \mu_Y, \qquad \sum_{i,j} x_i h(x_i, y_j) = \sum_i x_i f(x_i) = \mu_X, \qquad \sum_{i,j} h(x_i, y_j) = 1$$

Therefore

$$\sum_{i,j} (x_i - \mu_X)(y_j - \mu_Y) h(x_i, y_j)$$

$$= \sum_{i,j} (x_i y_j - \mu_X y_j - \mu_Y x_i + \mu_X \mu_Y) h(x_i, y_j)$$

$$= \sum_{i,j} x_i y_j h(x_i, y_j) - \mu_X \sum_{i,j} y_j h(x_i, y_j) - \mu_Y \sum_{i,j} x_i h(x_i, y_j) + \mu_X \mu_Y \sum_{i,j} h(x_i, y_j)$$

$$= \sum_{i,j} x_i y_j h(x_i, y_j) - \mu_X \mu_Y - \mu_X \mu_Y + \mu_X \mu_Y$$

$$= \sum_{i,j} x_i y_j h(x_i, y_j) - \mu_X \mu_Y$$

5.48. Prove Theorem 5.7. The standardized random variable Z has mean $\mu_Z = 0$ and standard deviation $\sigma_Z = 1$.

By definition $Z = \dfrac{X - \mu}{\sigma}$ where X has mean μ and standard deviation $\sigma > 0$. Using $E(X) = \mu$ and Theorem 5.2, we get

$$\mu_Z = E\left(\frac{X - \mu}{\sigma}\right) = E\left(\frac{X}{\sigma} - \frac{\mu}{\sigma}\right) = \frac{1}{\sigma}E(X) - \frac{\mu}{\sigma} = \frac{\mu}{\sigma} - \frac{\mu}{\sigma} = 0$$

Also, using Theorem 5.6, we get

$$\text{var}(Z) = \text{var}\left(\frac{X - \mu}{\sigma}\right) = \text{var}\left(\frac{X}{\sigma} - \frac{\mu}{\sigma}\right) = \frac{1}{\sigma^2}\text{var}(X) = \frac{\sigma^2}{\sigma^2} = 1$$

Therefore, $\sigma_Z = \sqrt{\text{var}(Z)} = \sqrt{1} = 1$.

5.49. Prove Theorem 5.8. Let X and Y be independent random variables on S. Then:

  (i) $E(XY) = E(X)E(Y)$.
  (ii) $\text{var}(X + Y) = \text{var}(X) + \text{var}(Y)$.
  (iii) $\text{cov}(X, Y) = 0$.

(Proof is given for the case when X and Y are discrete and finite.)
Since X and Y are independent, $h(x_i, y_j) = f(x_i)g(y_j)$. Thus

$$E(XY) = \sum_{i,j} x_i y_j h(x_i, y_j) = \sum_{i,j} x_i y_j f(x_i)g(y_j)$$

$$= \sum_i x_i f(x_i) \sum_j y_j g(y_j) = E(X)E(Y)$$

and

$$\text{cov}(X, Y) = E(XY) - \mu_X \mu_Y = E(X)E(Y) - \mu_X \mu_Y = 0$$

In order to prove (ii) we also need

$$\mu_{X+Y} = \mu_X + \mu_Y, \qquad \sum_{i,j} x_i^2 h(x_i, y_j) = \sum_i x_i^2 f(x_i), \qquad \sum_{i,j} y_j^2 h(x_i, y_j) = \sum_j y_j^2 g(y_j)$$

Hence

$$\text{var}(X + Y) = \sum_{i,j}(x_i + y_j)^2 h(x_i, y_j) - \mu_{X+Y}^2$$

$$= \sum_{i,j} x_i^2 h(x_i, y_j) + 2\sum_{i,j} x_i y_j h(x_i, y_j) + \sum_{i,j} y_j^2 h(x_i, y_j) - (\mu_X + \mu_Y)^2$$

$$= \sum_i x_i^2 f(x_i) + 2\sum_i x_i f(x_i) \sum_j y_j g(y_j) + \sum_j y_j^2 g(y_j) - \mu_X^2 - 2\mu_X \mu_Y - \mu_Y^2$$

$$= \sum_i x_i^2 f(x_i) - \mu_X^2 + \sum_j y_j^2 g(y_j) - \mu_Y^2 = \text{var}(X) + \text{var}(Y)$$

5.50. Prove Theorem 5.9. Let $X_1, X_2, \ldots, X_n$ be independent random variables on S. Then

$$\text{var}(X_1 + X_2 + \cdots + X_n) = \text{var}(X_1) + \text{var}(X_2) + \cdots + \text{var}(X_n)$$

(Proof is given for the case when $X_1, X_2, \ldots, X_n$ are all discrete and finite.)

We take for granted the analogs of Problem 5.49 and Theorem 5.11 for n random variables. Then

$$\text{var}(X_1 + \cdots + X_n) = E((X_1 + \cdots + X_n - \mu_{X_1 + \cdots + X_n})^2)$$

$$= \sum (x_1 + \cdots + x_n - \mu_{X_1 + \cdots + X_n})^2 h(x_1, \ldots, x_n)$$

$$= \sum (x_1 + \cdots + x_n - \mu_{X_1} - \cdots - \mu_{X_n})^2 h(x_1, \ldots, x_n)$$

$$= \sum \left\{ \sum_i \sum_j x_i x_j + \sum_i \sum_j \mu_{X_i} \mu_{X_j} - 2 \sum_i \sum_j \mu_{X_i} x_j \right\} h(x_1, \ldots, x_n)$$

where h is the joint distribution of $X_1, \ldots, X_n$, and $\mu_{X_1 + \cdots + X_n} = \mu_{X_1} + \cdots + \mu_{X_n}$. Since the X_i are pairwise independent, $\Sigma x_i x_j h(x_1, \ldots, x_n) = \mu_{X_i} \mu_{X_j}$ for $i \neq j$. Hence

$$\text{var}(X_1 + \cdots + X_n) = \sum_{i \neq j} \mu_{X_i} \mu_{Xj} + \sum_{i=1}^{n} E(X_i^2) + \sum_i \sum_j \mu_{X_i} \mu_{X_j} - 2 \sum_i \sum_j \mu_{X_i} \mu_{X_j}$$

$$= \sum_{i=1}^{n} E(X_i^2) - \sum_{i=1}^{n} (\mu_{X_i})^2 = \sum_{i=1}^{n} \text{var}(X_i) = \text{var}(X_1) + \cdots + \text{var}(X_n)$$

as required.

5.51. Prove Theorem 5.12 (Chebyshev's Inequality). For any $k > 0$,

$$P(\mu - k\sigma \leq X \leq \mu + k\sigma) \geq 1 - \frac{1}{k^2}$$

Note first that

$$P(|X - \mu| > k\sigma) = 1 - P(|X - \mu| \leq k\sigma) = 1 - P(\mu - k\sigma \leq X \leq \mu + k\sigma)$$

By definition

$$\sigma^2 = \text{var}(X) = \Sigma (x_i - \mu)^2 p_i$$

Delete all terms from the summation for which x_i is in the interval $[\mu - k\sigma, \mu + k\sigma]$, that is, delete all terms for which $|x_i - \mu| \leq k\sigma$. Denote the summation of the remaining terms by $\Sigma^* (x_i - \mu)^2 p_i$. Then

$$\sigma^2 \geq \Sigma^* (x_i - \mu)^2 p_i \geq \Sigma^* k^2 \sigma^2 p_i = k^2 \sigma^2 \Sigma^* p_i$$

$$= k^2 \sigma^2 P(|X - \mu| > k\sigma)$$

$$= k^2 \sigma^2 [1 - P(\mu - k\sigma \leq X \leq \mu + k\sigma)]$$

If $\sigma > 0$, then dividing by $k^2 \sigma^2$ gives

$$\frac{1}{k^2} \geq 1 - P(\mu - k\sigma \leq X \leq \mu + k\sigma)$$

or

$$P(\mu - k\sigma \leq X \leq \mu + k\sigma) \geq 1 - \frac{1}{k^2}$$

which proves Chebyshev's inequality for $\sigma > 0$. If $\sigma = 0$, then $x_i = \mu$ for all $p_i > 0$, and

$$P(\mu - k \cdot 0 \leq X \leq \mu + k \cdot 0) = P(X = \mu) = 1 > 1 - \frac{1}{k^2}$$

which completes the proof.

5.52. Let $X_1, X_2, \ldots, X_n$ be n independent and identically distributed random variables, each with mean μ and variance σ^2, and let $\bar{X}_n$ be the sample mean, that is,

$$\bar{X}_n = \frac{X_1 + X_2 + \cdots + X_n}{n}$$

(a) Prove that the mean of $\bar{X}_n$ is μ and the variance is σ^2/n.

(b) Prove Theorem 5.13 (weak law of large numbers): For any $\alpha > 0$,

$$P(\mu - \alpha \leq \bar{X}_n \leq \mu + \alpha) \to 1 \quad \text{as} \quad n \to \infty$$

(a) Using Theorems 5.2 and 5.3, we get

$$\mu_{\bar{X}_n} = E(\bar{X}_n) = E\left(\frac{X_1 + X_2 + \cdots + X_n}{n}\right) = \frac{1}{n} E(X_1 + X_2 + \cdots + X_n)$$

$$= \frac{1}{n}[E(X_1) + E(X_2) + \cdots + E(X_n)] = \frac{n\mu}{n} = \mu$$

Now using Theorems 5.3 and 5.9, we get

$$\text{var}(\bar{X}_n) = \text{var}\left(\frac{X_1 + X_2 + \cdots + X_n}{n}\right) = \frac{1}{n^2} \text{var}(X_1 + X_2 + \cdots + X_n)$$

$$= \frac{1}{n^2}[\text{var}(X_1) + \text{var}(X_2) + \cdots + \text{var}(X_n)] = \frac{n\sigma^2}{n^2} = \frac{\sigma^2}{n}$$

(b) The proof is based on an application of Chebyshev's inequality to the random variable $\bar{X}_n$. First note that by making the substitution $k\sigma = \alpha$, Chebyshev's inequality can be written as

$$P(\mu - \alpha \leq X \leq \mu + \alpha) \;\geq\; 1 - \frac{\sigma^2}{\alpha^2}$$

Applying Chebyshev's inequality in the form above, we get

$$P(\mu - \alpha \leq \bar{X}_n \leq \mu + \alpha) \;\geq\; 1 - \frac{\sigma^2}{n\alpha^2}$$

from which the desired result follows.

Supplementary Problems

RANDOM VARIABLES AND EXPECTED VALUE

5.53. Suppose a random variable X takes on the values $-4, 2, 3, 7$ with respective probabilities

$$\frac{k+2}{10}, \qquad \frac{2k-3}{10}, \qquad \frac{3k-4}{10}, \qquad \frac{k+1}{10}$$

Find the distribution and expected value of X.

5.54. A pair of dice is thrown. Let X denote the minimum of the two numbers which occur. Find the distribution and expectation of X.

5.55. A fair coin is tossed 4 times. Let Y denote the longest string of heads. Find the distribution and expectation of Y. (Compare with the random variable X in Problem 5.2.)

5.56. A coin, weighted so that $P(H) = 3/4$ and $P(T) = 1/4$, is tossed 3 times.   Let X denote the number of heads that appear.   (*a*) Find the distribution of X.   (*b*) Find $E(X)$.

5.57. A coin, weighted so that $P(H) = 1/3$ and $P(T) = 2/3$, is tossed until 1 head or 5 tails occur.   Find the expected number E of tosses of the coin.

5.58. The probability of team A winning any game is 1/2. Suppose A plays B in a tournament.   The first team to win 2 games in a row or 3 games wins the tournament.   Find the expected number E of games in the tournament.

5.59. A box contains 10 transistors of which 2 are defective. A transistor is selected from the box and tested until a nondefective one is chosen. Find the expected number E of transistors to be chosen.

5.60. Solve the preceding Problem 5.59 for the case when 3 of the 10 items are defective.

5.61. Five cards are numbered 1 to 5. Two cards are drawn at random (without replacement).   Let X denote the sum of the numbers drawn.   (*a*) Find the distribution of X.   (*b*) Find $E(X)$.

5.62. A lottery with 500 tickets gives 1 prize of \$100, 3 prizes of \$50 each, and 5 prizes of \$25 each.   (*a*) Find the expected winnings of a ticket.   (*b*) If a ticket costs \$1, what is the expected value of the game?

5.63. A player tosses 3 fair coins.   The player wins \$5 if 3 heads occur, \$3 if two heads occur, and \$1 if only 1 head occurs.   On the other hand, the player loses \$15 if 3 tails occur.   Find the value of the game to the player.

5.64. A player tosses 2 fair coins.   The player wins \$3 if 2 heads occur and \$1 if 1 head occurs.   For the game to be fair, how much should the player lose if no heads occur?

MEAN, VARIANCE, AND STANDARD DEVIATION

5.65. Find the mean μ, variance σ^2, and standard deviation σ of each distribution:

(*a*)

x	2	3	8
$f(x)$	1/4	1/2	1/4

(*b*)

x	-2	-1	7
$f(x)$	1/3	1/2	1/6

5.66. Find the mean μ, variance σ^2, and standard deviation σ of each distribution:

(*a*)

x	-1	0	1	2	3
$f(x)$	0.3	0.1	0.1	0.3	0.2

(*b*)

x	1	2	3	6	7
$f(x)$	0.2	0.1	0.3	0.1	0.3

5.67. Let X be a random variable with the following distribution:

x	1	3	4	5
$f(x)$	0.4	0.1	0.2	0.3

Find the mean μ, variance σ^2, and standard deviation σ of X.

5.68. Let X be the random variable in Problem 5.67.   Find the distribution, mean μ, variance σ^2, and standard deviation σ of each random variable Y: (*a*) $Y = 3X + 2$, (*b*) $Y = X^2$, (*c*) $Y = 2^X$.

5.69.   Let X be a random variable with the following distribution:

x	-1	1	2
$f(x)$	0.2	0.5	0.3

Find the mean μ, variance σ^2, and standard deviation σ of X.

5.70.   Let X be the random variable in Problem 5.69.   Find the distribution, mean μ, variance σ^2, and standard deviation σ of the random variable $Y = \Phi(X)$ where

$$(a) \quad \Phi(x) = x^4, \qquad (b) \quad \Phi(x) = 3^x, \qquad (c) \quad \Phi(x) = 2^{x+1}.$$

5.71.   Find the mean μ, variance σ^2, and standard deviation σ of the following two-point distribution where $p + q = 1$:

x	a	b
$f(x)$	p	q

5.72.   Show that $\sigma_X = 0$ if and only if X is a *constant function*, that is, $X(s) = k$ for every $s \in S$ or simply $X = k$.

5.73.   Two cards are selected from a box which contains 5 cards numbered 1, 1, 2, 2, and 3.   Let X denote the sum and Y the maximum of the 2 numbers drawn.   Find the distribution, mean, variance, and standard deviation of the random variables: (a) X, (b) Y, (c) $Z = X + Y$, (d) $W = XY$.

JOINT DISTRIBUTIONS, INDEPENDENT RANDOM VARIABLES

5.74.   Consider the joint distribution of X and Y in Fig. 5-23(a).   Find: (a) $E(X)$ and $E(Y)$, (b) cov(X, Y), (c) σ_X, σ_Y, and $\rho(X, Y)$.

5.75.   Consider the joint distribution of X and Y in Fig. 5-23(b).   Find: (a) $E(X)$ and $E(Y)$, (b) cov(X, Y), (c) σ_X, σ_Y, and $\rho(X, Y)$.

X \ Y	-4	2	7	Sum
1	1/8	1/4	1/8	1/2
5	1/4	1/8	1/8	1/2
Sum	3/8	5/8	1/4	

(a)

X \ Y	-2	-1	4	5	Sum
1	0.1	0.2	0	0.3	0.6
2	0.2	0.1	0.1	0	0.4
Sum	0.3	0.3	0.1	0.3	

(b)

Fig. 5-23

5.76.   Suppose X and Y are independent random variables with the following respective distributions:

x	1	2
$f(x)$	0.7	0.3

y	-2	5	8
$g(y)$	0.3	0.5	0.2

Find the joint distribution h of X and Y, and verify that cov$(X, Y) = 0$.

5.77. Consider the joint distribution of X and Y in Fig. 5-24(a). (a) Find $E(X)$ and $E(Y)$. (b) Determine whether X and Y are independent. (c) Find cov(X, Y).

5.78. Consider the joint distribution of X and Y in Fig. 5-24(b). (a) Find $E(X)$ and $E(Y)$. (b) Determine whether X and Y are independent. (c) Find the distribution, mean, and standard deviation of the random variable $Z = X + Y$.

X \ Y	2	3	4	Sum
1	0.06	0.15	0.09	0.30
2	0.14	0.35	0.21	0.70
Sum	3/8	5/8	1/4	

X \ Y	−2	−1	0	1	2	3	Sum
0	0.05	0.05	0.10	0	0.05	0.05	0.30
1	0.10	0.05	0.05	0.10	0	0.05	0.35
2	0.03	0.12	0.07	0.06	0.03	0.04	0.35
Sum	0.18	0.22	0.22	0.16	0.08	0.14	

(a)   (b)

Fig. 5-24

5.79. A fair coin is tossed 4 times. Let X denote the number of heads occurring, and let Y denote the longest string of heads occurring. (See Problems 5.2 and 5.55.)

(a) Determine the joint distribution of X and Y.

(b) Find cov(X, Y) and $\rho(X, Y)$.

5.80. Two cards are selected at random from a box which contains 5 cards numbered 1, 1, 2, 2, and 3. Let X denote the sum and Y the maximum of the 2 numbers drawn. (See Problem 5.73.) (a) Determine the joint distribution of X and Y. (b) Find cov(X, Y) and $\rho(X, Y)$.

CHEBYSHEV'S INEQUALITY

5.81. Let X be a random variable with mean μ and standard deviation σ. Use Chebyshev's inequality to estimate $P(\mu - 3\sigma \leq X \leq \mu + 3\sigma)$.

5.82. Let Z be the standard normal random variable with mean $\mu = 0$ and standard deviation $\sigma = 1$. Use Chebyshev's inequality to find a value b for which $P(-b \leq Z \leq b) \geq 0.9$.

5.83. Let X be a random variable with mean $\mu = 0$ and standard deviation $\sigma = 1.5$. Use Chebyshev's inequality to estimate $P(-3 \leq X \leq 3)$.

5.84. Let X be a random variable with mean $\mu = 70$. For what value of σ will Chebyshev's inequality yield $P(65 \leq X \leq 75) \geq 0.95$?

5.85. Let X be a random variable with mean $\mu = 100$ and standard deviation $\sigma = 10$. Use Chebyshev's inequality to estimate: (a) $P(X \geq 120)$ and (b) $P(X \leq 75)$.

MISCELLANEOUS PROBLEMS

5.86. Let X be a continuous random variable with the following distribution:

$$f(x) = \begin{cases} 1/8 & \text{if } 0 \leq x \leq 8 \\ 0 & \text{elsewhere} \end{cases}$$

Find: (a) $P(2 \leq X \leq 5)$, (b) $P(3 \leq X \leq 7)$, (c) $P(X \geq 6)$.

5.87. Determine and plot the graph of the cumulative distribution function F of the random variable X in Problem 5.86.

5.88. Let X be a continuous random variable with the following distribution:

$$f(x) = \begin{cases} kx & \text{if } 0 \le x \le 5 \\ 0 & \text{elsewhere} \end{cases}$$

Evaluate k and find: (a) $P(1 \le X \le 3)$, (b) $P(2 \le X \le 4)$, (c) $P(X \le 3)$.

5.89. Plot the graph of the cumulative distribution function F of the discrete random variable X with the following distribution:

x	-3	2	6
$f(x)$	1/4	1/2	1/4

5.90. Find the distribution function $f(x)$ of the continuous random variable X whose cumulative distribution function F follows:

(a) $F(x) = \begin{cases} 0 & \text{if } x < 0 \\ x^5 & \text{if } 0 \le x \le 1 \\ 1 & \text{if } x > 1 \end{cases}$ (b) $F(x) = \begin{cases} 0 & \text{if } x < 0 \\ \sin \pi x & \text{if } 0 \le x \le 1/2 \\ 1 & \text{if } x > 1/2 \end{cases}$

[Hint: $f(x) = F'(x)$, the derivative of $F(x)$, wherever it exists.]

5.91. Let X be a random variable for which $\sigma_X \neq 0$. Show that $\rho(X, X) = 1$ and $\rho(X, -X) = -1$.

5.92. Prove Theorem 5.11. Let X, Y, Z be random variables on S with $Z = \Phi(X, Y)$. Then

$$E(Z) = \sum_{i,j} \Phi(x_i, y_j) h(x_i, y_j)$$

where h is the joint distribution of X and Y.

Answers to Supplementary Problems

The following notation will be used:

$$[x_1, \ldots, x_n; f(x_1), \ldots, f(x_n)] \text{ for the distribution } f = \{(x_i, f(x_i)\};$$
$$[x_i; y_j; \text{ row by row}] \text{ for the joint distribution } h = \{[(x_i, y_j), h(x_i, y_j)]\}$$

5.53. $k = 2$; $[-4, 2, 3, 7; 0.4, 0.1, 0.2, 0.3]$; $E(X) = 1.3$.

5.54. $[1, 2, 3, 4, 5, 6; 11/36, 9/36, 7/36, 5/36, 3/36. 1/36]$; $E(X) = 91/36 \approx 2.5$.

5.55. $[0, 1, 2, 3, 4; 1/16, 7/16, 5/16, 2/16, 1/16]$; $E(X) = 27/16 \approx 1.7$.

5.56. (a) $[0, 1, 2, 3; 1/64, 9/64, 27/64, 27/64]$; (b) $E(X) = 2.25$.

5.57. $E = 211/81 \approx 2.6$.

5.58. $E = 23/8 \approx 2.9$.

5.59. $E = 11/9 \approx 1.2$.

5.60. $E = 11/8 \approx 1.4$.

5.61. (a) [3, 4, 5, …, 9; 0.1, 0.1, 0.2, 0.2, 0.2, 0.1, 0.1];   (b) $E(X) = 6$.

5.62. (a) 0.75; (b) -0.25.

5.63. 0.25.

5.64. \$5.

5.65. (a) $\mu = 4$, $\sigma^2 = 5.5$, $\sigma = 2.3$; (b) $\mu = 0$, $\sigma^2 = 10$, $\sigma = 3.2$.

5.66. (a) $\mu = 1$, $\sigma^2 = 2.4$, $\sigma = 1.5$; (b) $\mu = 4.0$, $\sigma^2 = 5.6$, $\sigma = 2.37$.

5.67. $\mu_X = 3$, $\sigma_X^2 = 3$, $\sigma_X = \sqrt{3} \approx 1.7$.

5.68. (a)   [5, 11, 14, 17; 0.4, 0.1, 0.2, 0.3], $\mu_Y = 11$, $\sigma_Y^2 = 27$, $\sigma_Y \approx 5.2$;
        (b)   [1, 9, 16, 25; 0.4, 0.1, 0.2, 0.3], $\mu_Y = 12$, $\sigma_Y^2 = 103.2$, $\sigma_Y \approx 10.2$;
        (c)   [2, 8, 16, 32; 0.4, 0.1, 0.2, 0.3], $\mu_Y = 14.4$, $\sigma_Y^2 = 159.0$, $\sigma_Y \approx 12.6$.

5.69. $\mu_X = 0.9$; $\sigma_X^2 = 1.09$; $\sigma_X = 1.04$.

5.70. (a)   [1, 1, 16; 0.2, 0.5, 0.3], $\mu_Y = 5.5$, $\sigma_Y^2 = 47.25$, $\sigma_Y = 6.87$;
        (b)   [1/3, 3, 9; 0.2, 0.5, 0.3], $\mu_Y = 4.67$, $\sigma_Y^2 = 5.21$, $\sigma_Y = 2.28$;
        (c)   [1, 2, 8; 0.2, 0.5, 0.3], $\mu_Y = 3.6$, $\sigma_Y^2 = 8.44$, $\sigma_Y = 2.91$.

5.71. $\sigma = ap + bq$; $\sigma^2 = pq(a - b)^2$; $\sigma = |a - b|\sqrt{pq}$.

5.73. (a)   [2, 3, 4, 5; 0.1, 0.4, 0.3, 0.2], $\mu_X = 3.6$, $\sigma_X^2 = 0.84$, $\sigma_X = 0.9$;
        (b)   [1, 2, 3; 0.1, 0.5, 0.4], $\mu_Y = 2.3$, $\sigma_Y^2 = 0.41$, $\sigma_Y = 0.64$;
        (c)   [3, 5, 6, 7, 8; 0.1, 0.4, 0.1, 0.2, 0.2], $\mu_Z = 5.9$, $\sigma_Z^2 = 2.3$, $\sigma_Z = 1.5$;
        (d)   [2, 6, 8, 12, 15; 0.1, 0.4, 0.1, 0.2, 0.2], $\mu_W = 8.8$, $\sigma_W^2 = 17.6$, $\sigma_W = 4.2$.

5.74. (a) $E(X) = 3$, $E(Y) = 1$; (b) $\text{cov}(X, Y) = 1.5$; (c) $\sigma_X = 2$, $\sigma_Y = 4.3$, $\rho(X, Y) = 0.17$.

5.75. (a) $E(X) = 1.4$, $E(Y) = 1$; (b) $\text{cov}(X, Y) = -0.5$; (c) $\sigma_X = 0.49$, $\sigma_Y = 3.1$, $\rho(X, Y) = -0.3$.

5.76. [1, 2; -2, 5, 8; 0.21, 0.35, 0.14; 0.09, 0.15, 0.06].

5.77. (a) $E(X) = 1.7$, $E(Y) = 3.1$; (b) yes; (c) must equal 0 since X and Y are independent.

5.78. (a) $E(X) = 1.05$, $E(Y) = 0.16$;
        (b)   no;
        (c)   [-2, -1, 0, 1, 2, 3, 4, 5; 0.05, 0.15, 0.18, 0.17, 0.22, 0.11, 0.08, 0.04], $\mu_Z = 1.21$, $\sigma_Z = \sqrt{3.21} \approx 1.79$.

5.79. (a)   [0, 1, 2, 3, 4; 0, 1, 2, 3, 4; 1/16, 0, 0, 0, 0; 0, 4/16, 0, 0, 0; 0, 3/16, 3/16, 0, 0; 0, 0, 2/16, 2/16;
              0, 0, 0, 0, 1/16];
        (b) $\text{cov}(X, Y) = 0.85$, $\rho(X, Y) = 0.89$.

5.80. (a) [2, 3, 4, 5; 1, 2, 3; 0.1, 0, 0; 0, 0.4, 0; 0, 0.1, 0.2; 0, 0, 0.2];   (b) $\text{cov}(X, Y) = 0.52$, $\rho(X, Y) = 0.9$.

5.81. $P \geq 1 - \frac{1}{32} \approx 0.89$.

5.82. $b = \sqrt{10} \approx 3.16$.

5.83. $P \geq 0.75$.

5.84. $\sigma = 5/\sqrt{20} \approx 1.12$.

5.85.   (a) $P \geq 0.75$; (b) $P \geq 0.84$.

5.86.   (a) 3/8; (b) 1/2; (c) 1/4.

5.87. $F(x)$ is equal to: 0 if $x < 0$, $x/8$ if $0 \leq x \leq 8$, and 1 if $x > 8$. See Fig. 5-25(a).

5.88. $k = 2/25$: (a) 8/25; (b) 12/25; (c) 9/25.

5.89.   See Fig. 5-25(b).

5.90.   (a) $f(x) = 5x^4$ between 0 and 1 and $f(x) = 0$ elsewhere;
    (b) $f(x) = \pi \cos x$ between 0 and 1/2 and $f(x) = 0$ elsewhere.

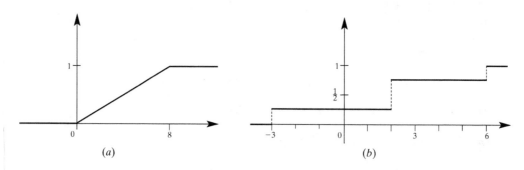

Fig. 5-25

CHAPTER 6

Binomial and Normal Distributions

6.1 INTRODUCTION

The last chapter defined a random variable X on a probability space S and its probability distribution $f(x)$. It was observed that one can discuss X and $f(x)$ without referring to the original probability space S. In fact, there are many applications of probability theory which give rise to the same probability distribution. This chapter mainly discusses two such important distributions in probability—the binomial distribution and the normal distribution. In addition, we will also briefly discuss other distributions, including the Poisson and multinomial distributions. Furthermore, we indicate how each distribution might be an appropriate probability model for some applications.

The central limit theorem, which plays a major role in probability theory, is also discussed in this chapter. This theorem may be viewed as a generalization of the fact that the discrete binomial distribution may be approximated by a continuous normal distribution.

6.2 BERNOULLI TRIALS, BINOMIAL DISTRIBUTION

Consider an experiment ε with only two outcomes, one called success (S) and the other called failure (F). (We will let p denote the probability of success in such an experiment and let $q = 1 - p$ denote the probability of failure.) Suppose the experiment ε is repeated and suppose the trials are independent, that is, suppose the outcome of any trial does not depend on any previous outcomes, such as tossing a coin. Such independent repeated trials of an experiment with two outcomes are called *Bernoulli trials*, named after the Swiss mathematician Jacob Bernoulli (1654–1705).

A *binomial experiment* consists of a fixed number, say n, of Bernoulli trials. (The term "binomial" comes from Theorem 6.1 below.) Such a binomial experiment will be denoted by

$$B(n, p)$$

That is, $B(n, p)$ denotes a binomial experiment with n trials and probability p of success.

Frequently, we are interested in the probability of a certain number of successes in a binomial experiment and not necessarily in the order in which they occur. The following theorem (proved in Problem 6.11) applies.

Theorem 6.1: The probability of exactly k success in a binomial experiment $B(n, p)$ is given by

$$P(k) = P(k \text{ successes}) = \binom{n}{k} p^k q^{n-k}$$

The probability of one or more successes is $1 - q^n$.

Here $\binom{n}{k}$ is the binomial coefficient which is defined and discussed in Chapter 2. Recall that $C(n, k)$ is also used for the binomial coefficient. Accordingly, we may alternately write

$$P(k) = C(n, k) p^k q^{n-k}$$

Observe that q^n denotes the probability of no successes, and hence $1 - q^n$ denotes the probability of one or more successes. Moreover, the probability of getting at least k successes, that is, k or more successes, is given by

$$P(k) + P(k + 1) + P(k + 2) + \cdots + P(n)$$

This follows from the fact that the events of getting k and k' successes are disjoint for $k \neq k'$.

EXAMPLE 6.1 The probability that Ann hits a target at any time is $p = \frac{1}{3}$; hence she misses with probability $q = 1 - p = \frac{2}{3}$. Suppose she fires at the target 7 times. This is a binomial experiment with $n = 7$ and $p = \frac{1}{3}$. Find the probability that she hits the target: (a) Exactly 3 times. (b) At least 1 time.

(a) Here $k = 3$ and hence $n - k = 4$. By Theorem 6.1, the probability that she hits the target 3 times is

$$P(3) = \binom{7}{3}\left(\frac{1}{3}\right)^3 \left(\frac{2}{3}\right)^4 = \frac{560}{2187} \approx 0.26$$

(b) The probability that she never hits the target, that is, all failures, is:

$$P(0) = q^7 = \left(\frac{2}{3}\right)^7 = \frac{128}{2187} \approx 0.06$$

Thus, the probability that she hit the target at least once is

$$1 - q^7 = \frac{2059}{2187} \approx 0.94 = 94\%$$

EXAMPLE 6.2 A fair coin is tossed 6 times; call heads a success. This is a binomial experiment with $n = 6$ and $p = q = \frac{1}{2}$. Find the probability that: (a) Exactly 2 heads occur. (b) At least 4 heads occur. (c) At least 1 head occurs.

(a) Here $k = 2$ and hence $n - k = 4$. Theorem 6.1 tells us that the probability that exactly 2 heads occur follows:

$$P(2) = \binom{6}{2}\left(\frac{1}{2}\right)^2 \left(\frac{1}{2}\right)^4 = \frac{15}{64} \approx 0.23$$

(b) The probability of getting at least 4 heads, that is, where $k = 4$, 5 or 6, follows:

$$P(4) + P(5) + P(6) = \binom{6}{4}\left(\frac{1}{2}\right)^4\left(\frac{1}{2}\right)^2 + \binom{6}{5}\left(\frac{1}{2}\right)^5\left(\frac{1}{2}\right) + \binom{6}{6}\left(\frac{1}{2}\right)^6$$

$$= \frac{15}{64} + \frac{6}{64} + \frac{1}{64} = \frac{22}{64} \approx 0.34$$

(c) The probability of getting no heads (that is, all failures) is $q^6 = (1/2)^6 = 1/64$, so the probability of 1 or more heads is

$$1 - q^n = 1 - \tfrac{1}{64} = \tfrac{63}{64} \approx 0.94$$

Binomial Distribution

Consider a binomial experiment $B(n, p)$. That is, $B(n, p)$ consists of n independent repeated trials with two outcomes, success or failure, and p is the probability of success and $q = 1 - p$ is the probability of failure. Let X denote the number of successes in such an experiment. Then X is a random variable with the following distribution:

k	0	1	2	$\cdots$	n
$P(k)$	q^n	$\binom{n}{1}q^{n-1}p$	$\binom{n}{2}q^{n-2}p^2$	$\cdots$	p^n

This distribution for a binomial experiment $B(n, p)$ is called the *binomial distribution* since it corresponds to the successive terms of the following binomial expansion:

$$(q + p)^n = q^n + \binom{n}{1}q^{n-1}p + \binom{n}{2}q^{n-2}p^2 + \cdots + p^n$$

Thus, $B(n, p)$ will also be used to denote the above binomial distribution.

EXAMPLE 6.3 Suppose a fair coin is tossed 6 times, and heads is called a success. This is a binomial experiment with $n = 6$ and $p = q = \frac{1}{2}$. By Example 6.2,

$$P(2) = \tfrac{15}{64}, \qquad P(4) = \tfrac{15}{64}, \qquad P(5) = \tfrac{6}{64}, \qquad P(6) = \tfrac{1}{54}$$

Similarly,

$$P(0) = \tfrac{1}{64}, \qquad P(1) = \tfrac{6}{54}, \qquad P(3) = \tfrac{20}{64}$$

Thus, the binomial distribution $B(6, \frac{1}{2})$ follows:

k	0	1	2	3	4	5	6
$P(k)$	$\tfrac{1}{64}$	$\tfrac{6}{64}$	$\tfrac{15}{64}$	$\tfrac{20}{64}$	$\tfrac{15}{64}$	$\tfrac{6}{64}$	$\tfrac{1}{64}$

Properties of the binomial distribution follow:

Theorem 6.2:

Binomial distribution $B(n, p)$	
Mean or expected number of successes	$\mu = np$
Variance	$\sigma^2 = npq$
Standard deviation	$\sigma = \sqrt{npq}$

This theorem is proved in Problem 6.18.

EXAMPLE 6.4

(a) The probability that John hits a target is $p = 1/4$. He fires 100 times. Find the expected number μ of times he will hit the target and the standard deviation σ.

Here $p = 1/4$ and so $q = 3/4$. Hence

$$\mu = np = 100 \cdot \tfrac{1}{4} = 25 \quad \text{and} \quad \sigma = \sqrt{npq} = \sqrt{100 \cdot \tfrac{1}{4} \cdot \tfrac{3}{4}} = 2.5$$

(b) A fair die is tossed 180 times. Find the expected number μ of times the face 6 will appear and the standard deviation σ.

Here $p = 1/6$ and so $q = 5/6$. Hence

$$\mu = np = 180 \cdot \tfrac{1}{6} = 30 \quad \text{and} \quad \sigma = \sqrt{npq} = \sqrt{180 \cdot \tfrac{1}{6} \cdot \tfrac{5}{6}} = 5$$

(c) Find the expected number $E(X)$ of correct answers obtained by guessing in a 30-question true-false test.

Here $p = \tfrac{1}{2}$. Hence $E(X) = np = 30 \cdot \tfrac{1}{2} = 15$.

6.3 NORMAL DISTRIBUTION

Let X be a random variable on an infinite sample space S where, by definition, $\{a \leq X \leq b\}$ is an event in S. Recall (Section 5.10) that X is said to be *continuous* if there is a function $f(x)$ defined on the real line $\mathbf{R} = (-\infty, \infty)$ such that

(i) f is nonnegative.

(ii) The area under the curve of f is one.

(iii) The probability that X lies in the interval $[a, b]$ is equal to the area under f between $x = a$ and $x = b$.

These properties may be restated as follows where we use the language of calculus for the area under a curve:

$$\text{(i)} \quad f(x) \geq 0, \quad \text{(ii)} \quad \int_{-\infty}^{\infty} f(x)\, dx = 1, \quad \text{(iii)} \quad P(a \leq X \leq b) = \int_{a}^{b} f(x)\, dx$$

The function $f(x)$ is called the *probability density function* or, simply, *distribution* of X.

Furthermore, the expected value $\mu = E(X)$ and the variance $\mathrm{var}(X)$ of a continuous random variable X, with density function $f(x)$, is defined by the integrals

$$E(X) = \int_{-\infty}^{\infty} xf(x)\, dx \quad \text{and} \quad \mathrm{var}(X) = \int_{-\infty}^{\infty} (x - \mu)^2 f(x)\, dx$$

As in the case of a discrete random variable, the standard deviation σ of X is the nonnegative square root of $\mathrm{var}(X)$.

Normal Random Variable

The most important example of a continuous random variable X is the *normal* random variable, whose density function has the familar bell-shaped curve. This distribution was discovered by De Moive in 1733 as the limiting form of the binomial distribution. Although the normal distribution is sometimes called the "Gaussian distribution" after Gauss who discussed it in 1809, it was actually already known in 1774 by LaPlace.

Formally, a random variable X is said to be *normally distributed* if its density function $f(x)$ has the following form:

$$f(x) = \frac{1}{\sqrt{2\pi}\,\sigma} \exp\left[-\frac{1}{2}\left(\frac{x - \mu}{\sigma}\right)^2 \right]$$

where μ is any real number and σ is any positive number.

The above distribution, which depends on the *parameters* μ and σ, will be denoted by

$$N(\mu, \sigma^2)$$

Thus, we say that X is $N(\mu, \sigma^2)$ if the above function $f(x)$ is its distribution.

The two diagrams in Fig. 6-1 show the changes in the bell-shaped normal curves as μ and σ vary. Specifically, Fig. 6-1(a) shows the distribution for three values of μ and a constant value of σ. In Fig. 6-1(b) μ is constant and three values of σ are used.

Observe that each curve in Fig. 6-1 reaches its highest point at $x = \mu$, and that the curve is symmetric about $x = \mu$. The *inflection* points, where the direction of the bend of the curve changes, occur when $x = \mu + \sigma$ and $x = \mu - \sigma$. Furthermore, although the distribution is defined for all real numbers, the probability of any large deviation from the mean μ is extremely small and hence may be neglected in most practical applications.

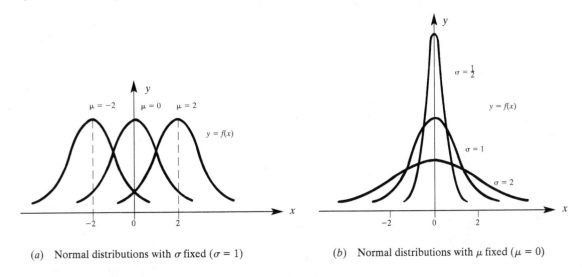

(a) Normal distributions with σ fixed ($\sigma = 1$)   (b) Normal distributions with μ fixed ($\mu = 0$)

Fig. 6-1

Properties of the normal distribution follow:

Theorem 6.3:

Normal distribution $N(\mu, \sigma^2)$	
Mean or expected value	μ
Variance	σ^2
Standard deviation	σ

That is, the mean, variance, and standard deviation of the normal distribution $N(\mu, \sigma^2)$ are μ, σ^2, and σ, respectively. This is why the symbols μ and σ are used as the parameters in the definition of the above density function $f(x)$.

Standardized Normal Distribution

Suppose X is any normal distribution $N(\mu, \sigma^2)$. Recall that the standardized random variable corresponding to X is defined by

$$Z = \frac{X - \mu}{\sigma}$$

We note that Z is also a normal distribution and that $\mu = 0$ and $\sigma = 1$, that is, Z is $N(0, 1)$.   The density function for Z, obtained by setting $z = (x - \mu)/\sigma$ in the above formula for $N(\mu, \sigma^2)$, follows:

$$\phi(z) = \frac{1}{\sqrt{2\pi}}\, e^{-z^2/2}$$

The graph of this function is shown in Fig. 6-2.

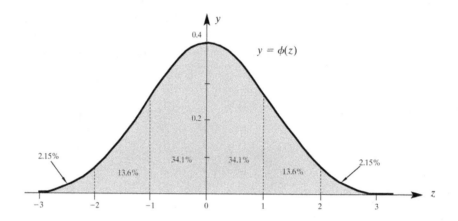

Fig. 6-2.   Normal distribution $N(0, 1)$.

Figure 6-2 also tells us that the percentage of area under the standardized normal curve $\phi(z)$ and hence also under any normal distribution X is as follows:

68.2%	for	$-1 \leq z \leq 1$	and for	$\mu - \sigma \leq x \leq \mu + \sigma$
95.4%	for	$-2 \leq z \leq 2$	and for	$\mu - 2\sigma \leq x \leq \mu + 2\sigma$
99.7%	for	$-3 \leq z \leq 3$	and for	$\mu - 3\sigma \leq x \leq \mu + 3\sigma$

This gives rise to the so-called

> 68–95–99.7 rule

This rule says that, in a normally distributed population, 68 percent (approximately) of the population falls within one standard deviation of the mean, 95 percent falls within two standard deviations of the mean, and 99.7 percent falls within three standard deviations of the mean.

6.4   EVALUATING NORMAL PROBABILITIES

Consider any continuous random variable X with density function $f(x)$.   Recall that the probability $P(a \leq X \leq b)$ is equal to the area under the curve f between $x = a$ and $x = b$.   In the language of calculus,

$$P(a \leq X \leq b) = \int_a^b f(x)\, dx$$

However, if X is a normal distribution, then we are able to evaluate such areas without calculus.   We show how in this section in two steps: first with the standard normal distribution Z, and then with any normal distribution X.

Evaluating Standard Normal Probabilities

Table 6-1 gives the area under the standard normal curve ϕ between 0 and z, where $0 \leq z < 4$ and z is given in steps of 0.01. This area is denoted by $\Phi(z)$, as indicated by the illustration in the table.

EXAMPLE 6.5 Find: (a) $\Phi(1.72)$, (b) $\Phi(0.34)$, (c) $\Phi(2.3)$, (d) $\Phi(4.3)$.

(a) To find $\Phi(1.72)$, look down on the left for the row labeled 1.7, and then continue right for the column labeled 2. The entry in the table corresponding to row 1.7 and column 2 is 0.457 3. Thus, $\Phi(1.72) = 0.457\ 3$.

(b) To find $\Phi(0.46)$, look down on the left for the row labeled 0.4, and then continue right for the column labeled 6. The entry corresponding to row 0.4 and column 6 is 0.177 2. Thus, $\Phi(0.46) = 0.177\ 2$.

(c) To find $\Phi(2.3)$, look on the left for the row labeled 2.3. The first entry 0.489 3 in the row corresponds to $2.3 = 2.30$. Thus, $\Phi(2.3) = 0.489\ 3$.

(d) The value of $\Phi(z)$ for any $z \geq 3.9$ is 0.500 0. Thus, $\Phi(4.3) = 0.500\ 0$, even though 4.3 is not in the table.

Using Table 6-1 and the symmetry of the curve, we can find $P(z_1 \leq Z \leq z_2)$, the area under the curve between any two values z_1 and z_2, as follows:

$$P(z_1 \leq Z \leq z_2) = \begin{cases} \Phi(z_2) + \Phi(|z_1|) & \text{if } z_1 \leq 0 \leq z_2 \\ \Phi(z_2) - \Phi(z_1) & \text{if } 0 \leq z_1 \leq z_2 \\ \Phi(|z_1|) - \Phi(|z_2|) & \text{if } z_1 \leq z_2 \leq 0 \end{cases}$$

These cases are pictured in Fig. 6-3.

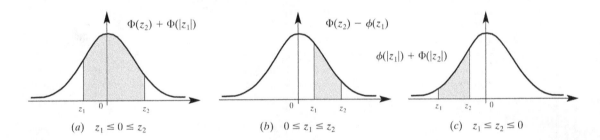

(a) $z_1 \leq 0 \leq z_2$ (b) $0 \leq z_1 \leq z_2$ (c) $z_1 \leq z_2 \leq 0$

Fig. 6-3

EXAMPLE 6.6 Find the following probabilities for the standard normal distribution Z:

(a) $P(-0.5 \leq Z \leq 1.1)$ (c) $P(0.2 \leq Z \leq 1.4)$

(b) $P(-0.38 \leq Z \leq 1.72)$ (d) $P(-1.5 \leq Z \leq -0.7)$

(a) Referring to Fig. 6-3(a),

$$P(-0.5 \leq Z \leq 1.1) = \Phi(1.1) + \Phi(0.5) = 0.364\ 3 + 0.191\ 5 = 0.555\ 8$$

(b) Referring to Fig. 6-3(a),

$$P(-0.38 \leq Z \leq 1.72) = \Phi(1.72) + \Phi(0.38) = 0.457\ 3 + 0.148\ 0 = 0.605\ 3$$

(c) Referring to Fig. 6-3(b),

$$P(0.2 \leq Z \leq 1.4) = \Phi(1.4) - \Phi(0.2) = 0.419\ 2 - 0.079\ 3 = 0.339\ 9$$

(d) Referring to Fig. 6-3(c),

$$P(-1.5 \leq Z \leq -0.7) = \Phi(1.5) - \Phi(0.7) = 0.433\ 2 - 0.258\ 0 = 0.175\ 2$$

Table 6-1   Standard Normal Curve Areas

This table gives areas $\Phi(z)$ under the standard normal distribution ϕ between 0 and $z \geq 0$ in steps of 0.01.

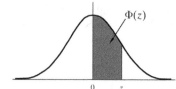

z	0	1	2	3	4	5	6	7	8	9
0.0	0.0000	0.0040	0.0080	0.0120	0.0160	0.0199	0.0239	0.0279	0.0319	0.0359
0.1	0.0398	0.0438	0.0478	0.0517	0.0557	0.0596	0.0636	0.0675	0.0714	0.0754
0.2	0.0793	0.0832	0.0871	0.0910	0.0948	0.0987	0.1026	0.1064	0.1103	0.1141
0.3	0.1179	0.1217	0.1255	0.1293	0.1331	0.1368	0.1406	0.1443	0.1480	0.1517
0.4	0.1554	0.1591	0.1628	0.1664	0.1700	0.1736	0.1772	0.1808	0.1844	0.1879
0.5	0.1915	0.1950	0.1985	0.2019	0.2054	0.2088	0.2123	0.2157	0.2190	0.2224
0.6	0.2258	0.2291	0.2324	0.2357	0.2389	0.2422	0.2454	0.2486	0.2518	0.2549
0.7	0.2580	0.2612	0.2642	0.2673	0.2704	0.2734	0.2764	0.2794	0.2823	0.2852
0.8	0.2881	0.2910	0.2939	0.2967	0.2996	0.3023	0.3051	0.3078	0.3106	0.3133
0.9	0.3159	0.3186	0.3212	0.3238	0.3264	0.3289	0.3315	0.3340	0.3365	0.3389
1.0	0.3413	0.3438	0.3461	0.3485	0.3508	0.3531	0.3554	0.3577	0.3599	0.3621
1.1	0.3643	0.3665	0.3686	0.3708	0.3729	0.3749	0.3770	0.3790	0.3810	0.3830
1.2	0.3849	0.3869	0.3888	0.3907	0.3925	0.3944	0.3962	0.3980	0.3997	0.4015
1.3	0.4032	0.4049	0.4066	0.4082	0.4099	0.4115	0.4131	0.4147	0.4162	0.4177
1.4	0.4192	0.4207	0.4222	0.4236	0.4251	0.4265	0.4279	0.4292	0.4306	0.4319
1.5	0.4332	0.4345	0.4357	0.4370	0.4382	0.4394	0.4406	0.4418	0.4429	0.4441
1.6	0.4452	0.4463	0.4474	0.4484	0.4495	0.4505	0.4515	0.4525	0.4535	0.4545
1.7	0.4554	0.4564	0.4573	0.4582	0.4591	0.4599	0.4608	0.4616	0.4625	0.4633
1.8	0.4641	0.4649	0.4656	0.4664	0.4671	0.4678	0.4686	0.4693	0.4699	0.4706
1.9	0.4713	0.4719	0.4726	0.4732	0.4738	0.4744	0.4750	0.4756	0.4761	0.4767
2.0	0.4772	0.4778	0.4783	0.4788	0.4793	0.4798	0.4803	0.4808	0.4812	0.4817
2.1	0.4821	0.4826	0.4830	0.4834	0.4838	0.4842	0.4846	0.4850	0.4854	0.4857
2.2	0.4861	0.4864	0.4868	0.4871	0.4875	0.4878	0.4881	0.4884	0.4887	0.4890
2.3	0.4893	0.4896	0.4898	0.4901	0.4904	0.4906	0.4909	0.4911	0.4913	0.4916
2.4	0.4918	0.4920	0.4922	0.4925	0.4927	0.4929	0.4931	0.4932	0.4934	0.4936
2.5	0.4938	0.4940	0.4941	0.4943	0.4945	0.4946	0.4948	0.4949	0.4951	0.4952
2.6	0.4953	0.4955	0.4956	0.4957	0.4959	0.4960	0.4961	0.4962	0.4963	0.4964
2.7	0.4965	0.4966	0.4967	0.4968	0.4969	0.4970	0.4971	0.4972	0.4973	0.4974
2.8	0.4974	0.4975	0.4976	0.4977	0.4977	0.4978	0.4979	0.4979	0.4980	0.4981
2.9	0.4981	0.4982	0.4982	0.4983	0.4984	0.4984	0.4985	0.4985	0.4986	0.4986
3.0	0.4987	0.4987	0.4987	0.4988	0.4988	0.4989	0.4989	0.4989	0.4990	0.4990
3.1	0.4990	0.4991	0.4991	0.4991	0.4992	0.4992	0.4992	0.4992	0.4993	0.4993
3.2	0.4993	0.4993	0.4994	0.4994	0.4994	0.4994	0.4994	0.4995	0.4995	0.4995
3.3	0.4995	0.4995	0.4995	0.4996	0.4996	0.4996	0.4996	0.4996	0.4996	0.4997
3.4	0.4997	0.4997	0.4997	0.4997	0.4997	0.4997	0.4997	0.4997	0.4997	0.4998
3.5	0.4998	0.4998	0.4998	0.4998	0.4998	0.4998	0.4998	0.4998	0.4998	0.4998
3.6	0.4998	0.4998	0.4999	0.4999	0.4999	0.4999	0.4999	0.4999	0.4999	0.4999
3.7	0.4999	0.4999	0.4999	0.4999	0.4999	0.4999	0.4999	0.4999	0.4999	0.4999
3.8	0.4999	0.4999	0.4999	0.4999	0.4999	0.4999	0.4999	0.4999	0.4999	0.4999
3.9	0.5000	0.5000	0.5000	0.5000	0.5000	0.5000	0.5000	0.5000	0.5000	0.5000

The "tail end" of a one-sided probability for the standard normal distribution Z can also be obtained from Table 6-1 by using the fact that the total area under the normal curve is 1 and hence half the area is 1/2. There are two cases, the probability that $Z \leq z_1$ and the probability that $Z \geq z_1$.

The probability in the first case follows:

$$P(Z \leq z_1) = \begin{cases} 0.500\,0 + \Phi(z_1) & \text{if } 0 \leq z_1 \\ 0.500\,0 - \Phi(|z_1|) & \text{if } z_1 \leq 0 \end{cases}$$

These two possibilities are pictured in Fig. 6-4(a).

The probability in the second case follows:

$$P(Z \geq z_1) = \begin{cases} 0.500\,0 - \Phi(z_1) & \text{if } 0 \leq z_1 \\ 0.500\,0 + \Phi(|z_1|) & \text{if } z_1 \leq 0 \end{cases}$$

These two possibilities are pictured in Fig. 6-4(b).

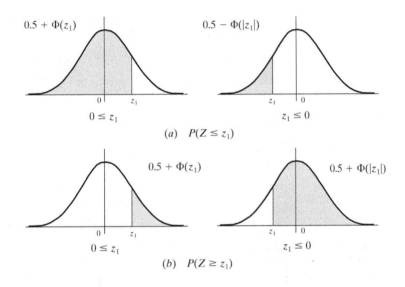

Fig. 6-4

EXAMPLE 6.7   Find the following one-sided probabilities for the standard normal distribution Z:

(a) $P(Z \leq 0.75)$ (b) $P(Z \leq -1.2)$ (c) $P(Z \geq 0.60)$ (d) $P(Z \geq -0.45)$

(a)   Referring to Fig. 6-4(a),

$$P(Z \leq 0.75) = 0.5 + \Phi(0.75) = 0.500\,0 + 0.242\,2 = 0.742\,2$$

(b)   Referring to Fig. 6-4(a),

$$P(Z \leq -1.2) = 0.5 - \Phi(1.2) = 0.500\,0 - 0.384\,9 = 0.115\,1$$

(c)   Referring to Fig. 6-4(b),

$$P(Z \geq 0.60) = 0.5 - \Phi(0.60) = 0.500\,0 - 0.225\,8 = 0.274\,2$$

(d)   Referring to Fig. 6-4(b),

$$P(Z \geq -0.45) = 0.5 + \Phi(-0.45) = 0.500\,0 + 0.173\,6 = 0.673\,6$$

Evaluating Arbitrary Normal Probabilities

Suppose X is a normal distribution, say X is $N(\mu, \sigma^2)$. We evaluate $P(a \leq X \leq b)$ by first changing a and b into corresponding standard units as follows:

$$z_1 = \frac{a - \mu}{\sigma} \quad \text{and} \quad z_2 = \frac{b - \mu}{\sigma}$$

Then
$$P(a \leq X \leq b) = P(z_1 \leq Z \leq z_2)$$

This is the area under the standard normal curve between z_1 and z_2 which can be found, as above, using Table 6-1 on page 184.

One-sided probabilities are obtained similarly. Namely,

$$P(X \leq a) = P(Z \leq z) \quad \text{and} \quad P(X \geq a) = P(Z \geq z)$$

Here again a is changed into its corresponding standard unit using $z = (a - \mu)/\sigma$.

EXAMPLE 6.8 Suppose X is the normal distribution $N(70, 4)$. Find:

(a) $P(68 \leq X \leq 74)$ (b) $P(72 \leq X \leq 75)$ (c) $P(63 \leq X \leq 68)$ (d) $P(X \geq 73)$

X has mean $\mu = 70$ and standard deviation $\sigma = \sqrt{4} = 2$. With reference to Figs. 6-3 and 6-4, we make the following computations:

(a) Transform $a = 68$, $b = 74$ into standard units as follows:

$$z_1 = \frac{68 - \mu}{\sigma} = \frac{68 - 70}{2} = -1, \qquad z_2 = \frac{74 - \mu}{\sigma} = \frac{74 - 70}{2} = 2$$

Therefore [Fig. 6-3(a)]

$$P(68 \leq X \leq 74) = P(-1 \leq Z \leq 2) = \Phi(2) + \Phi(1)$$
$$= 0.477\,2 + 0.341\,3 = 0.818\,4$$

(b) Transform $a = 72$, $b = 75$ into standard units:

$$z_1 = \frac{72 - 70}{2} = 1, \qquad z_2 = \frac{75 - 70}{2} = 2.5$$

Accordingly [Fig. 6-3(b)]:

$$P(72 \leq X \leq 75) = P(1 \leq Z \leq 2.5) = \Phi(2.5) - \Phi(1)$$
$$= 0.493\,8 - 0.341\,3 = 0.152\,5$$

(c) Transform $a = 63$, $b = 68$ into standard units:

$$z_1 = \frac{63 - 70}{2} = -3.5, \qquad z_2 = \frac{68 - 70}{2} = -1$$

Therefore [Fig. 6-3(c)]

$$P(63 \leq X \leq 68) = P(-3.5 \leq Z \leq -1) = \Phi(3.5) - \Phi(1)$$
$$= 0.499\,8 - 0.341\,3 = 0.158\,5$$

(d) Transform $a = 73$ into the standard unit $z = (73 - 70)/2 = 1.5$. Thus [Fig. 6.4(b)]

$$P(X \geq 73) = P(Z \geq 1.5) = 0.5 - \Phi(1.5)$$
$$= 0.500\,0 - 0.433\,2 = 0.066\,8$$

EXAMPLE 6.9 Verify the above 68–95–99.7 rule, that is, for a normal random variable X, show that:

(a) $P(\mu - \sigma \leq X \leq \mu + \sigma) \approx 0.68,$ (b) $P(\mu - 2\sigma \leq X \leq \mu + 2\sigma) \approx 0.95,$

(c) $P(\mu - 3\sigma \leq X \leq \mu + 3\sigma) \approx 0.997$

In each case, change to standard units and then use Table 6-1:

(a) $P(\mu - \sigma \leq X \leq \mu + \sigma) = P(-1 \leq Z \leq 1) = 2\Phi(1) = 2(0.341\,3) \approx 0.68$

(b) $P(\mu - 2\sigma \leq X \leq \mu + 2\sigma) = P(-2 \leq Z \leq 2) = 2\Phi(2) = 2(0.477\,2) \approx 0.95$

(c) $P(\mu - 3\sigma \leq X \leq \mu + 3\sigma) = P(-3 \leq Z \leq 3) = 2\Phi(3) = 2(0.498\,7) \approx 0.997$

Remark: Let X be any continuous random variable, which includes the normal random variables. Then X has the property that

$$P(X = a) \equiv P(a \leq X \leq a) = 0$$

Accordingly, for continuous data, such as heights, weights, and temperatures (whose measurements are really approximations), we usually ask for the probability that X lies in some interval $[a, b]$. On the other hand, we may sometimes ask for the probability that "$X = a$", where we mean the probability that X lies in some small interval $[a - \varepsilon, a + \varepsilon]$ centered at a. (Here the ε corresponds to the accuracy of the measurement.) This is illustrated in the next example.

EXAMPLE 6.10 Suppose the heights of American men are (approximately) normally distributed with mean $\mu = 68$ and standard deviation $\sigma = 2.5$. Find the percentage of American men who are:

(a) Between $a = 66$ and $b = 71$ in tall,   (b) (Approximately) 6 ft tall

(a) Transform a and b into standard units obtaining

$$z_1 = \frac{66 - 68}{2.5} = -0.80 \quad \text{and} \quad z_2 = \frac{71 - 68}{2.5} = 1.20$$

Here $z_1 < 0 < z_2$. Hence

$$P(66 \leq X \leq 71) = P(-0.8 \leq Z \leq 1.2) = \Phi(1.2) + \Phi(0.8)$$

$$= 0.384\,9 + 0.288\,1 = 0.673\,0$$

That is, approximately 67.3 percent of American men are between 66 and 71 in tall.

(b) Assuming heights are rounded off to the nearest inch, we are really asking the percentage of American men who are between $a = 71.5$ and $b = 72.5$ inches tall. Transform a and b into standard units obtaining

$$z_1 = \frac{71.5 - 68}{2.5} = 1.4 \quad \text{and} \quad z_2 = \frac{72.5 - 68}{2.5} = 1.8$$

Here $0 < z_1 < z_2$. Therefore

$$P(71.5 \leq X \leq 72.5) = P(1.4 \leq Z \leq 1.8) = \Phi(1.8) - \Phi(1.4)$$

$$= 0.464\,1 - 0.419\,2 = 0.044\,9$$

That is, about 4.5 percent of American men are (approximately) 6 ft tall.

6.5 NORMAL APPROXIMATION OF THE BINOMIAL DISTRIBUTION

The binomial probabilities $P(k) = \binom{n}{k} p^k q^{n-k}$ become increasingly difficult to compute as n gets larger. However, there is a way to approximate $P(k)$ by means of a normal distribution when an exact computation is impractical. This is the topic in this section.

Probability Histogram for $B(n,p)$

The probability histograms for $B(10, 0.1)$, $B(10, 0.5)$, $B(10, 0.7)$ are pictured in Fig. 6-5. (Rectangles whose heights are less than 0.01 have been omitted.) Generally speaking, the histogram of a binomial distribution $B(n,p)$ rises as k approaches the mean $\mu = np$ and falls off as k moves away from μ. Furthermore:

(1) For $p = 0.5$, the histogram is symmetric about the mean μ as in Fig. 6-5(b).

(2) For $p < 0.5$, the graph is skewed to the right as in Fig. 6-5(a).

(3) For $p > 0.5$, the graph is skewed to the left as in Fig. 6-5(c).

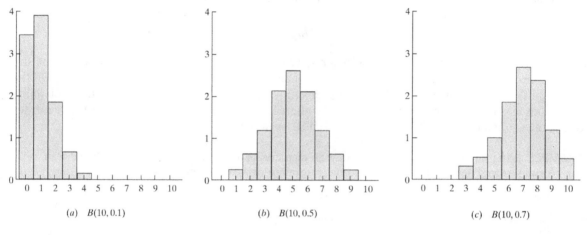

    (a) $B(10, 0.1)$ (b) $B(10, 0.5)$ (c) $B(10, 0.7)$

Fig. 6-5

Consider now the following distribution for $B(20, 0.7)$ where an asterisk (*) indicates that $P(k)$ is less than 0.01:

k	0	1	$\cdots$	8	9	10	11	12	13	14	15	16	17	18	19	20
$P(k)$	*	*	$\cdots$	*	0.01	0.03	0.07	0.11	0.16	0.19	0.18	0.13	0.07	0.03	0.01	*

The probability histogram for $B(20, 0.7)$ appears in Fig. 6-6.

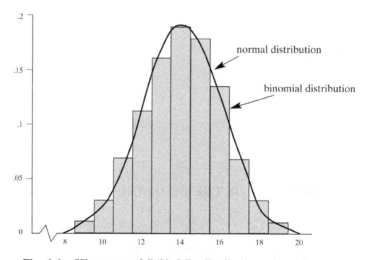

Fig. 6-6.   Histogram of $B(20, 0.7)$; distribution of $N(14, 4.2)$.

Although $p \neq 0.5$, observe that the histogram for $B(20, 0.7)$ is still nearly symmetric about the mean $\mu = np = 20(0.7) = 14$ for values of k between 8 and 20. Also, for k outside that range, $P(k)$ is practically 0. Furthermore, the standard deviation for $B(20, 0.7)$ is (approximately) $\sigma = 2$. Accordingly, the interval $[8, 20]$ is approximately $[\mu - 3\sigma, \mu + 3\sigma]$. These results are typical for binomial distributions $B(n, p)$ in which both np and nq are at least 5. We state these results more formally:

Basic Property of the Binomial Probability Histogram: For $np \geq 5$ and $nq \geq 5$, the probability histogram for $B(n, p)$ is nearly symmetric about $\mu = np$ over the interval $[\mu - 3\sigma, \mu + 3\sigma]$, where $\sigma = \sqrt{npq}$, and outside this interval $P(k) \approx 0$.

Normal Approximation, Central Limit Theorem

The density curve for the normal distribution $N(14, 4.2)$ is superimposed on the probability histogram for the binomial distribution $B(20, 0.7)$ in Fig. 6-6. Here $\mu = 14$ and $\sigma = \sqrt{4.2}$ for both distributions. The following is the fundamental relationship between any two such distributions:

For any integer value of k between $\mu - 3\sigma$ and $\mu + 3\sigma$, the area under the normal curve between $k - 0.5$ and $k + 0.5$ is approximately equal to $P(k)$, the area of the rectangle at k.

In other words:

The binomial probability $P(k)$ can be approximated by the normal probability
$$P(k - 0.05 \leq X \leq k + 0.5)$$

The following fundamental central limit theorem is the theoretical justification for the above approximation of $B(n, p)$ by $N(np, npq)$.

Central Limit Theorem 6.4: Let $X_1, X_2, X_3, \ldots$ be a sequence of independent random variables with the same distribution and with mean μ and variance σ^2. Let

$$Z_n = \frac{\bar{X}_n - \mu}{\sigma/\sqrt{n}}$$

where $\bar{X}_n = (X_1 + X_2 + \cdots + X_n)/n$. Then for large n and any interval $\{a \leq x \leq b\}$,

$$P(a \leq Z_n \leq b) \approx P(a \leq \phi \leq b)$$

where ϕ is the standard normal distribution.

Recall that $\bar{X}_n$ was called the sample mean of the random variables $X_1, \ldots, X_n$. Thus, Z_n in the above theorem is the standardized sample mean. Roughly speaking, the central limit theorem says that in any sequence of repeated trials the standardized sample mean approaches the standard normal curve as the number of trials increases.

6.6 CALCULATIONS OF BINOMIAL PROBABILITIES USING THE NORMAL APPROXIMATION

Let BP denote the binomial probability and let NP denote the normal probability. As noted above, for any integer value of k between $\mu - 3\sigma$ and $\mu + 3\sigma$, we have

$$BP(k) \approx NP(k - 0.5 \leq X \leq k + 0.5)$$

Accordingly, for nonnegative integers n_1 and n_2,

$$BP(n_1 \leq k \leq n_2) \approx NP(n_1 - 0.5 \leq X \leq n_2 + 0.5)$$

These formulas are used in the following examples.

EXAMPLE 6.11   A fair coin is tossed 100 times. Find the probability P that heads occur exactly 60 times.
This is a binomial experiment $B(n, p)$ with $n = 100$, $p = 0.5$, and $q = 1 - p = 0.5$. First we find

$$\mu = np = 100(0.5) = 50, \qquad \sigma^2 = npq = 100(0.5)(0.5) = 25, \qquad \text{and so} \qquad \sigma = 5$$

We use the normal distribution to approximate the binomial probability $P(60)$. We have

$$BP(60) \approx NP(59.5 \leq X \leq 60.5)$$

Transforming $a = 59.5$ and $b = 60.5$ into standard units yields

$$z_1 = \frac{59.5 - 50}{5} = 1.9 \qquad \text{and} \qquad z_2 = \frac{60.5 - 50}{5} = 2.1$$

Here $0 < z_1 < z_2$. Therefore [Fig. 6-3(b)]

$$P = BP(60) \approx NP(59.5 \leq X \leq 60.5) = NP(1.9 \leq Z \leq 2.1)$$

$$= \Phi(2.1) - \Phi(1.9) = 0.482\,1 - 0.471\,3 = 0.010\,8$$

Thus, 60 heads will occur approximately 1 percent of the time.

Remark: The above result agrees with the exact value of $BP(60)$ to four decimal places. That is, to four decimal places:

$$BP(60) = \binom{100}{60} (0.5)^{60} (0.5)^{30} = 0.010\,8$$

However, calculating $BP(60)$ directly is difficult.

EXAMPLE 6.12   A fair coin is tossed 100 times (as in Example 6.11). Find the probability P that heads occur between 48 and 53 times inclusive.
Again, we have the binomial experiment $B(n, p)$ with $n = 100$, $p = 0.5$, and $q = 0.5$; and again we have

$$\mu = np = 100(0.5) = 50 \qquad \text{and} \qquad \sigma = \sqrt{npq} = \sqrt{25} = 5$$

We seek $BP(48 \leq k \leq 53)$ or, assuming the data are continuous, $NP(47.5 \leq X \leq 53.5)$. Transforming $a = 47.5$ and $b = 53.5$ into standard units yields

$$z_1 = \frac{47.5 - 50}{5} = -0.5 \qquad \text{and} \qquad z_2 = \frac{53.5 - 50}{5} = 0.7$$

Here, $z_1 < 0 < z_2$. Accordingly [Fig. 6-3(a)]

$$P = BP(48 \leq k \leq 53) \approx NP(47.5 \leq X \leq 53.5) = NP(-0.5 \leq Z \leq 0.7)$$

$$= \Phi(0.7) + \Phi(0.5) = 0.258\,0 + 0.191\,5 = 0.449\,5$$

Thus, 48 to 53 heads will occur approximately 45 percent of the time.

Approximation of One-Sided Binomial Probabilities

The following formulas are used for the normal approximation to one-sided binomial probabilities:

$$BP(k \le n_1) \approx NP(X \le n_1 + 0.5) \quad \text{and} \quad BP(k \ge n_1) \approx NP(X \ge n_1 - 0.5)$$

The following remark justifies these one-sided approximations.

Remark: For the binomial distribution $B(n, p)$, the binomial variable k lies between 0 and n. Thus we should actually replace:

$$BP(k \le n_1) \text{ by } BP(0 \le k \le n_1) \quad \text{and} \quad BP(k \ge n_1) \text{ by } BP(n_1 \le k \le n)$$

This would yield the following approximations:

$$BP(0 \le k \le n_1) \approx NP(-0.5 \le X \le n_1 + 0.5) = NP(X \le n_1 + 0.5) - NP(X \le -0.5)$$

$$BP(n_1 \le k \le n) \approx NP(n_1 - 0.5 \le X \le n + 0.5) = NP(X \ge n_1 - 0.5) - NP(X \ge n + 0.5)$$

However, $NP(X \le -0.5)$ and $NP(X \ge 0.5)$ are very small and can be neglected. This is the reason for the above one-sided approximations.

EXAMPLE 6.13 A fair coin is tossed 100 times (as in Example 6.11). Find the probability P that heads occur less than 45 times.

Again, we have the binomial experiment $B(n, p)$ with $n = 100$, $p = 0.5$, and $q = 0.5$; and again we have

$$\mu = np = 100(0.5) = 50 \quad \text{and} \quad \sigma = \sqrt{npq} = \sqrt{25} = 5$$

We seek $BP(k < 45) = BP(k \le 44)$ or, approximately, $NP(X \le 44.5)$. Transforming $a = 44.5$ into standard units yields

$$z_1 = (44.5 - 50)/5 = -1.1$$

Here $z_1 < 0$. Accordingly [Fig. 6-4(a)]

$$P = BP(k \le 44) \approx NP(X \le 44.5) = NP(Z \le 1.1)$$

$$= 0.5 - \Phi(1.1) = 0.5 - 0.364\,3 = 0.135\,7$$

Thus, less than 45 heads will occur approximately 13.6 percent of the time.

6.7 POISSON DISTRIBUTION

A discrete random variable X is said to have the *Poisson distribution* with parameter $\lambda > 0$ if X takes on nonnegative integer values $k = 0, 1, 2, \ldots$ with respective probabilities

$$P(k) = f(k; \lambda) = \frac{\lambda^k e^{-\lambda}}{k!}$$

Such a distribution will be denoted by POI(λ). (This distribution is named after S. D. Poisson who discovered it in the early part of the nineteenth century.)

The values of $f(k; \lambda)$ can be obtained by using Table 6-2 which gives values of $e^{-\lambda}$ for various values of λ, or by using logarithms.

Table 6-2 Values of $e^{-\lambda}$

λ	0.0	0.1	0.2	0.3	0.4	0.5	0.6	0.7	0.8	0.9
$e^{-\lambda}$	1.000	0.905	0.819	0.741	0.670	0.607	0.549	0.497	0.449	0.407
λ	1	2	3	4	5	6	7	8	9	10
$e^{-\lambda}$	0.368	0.135	0.049 8	0.018 3	0.006 74	0.002 48	0.000 912	0.000 335	0.000 123	0.000 045

The Poisson distribution appears in many natural phenomena, such as the number of telephone calls per minute at some switchboard, the number of misprints per page in a large text, and the number of α particles emitted by a radioactive substance. Bar charts of the Poisson distribution for various values of λ appear in Fig. 6-7.

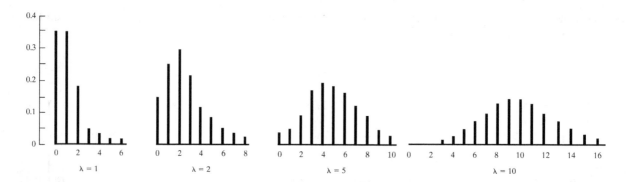

Fig. 6-7. Poisson distribution for selected values of λ.

Properties of the Poisson distribution follow:

Theorem 6.5:

Poisson distribution with parameter λ	
Mean or expected value	$\mu = \lambda$
Variance	$\sigma^2 = \lambda$
Standard deviation	$\sigma = \sqrt{\lambda}$

Although the Poisson distribution is of independent interest, it also provides us with a close approximation of the binomial distribution for small k provided p is small and $\lambda = np$ (Problem 6.43). This property is indicated in Table 6-3 which compares the binomial and Poisson distributions for small values of k with $n = 100$, $p = 1/100$, and $\lambda = np = 1$.

Table 6-3   Comparison of Binomial and Poisson Distributions with $n = 100$, $p = 1/100$, and $\lambda = np = 1$

k	0	1	2	3	4	5
Binomial	0.366	0.370	0.185	0.061 0	0.014 9	0.002 9
Poisson	0.368	0.368	0.184	0.061 3	0.015 3	0.003 07

EXAMPLE 6.14   Suppose 2 percent of the items produced by a factory are defective.   Find the probability P that there are 3 defective items in a sample of 100 items.

The binomial distribution with $n = 100$ and $p = 0.2$ applies.   However, since p is small, we can use the Poisson approximation with $\lambda = np = 2$.   Thus

$$P = f(3, 2) = \frac{2^{3e} - 2}{3!} = 8(0.135)/6 = 0.180$$

On the other hand, using the binomial distribution, we would need to calculate

$$P(3) = C(100, 3)(0.02)^3 (0.98)^{97} = (161,700)(0.02)^3 (0.98)^{97} \approx 0.182$$

Thus, the difference is only about 2 percent.

6.8 MISCELLANEOUS DISCRETE RANDOM VARIABLES

This section discusses a number of miscellaneous discrete random variables.   Recall that a random variable X is discrete if its range space is finite or countably infinite.

(a) Multinomial Distribution:

The binomial distribution is generalized as follows.   Suppose the sample space S of an experiment ε is partitioned into, say, s mutually exclusive events $A_1, A_2, \ldots, A_s$ with respective probabilities $p_1, p_2, \ldots, p_s$.   (Hence $p_1 + p_2 + \cdots + p_s = 1$.) Then

Theorem 6.6:   In n repeated trials, the probability that A_1 occurs k_1 times,

$$A_2 \text{ occurs } k_2 \text{ times}, \ldots, A_s \text{ occurs } k_s \text{ times}$$

is equal to

$$\frac{n!}{k_1! \, k_2! \cdots k_s!} p_1^{k_1} p_2^{k_2} \cdots p_s^{k_s}$$

where $k_1 + k_2 + \cdots + k_s = 1$.

The above numbers form the so-called *multinomial distribution* since they are precisely the terms in the expansion of the expression $(p_1 + p_2 + \cdots + p_s)^n$.   Observe that when $s = 2$ we obtain the binomial distribution discussed at the beginning of the chapter.

The process of repeated trials of the above experiment ε implicitly defines s discrete random variables $X_1, X_2, \ldots, X_s$.   Specifically, define X_1 to be the number of times A_1 occurs when ε is repeated n times.   Define X_2 to be the number of times A_2 occurs when ε is repeated n times.   And so on.   (Observe that the random variables are not independent since the knowledge of any $s - 1$ of them gives the remaining one.)

EXAMPLE 6.15   A fair die is tossed 8 times.   Find the probability p of obtaining 5 and 6 exactly twice and the other numbers exactly once.

Here we use the multinomial distribution to obtain

$$p = \frac{8!}{2! \, 2! \, 1! \, 1! \, 1!} \left(\frac{1}{6}\right)^2 \left(\frac{1}{6}\right)^2 \left(\frac{1}{6}\right) \left(\frac{1}{6}\right) \left(\frac{1}{6}\right) \left(\frac{1}{6}\right) = \frac{35}{5832} \approx 0.006$$

(b) *Hypergeometric Distribution:*

Consider a binomial experiment $B(n, p)$, that is, the experiment is repeated n times where each time the probability of success is p and the probability of failure is $q = 1 - p$. This experiment may be modeled as follows:

(i)   Choose a population of N elements where pN of the elements are designated as success and the remaining qN elements are designated as failures.

(ii) Choose a sample S of n elements *with replacement*, that is, each time a sample element is chosen, it is replaced in the population before the next sample element is chosen.

Then Theorem 6.1 tells us that the probability $P(k)$ that the sample S has k success elements is as follows:

$$P(k) = C(n, k)\, p^k\, q^{n-k}$$

EXAMPLE 6.16   Consider the binomial experiment $B(5, 0.6)$. Here $p = 0.6$ and $q = 0.4$. One model would be a box with $N = 10$ marbles of which $Np = 6$ have the color white (success) and $Nq = 4$ have the color red (failure). The probability of choosing a white marble is $p = 6/10 = 0.6$, as required. Suppose a random sample S of size $n = 5$ is chosen, with replacement. The probability of success for each of the $n = 5$ choices will be $p = 0.6$, and hence we have a model of the binomial experiment $B(5, 0.2)$.

The probability $P(3)$ that our sample has exactly 3 white (and hence 2 red) marbles follows:

$$P(3) = C(5, 3)(0.6)^3\, (0.4)^2 = 10(0.6)^3\, (0.4)^2 = 0.345$$

The hypergeometric distribution applies to sampling when there is no replacement. We illustrate with an example.

EXAMPLE 6.17   A class with $N = 10$ students has $M = 6$ men. Hence there are $N - M = 4$ women. Suppose a random sample of $n = 5$ students are selected. Find the probability $p = P(3)$ that exactly $k = 3$ men (and hence $n - k = 2$ women) are selected.

The probability follows:

$$p = P(3) = \frac{C(M, k)\, C(N - M, n - k)}{C(N, n)} = \frac{C(6, 3)\, C(4, 2)}{C(10, 5)} = \frac{(20)(6)}{252} \approx 0.476$$

The denominator $C(10, 5)$ denotes the number of possible ways of selecting a sample of $n = 5$ from the 10 students. The $C(6, 3)$ denotes the number of possible ways of selecting 3 men from the 6 men, and the $C(4, 2)$ denotes the number of possible ways of selecting 2 women from the 4 women.

The following theorem applies where $\min(M, n)$ means the minimum of the two numbers.

Theorem 6.7:   Suppose positive integers N, M, n are given with $M, n \leq N$. Then the following is a discrete probability distribution:

$$P(k) = \frac{C(M, k)\, C(N - M, n - k)}{C(N, n)} \qquad \text{for} \qquad k = 1, 2, \ldots, \min(M, n)$$

The above numbers form the so-called *hypergeometric distribution*; it is characterized by three parameters, n, N, M, and it is sometimes denoted by $\text{HYP}(n, N, M)$. A random variable X with this distribution is called a *hypergeometric* random variable.

If n is much smaller than M and N, then the hypergeometric distribution approaches the binomial distribution. Roughly speaking, with a large population, sampling with or without replacement is almost identical.

(c) *Geometric Distribution:*

Consider repeated trials of a Bernoulli experiment ε with probability p of success and $q = 1 = p$ of failure. Let X denote the number of times ε must be repeated until finally obtaining a success. (For example, one may continually fire at a target until finally hitting the target.) Then X is a random variable with the following distribution:

k	1	2	3	4	5	$\cdots$
$P(k)$	p	qp	$q^2 p$	$q^3 p$	$q^4 p$	$\cdots$

In other words, ε will be repeated k times only in the case that there is a sequence of $k - 1$ failures followed by a success. Thus, $P(k) = q^{k-1} p = pq^{k-1}$, as indicated by the above distribution table.

First we show that the above is a probability distribution, that is, that $\Sigma q^{k-1} p = 1$. Recall that the geometric series

$$1 + q + q^2 + \cdots = \frac{1}{1 - q}$$

Hence, using $p = 1 - q$, we have

$$\sum q^{k-1} p = p(1 + q + q^2 + \cdots) = \frac{p}{1 - q} = \frac{p}{p} = 1$$

as required.

The above distribution is called the *geometric distribution*; it is characterized by a single parameter p since $q = p - 1$, and it is sometimes denoted by GEO(p). A random variable X with such a distribution is called *geometric* random variable.

The following theorem applies.

Theorem 6.8:   Let X be a geometric random variable with distribution GEO(p). Then

   (i)   Expectation $E(X) = 1/p$.

   (ii) Variance $\text{var}(X) = q/p^2$.

   (iii) Cumulative distribution $F(k) = 1 - q^k$.

   (iv) $P(k > r) = q^r$.

   (See Problems 6.78, 6.90, and 6.91.)

EXAMPLE 6.18 Suppose the probability of a rocket hitting a target is $p = 0.2$, and a rocket is repeatedly fired until the target is hit.

(a) Find the expected number E of rockets which will be fired.

(b) Find the probability P that 4 or more rockets will be needed to finally hit the target.

(a) By Theorem 6.8, $E = 1/p = 1/(0.2) = 5$.

(b) First find $q = 1 - 0.2 = 0.8$. Then, by Theorem 6.8,

$$P(k > 3) = q^3 = (0.8)^3 = 0.512$$

That is, there is about a 50–50 chance of hitting the target with less than 4 rockets.

6.9 MISCELLANEOUS CONTINUOUS RANDOM VARIABLES

This section discusses another two continuous random variables, one with a uniform distribution on an interval and the other with an exponential distribution. Recall that a random variable X is continuous if its range space is noncountable and if there exists a density function $f(x)$ defined on $\mathbf{R} = (-\infty, \infty)$ such that

(i) $f(x) \geq 0,$ (ii) $\displaystyle\int_{-\infty}^{\infty} f(x)\, dx = 1,$ (iii) $P(a \leq X \leq b) = \displaystyle\int_{a}^{b} f(x)\, dx$

That is,

(i) f is nonnegative.

(ii) The area under its curve is one.

(iii) The probability that X lies in the interval $[a, b]$ is equal to the area under f between $x = a$ and $x = b$.

The cumulative distribution function $F(x)$ for the density function $f(x)$ is defined by

$$F(x) = \int_{-\infty}^{x} f(t)\, dt$$

Frequently, a continuous random variable X is defined by simply giving its density function $f(x)$. Also, if $f(x)$ is explicitly given for only certain values of x, then we assume $f(x) = 0$ for the remaining values of x in $\mathbf{R}$.

(a) *Uniform Distribution on an Interval:*

A continuous random variable X is said to have a *uniform distribution* on an interval $I = [a, b]$, where $a < b$, if its density function $f(x)$ has the constant value k on the interval and zero elsewhere. Since the area under f must be 1, we easily get that $k(b - a) = 1$ or that $k = 1/(b - a)$. That is,

$$f(x) = \begin{cases} \dfrac{1}{b - a} & \text{for } a \leq x \leq b \\ 0 & \text{elsewhere} \end{cases}$$

The notation UNIF(a, b) is sometimes used to denote this distribution; its graph is exhibited in Fig. 6-8(a).

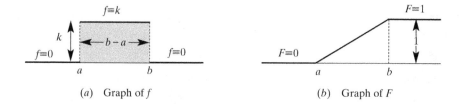

(a) Graph of f (b) Graph of F

Fig. 6-8

The following theorem, proved in Problem 6.48, applies.

Theorem 6.9:   Let X be a continuous random variable with distribution UNIF(a, b).   Then

(i)   Expectation $E(X) = \dfrac{a + b}{2}$

(ii)   Variance $\text{var}(X) = \dfrac{(b - a)^2}{12}$

(iii) Cumulative distribution function:

$$F(x) = \begin{cases} 0 & \text{for } x < a \\ \dfrac{x - a}{b - a} & \text{for } a \le x \le b \\ 1 & \text{for } x > b \end{cases}$$

The cumulative distribution function $F(x)$ of UNIF(a, b) is exhibited in Fig. 6-8(b).   Observe that $F(x)$ is 0 before the interval $[a, b]$, increases linearly from 0 to 1 on the interval $[a, b]$, and then remains at 1 after the interval $[a, b]$.

(b)   *Exponential Distribution:*

A continuous random variable X is said to have an *exponential distribution* with parameter β (where $\beta > 0$) if its density function $f(x)$ has the form

$$f(x) = \frac{1}{\beta} e^{-x/\beta}, \qquad x > 0$$

and 0 elsewhere.   The notation EXP(β) is sometimes used to denote such a distribution.   A picture of this distribution for various values of β appears in Fig. 6-9.

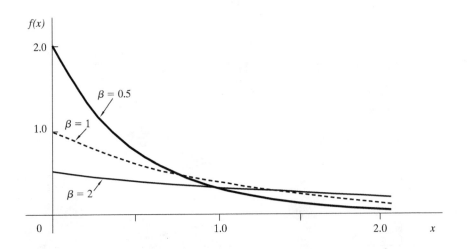

Fig. 6-9.   Exponential distribution for various values of β.

The following theorem applies:

Theorem 6.10:   Let X be a continuous random variable with distribution EXP(β).   Then

(i)   Expectation $E(X) = \beta$

(ii)   Variance $\text{var}(X) = \beta^2$

(iii) Cumulative distribution function:

$$F(x) = 1 - e^{-x/\beta}, \qquad x > 0$$

The exponential distribution also has the following important "no-memory property".

Theorem 6.11:   Let X have an exponential distribution.   Then

$$P(X > a + t \mid X > a) = P(X > t)$$

That is, suppose the lifetime of a certain solid-state component X is exponential.   Theorem 6.11 states that the probability that X will last t units after it has already lasted a units is the same as the probability that X will last t units when X was new.   In other words, a used component is just as reliable as a new component.

EXAMPLE 6.19   Suppose the lifetime X (in days) of a certain component C is exponential with $\beta = 120$.   Find the probability that the component C will last:

(*a*)   less than 60 days,   (*b*)   more than 240 days.

The following are the distribution $f(x)$ and cumulative distribution $F(x)$ with $\beta = 120$:

$$f(x) = \frac{1}{120} e^{-x/120} \quad \text{and} \quad F(x) = 1 - e^{-x/120}$$

(*a*)   The probability that C will last less than 60 days is

$$P(X < 60) = F(60) = 1 - e^{-0.5} = 0.393$$

(*b*)   The probability that C will last less than 240 days is

$$P(X < 240) = F(240) = 1 - e^{-2} = 0.865$$

Hence $$P(X > 240) = 1 - F(240) = 1 - 0.865 = 0.135$$

These probabilities are pictured in Fig. 6-10.

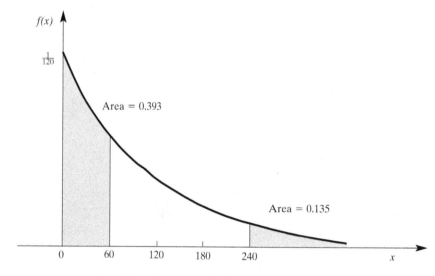

Fig. 6-10.   Exponential distribution with $\beta = 120$.

EXAMPLE 6.20   Consider the component C in Example 6.19.   If C is still working after 100 days, find the probability that it will last more than 340 days.

By the "no-memory property" Theorem 6.11 of the exponential distribution:

$$P(X > 340 \mid X > 100) = P(X > 240) = 0.135$$

That is, after working 100 days, the life expectancy of the used component C is the same as a new component.

Solved Problems

BINOMIAL DISTRIBUTION

6.1. Compute $P(k)$ for the binomial distribution $B(n,p)$ where

(a) $n = 5, p = \frac{1}{3}, k = 2$;   (b) $n = 10, p = \frac{1}{2}, k = 7$;   (c) $n = 4, p = \frac{1}{4}, k = 3$

Use Theorem 6.1 that $P(k) = \binom{n}{k} p^k q^{n-k}$ where $q = 1 - p$.

(a) Here $q = \frac{2}{3}$, so $P(2) = \binom{5}{2}\left(\frac{1}{3}\right)^2\left(\frac{2}{3}\right)^3 = \frac{5 \cdot 4}{2 \cdot 1}\left(\frac{1}{9}\right)\left(\frac{8}{27}\right) = \frac{80}{243} \approx 0.329$

(b) Here $q = \frac{1}{2}$, so $P(7) = \binom{10}{7}\left(\frac{1}{2}\right)^7\left(\frac{1}{2}\right)^3 = 120\left(\frac{1}{128}\right)\left(\frac{1}{8}\right) = \frac{15}{128} \approx 0.117$

(c) Here $q = \frac{3}{4}$, so $P(3) = \binom{4}{3}\left(\frac{1}{4}\right)^3\left(\frac{3}{4}\right) = 4\left(\frac{1}{64}\right)\left(\frac{3}{4}\right) = \frac{3}{64} \approx 0.047$

6.2. A fair coin is tossed 3 times. Find the probability that there will appear:

(a) 3 heads, (b) exactly 2 heads, (c) exactly 1 head, (d) no heads.

Method 1: We obtain the following equiprobable space with 8 elements:

$$S = \{HHH, HHT, HTH, HTT, THH, THT, TTH, TTT\}$$

Then we simply count the number of ways the event can occur.

(a) 3 heads *HHH* occurs only once; hence

$$P(3 \text{ heads}) = 1/8.$$

(b) 2 heads occurs 3 times, *HHT, HTH, THH*; hence

$$P(\text{exactly 2 heads}) = 3/8.$$

(c) 1 head occurs 3 times, *HTT, THT, TTH*; hence

$$P(\text{exactly 1 head}) = 3/8.$$

(d) No heads, that is, 3 tails *TTT*, occurs only once; hence

$$P(\text{no heads}) = P(3 \text{ tails}) = 1/8.$$

Method 2: Use Theorem 6.1 with $n = 3$ and $p = q = 1/2$.

(a) Here $k = 3$, so $P = P(3) = \left(\frac{1}{2}\right)^3 = \frac{1}{8} = 0.125$.

(b) Here $k = 2$, so $P = P(2) = \binom{3}{2}\left(\frac{1}{2}\right)^2\left(\frac{1}{2}\right)^1 = 3\left(\frac{1}{4}\right)\left(\frac{1}{2}\right) = \frac{3}{8} = 0.375$.

(c) Here $k = 1$, so $P = P(1) = \binom{3}{1}\left(\frac{1}{2}\right)^1\left(\frac{1}{2}\right)^2 = 3\left(\frac{1}{2}\right)\left(\frac{1}{4}\right) = \frac{3}{8} = 0.375$.

(d) Here $k = 0$, so $P = P(0) = \left(\frac{1}{2}\right)^3 = \frac{1}{8} = 0.125$.

6.3. The probability that John hits a target is $p = 1/4$. He fires $n = 6$ times. Find the probability that he hits the target: (*a*) exactly 2 times, (*b*) more than 4 times, (*c*) at least once.

This is a binomial experiment with $n = 6$, $p = 1/4$, and $q = 1 - p = 3/4$; hence use Theorem 6.1.

(*a*) $P(2) = \binom{6}{2} (1/4)^2 (3/4)^4 = 15(3^4)/(4^6) = 1215/4096 \approx 0.297$.

(*b*) John hits the target more than 4 times if he hits it 5 or 6 times. Hence

$$P(X > 4) = P(5) + P(6) = \binom{6}{5} (1/4)^5 (3/4)^1 + (1/4)^6$$

$$= 18/4^6 + 1/4^6 = 19/4^6 = 19/4096 \approx 0.004\,6$$

(*c*) Here $q^6 = (3/4)^6 = 729/4096$ is the probability that John misses all 6 times; hence

$$P(\text{one or more}) = 1 - 729/4096 = 3367/4096 \approx 0.82$$

6.4. Suppose 20 percent of the items produced by a factory are defective. Suppose 4 items are chosen at random. Find the probability that:

(*a*) 2 are defective, (*b*) 3 are defective, (*c*) none is defective.

This is a binomial experiment with $n = 4$, $p = 0.2$ and $q = 1 - p = 0.8$, that is, $B(4, 0.2)$. Hence use Theorem 6.1.

(*a*) Here $k = 2$ and $P(2) = \binom{4}{2} (0.2)^2 (0.8)^2 \approx 0.153\,6$.

(*b*) Here $k = 3$ and $P(3) = \binom{4}{3} (0.2)^3 (0.8) \approx 0.025\,6$.

(*c*) Here $P(0) = q^4 = (0.8)^4 = 0.409\,5$. Hence

$$P(X > 0) = 1 - P(0) = 1 - 0.409\,5 = 0.590\,4$$

6.5. Team A has probability 2/3 of winning whenever it plays. Suppose A plays 4 games. Find the probability that A wins more than half of its games.

This is a binomial experiment with $n = 4$, $p = 2/3$ and $q = 1 - p = 1/3$. A wins more than half its games if it wins 3 or 4 games. Hence

$$P(X > 2) = P(3) + P(4) = C(4,3)(2/3)^3 (1/3) + (2/3)^4$$

$$= 32/81 + 16/81 = 48/81 \approx 0.593$$

6.6. A family has 6 children. Find the probability P that there are: (*a*) 3 boys and 3 girls, (*b*) fewer boys than girls. (Assume that the probability of any particular child being a boy is 1/2.)

This is a binomial experiment with $n = 6$ and $p = q = 1/2$.

(*a*) $P = P(3 \text{ boys}) = C(6,3)(1/2)^3 (1/2)^3 = 20/64 = 5/16$.

(*b*) There are fewer boys than girls if there are 0, 1 or 2 boys. Hence

$$P = P(0) + P(1) + P(2)$$

$$= (1/2)^6 + C(6,1)(1/2)^5 (1/2) + C(6,2)(1/2)^4 (1/2)^2$$

$$= 22/64 = 11/32$$

Alternatively, the probability of different numbers of boys and girls is $1 - 5/16 = 11/16$. Half of these cases will have fewer boys than girls; hence

$$P = (1/2)(11/32) = 11/64$$

6.7. Find the number of dice that must be thrown so that there is a better-than-even chance of obtaining at least one six.

The probability of not obtaining a six on n dice is $q = (5/6)^n$. Thus, we seek the smallest n such that q is less than 1/2. Compute:

$$(5/6)^1 = 5/6, \qquad (5/6)^2 = 25/36, \qquad (5/6)^3 = 125/216, \qquad (5/6)^4 = 6255/1296 < 1/2$$

Thus, 4 dice must be thrown.

6.8. A certain type of missile hits its target with probability $p = 0.3$. Find the number of missiles that should be fired so that there is at least a 90 percent probability of hitting the target.

The probability of missing the target is $q = 1 - p = 0.7$. Hence the probability that n missiles miss the target is $(0.7)^n$. Thus, we seek the smallest n for which

$$1 - (0.7)^n > 0.90 \text{ or equivalently } (0.7)^n < 0.10$$

Compute

$$(0.7)^1 = 0.7, \ (0.7)^2 = 0.49, \ (0.7)^3 = 0.343, \ (0.7)^4 = 0.240, \ (0.7)^5 = 0.168, \ (0.7)^8 = 0.118, \ (0.7)^9 = 0.0823$$

Thus, at least 9 missiles should be fired.

6.9. The mathematics department has 8 graduate assistants who are assigned to the same office. Each assistant is just as likely to study at home as in the office. Find the minimum number m of desks that should be put in the office so that each assistant has a desk at least 90 percent of the time.

This problem can be modeled as a binomial experiment where

$n = 8 =$ number of assistants assigned to the office

$p = \frac{1}{2} =$ probability that an assistant will study in the office

$X =$ number of assistants studying in the office

Suppose there are k desks in the office, where $k \leq 8$. Then a graduate student will not have a desk if $X > k$. Note that

$$P(X > k) = P(k + 1) + P(k + 2) + \cdots + P(8)$$

We seek the smallest value of k for which $P(X > k)$ is less than 10 percent.

Compute $P(8), P(7), P(6), \ldots$ until the sum exceeds 10 percent. Using Theorem 6.1 with $n = 8$ and $p = q = 1/2$, we obtain

$$P(8) = (1/2)^8 = 1/256, \quad P(7) = 8(1/2)^7 (1/2) = 8/256, \quad P(6) = 28(1/2)^6 (1/2)^2 = 28/256$$

Now $P(8) + P(7) + P(6) = 37/256 > 10$ percent but $P(8) + P(7) < 10$ percent. Thus, $m = 6$ desks are needed.

6.10. A man fires at a target $n = 6$ times and hits it $k = 2$ times. (*a*) List the different ways that this can happen. (*b*) How many ways are there?

(*a*) List all sequences with 2 S's (successes) and 4 F's (failures):

SSFFFF, SFSFFF, SFFSFF, SFFFSF, SFFFFS, FSSFFF, FSFSFF, FSFFSF,
FSFFFS, FFSSFF, FFSFSF, FFSFFS, FFFSSF, FFFSFS, FFFFSS

(*b*) There are 15 different ways as indicated by the list.

Observe that this is equal to $C(6, 2) = \binom{6}{2}$ since we are distributing $k = 2$ letters S among the

$n = 6$ positions in the sequence.

6.11. Prove Theorem 6.1. The probability of exactly k successes in a binomial experiment $B(n, p)$ is

$$P(k) = P(k \text{ successes}) = \binom{n}{k} p^k q^{n-k}$$

The probability of one or more successes is $1 - q^n$.

The sample space of the n repeated trials consists of all n-tuples (that is, n-element sequences) whose components are either S (success) or F (failure). Let A be the event of exactly k successes. Then A consists of all n-tuples of which k components are S and $n - k$ components are F. The number of such n-tuples in the event A is equal to the number of ways that k letters S can be distributed among the n components of an n-tuple; hence A consists of $C(n, k) = \binom{n}{k}$ sample points.

The probability of each point in A is $p^k q^{n-k}$; hence

$$P(A) = \binom{n}{k} p^k q^{n-k}$$

In particular, the probability of no successes is

$$P(0) = \binom{n}{0} p^0 q^n = q^n$$

Thus, the probability of one or more successes is $1 - q^n$.

EXPECTED VALUE AND STANDARD DEVIATION

6.12. Four fair coins are tossed. Let X denote the number of heads occurring. Calculate the expected value of X directly and compare with Theorem 6.2.

X is binomially distributed with $n = 4$ and $p = q = \frac{1}{2}$. We have

$$P(0) = \frac{1}{16}, \qquad P(1) = \frac{4}{16}, \qquad P(2) = \frac{6}{16}, \qquad P(3) = \frac{4}{16}, \qquad P(4) = \frac{1}{16}$$

Thus, the expected value is

$$E(X) = 0\left(\frac{1}{16}\right) + 1\left(\frac{4}{16}\right) + 2\left(\frac{6}{16}\right) + 3\left(\frac{4}{16}\right) + 4\left(\frac{1}{16}\right) = \frac{32}{16} = 2$$

This agrees with Theorem 6.2, which states that

$$E(X) = np = 4\left(\frac{1}{2}\right) = 2$$

6.13. A family has 8 children. [We assume male and female children are equally probable.] (a) Determine the expected number E of girls. (b) Find the probability P that the expected number of girls does occur.

(a) The number of girls is binomially distributed with $n = 8$ and $p = q = 0.5$. By Theorem 6.2,

$$E = np = 8(0.5) = 4$$

(b) We seek the probability of 4 girls. By Theorem 6.1, with $k = 4$,

$$P = P(4 \text{ girls}) = \binom{8}{4} (0.5)^4 (0.5)^4 \approx 0.27 = 27\%$$

6.14. The probability that a man hits a target is $p = 1/10 = 0.1$.   He fires $n = 100$ times.   Find the expected number E of times he will hit the target and the standard deviation σ.

This is a binomial experiment $B(n,p)$ where $n = 100$, $p = 0.1$, and $q = 1 - p = 0.9$.   Thus, apply Theorem 6.2 to obtain

$$E = np = 100(0.1) = 10, \qquad \sigma^2 = npq = 100(0.1)(0.9) = 9, \qquad \sigma = \sqrt{9} = 3$$

6.15. The probability is 0.02 that an item produced by a factory is defective.   A shipment of 10,000 items is sent to a warehouse.   Find the expected number E of defective items and the standard deviation σ.

This is a binomial experiment $B(n,p)$ with $n = 10{,}000$, $p = 0.02$, and $q = 1 - p = 0.98$.   By Theorem 6.2,

$$E = np = (10{,}000)(0.02) = 200, \qquad \sigma^2 = npq = (10{,}000)(0.02)(0.98) = 196, \qquad \sigma = \sqrt{196} = 14$$

6.16. A student takes an 18-question multiple-choice exam, with 4 choices per question.   Suppose one of the choices is obviously incorrect, and the student makes an "educated" guess of the remaining choices.   Find the expected number E of correct answers and the standard deviation σ.

This is a binomial experiment $B(n,p)$ where $n = 18$, $p = 1/3$, and $q = 1 - p = 2/3$.   Thus, apply Theorem 6.2 to obtain

$$E = np = 18(1/3) = 6, \qquad \sigma^2 = npq = 18(1/3)(2/3) = 4, \qquad \sigma = \sqrt{4} = 2$$

6.17. A fair die is tossed 300 times.   Find the expected number E and the standard deviation σ of the number of 2's.

The number of 2's is binomially distributed with $n = 300$ and $p = 1/6$.   Also, $q = 1 - p = 5/6$.   By Theorem 6.2,

$$E = np = 300(1/6) = 50, \qquad \sigma^2 = npq = 300(1/6)(5/6) = 41.67, \qquad \sigma = \sqrt{41.67} \approx 6.45$$

6.18. Prove Theorem 6.2.   Let X be the binomial random variable $B(n,p)$.   Then:

(i) $\mu = E(X) = np$, (ii) $\mathrm{var}(X) = npq$.

On the sample space of n Bernoulli trials, let X_i (for $i = 1, 2, \ldots, n$) be the random variable which has the value 1 or 0 according as the ith trial is a success or a failure.   Then each X_i has the following distribution:

x	0	1
$P(x)$	q	p

and the total number of successes is $X = X_1 + X_2 + \cdots + X_n$.

(i)   For each i, we have

$$E(X_i) = 0(q) + 1(p) = p$$

Using the linearity property of E [Theorem 5.3 and Corollary 5.4], we have

$$E(X) = E(X_1 + X_2 + \cdots + X_n)$$
$$= E(X_1) + E(X_2) + \cdots + E(X_n)$$
$$= p + p + \cdots + p = np$$

(ii)   For each i, we have

$$E(X_i^2) = 0^2(q) + 1^2(p) = p$$

and

$$\text{var}(X_i) = E(X_i^2) - [E(X)_i]^2 = p - p^2 = p(1-p) = pq$$

The n random variables X_i are independent.   Therefore, by Theorem 5.9,

$$\text{var}(X) = \text{var}(X_1 + X_2 + \cdots + X_n)$$

$$= \text{var}(X_1) + \text{var}(X_2) + \cdots + \text{var}(X_n)$$

$$= pq + pq + \cdots + pq = npq$$

6.19. Give a direct proof of Theorem 6.2(i).   Let X be the binomial random variable $B(n, p)$.   Then $\mu = E(X) = np$.

Using the notation $P(k) = \binom{n}{k} p^k q^{n-k}$, we obtain the following where the last expression is obtained by dropping the term with $k = 0$, since its value is 0, and by factoring out np from each term:

$$E(X) = \sum_{k=0}^{n} kP(k) = \sum_{k=0}^{n} k\frac{n!}{k!(n-k)!} p^k q^{n-k}$$

$$= np \sum_{k=1}^{n} \frac{(n-1)!}{(k-1)!(n-k)!} p^{k-1} q^{n-k}$$

Let $s = k - 1$ in the above sum.   As k runs through the values 1 to n, s runs through the values 0 to $n - 1$.   Therefore

$$E(X) = np \sum_{s=0}^{n-1} \frac{(n-1)}{s!(n-1-s)!} p^s q^{n-1-s} = np$$

where, by the binomial theorem,

$$\sum_{s=0}^{n-1} k\frac{(n-1)!}{s!(n-1-s)!} p^s q^{n-1-s} = (p + q)^{n-1} = 1^{n-1} = 1$$

Thus, the theorem is proved.

6.20. Give a direct proof of Theorem 6.2(ii).   Let X be the binomial random variable $B(n, p)$.   Then $\text{var}(X) = npq$.

We first compute $E(X^2)$.   We have

$$E(X^2) = \sum_{k=0}^{n} k^2 P(k) = \sum_{k=0}^{n} k^2 \frac{n!}{k!(n-k)!} p^k q^{n-k}$$

$$= np \sum_{k=1}^{n} k\frac{(n-1)!}{(k-1)!(n-k)!} p^{k-1} q^{n-k}$$

Again we let $s = k - 1$ in the above sum and obtain

$$E(X^2) = np \sum_{s=0}^{n-1} (s + 1) \frac{(n-1)!}{s!(n-1-s)!} p^s q^{n-1-s}$$

$$= np \sum_{s=0}^{n-1} s\frac{(n-1)!}{s!(n-1-s)!} p^s q^{n-1-s} + np \sum_{s=0}^{n-1} \frac{(n-1)!}{s!(n-1-s)!} p^s q^{n-1-s}$$

Using Theorem 6.2(i), the first sum in the last expression is equal to $(n-1)p$; and, by the binomial theorem, the second sum is equal to 1.   Thus

$$E(X^2) = np(n-1)p + np = (np)^2 + np(1-p) = (np)^2 + npq$$

Hence,

$$\text{var}(X) = E(X^2) - \mu_X^2 = (np)^2 + npq - (np)^2 = npq$$

Thus, the theorem is proved.

NORMAL DISTRIBUTION

6.21. The mean and standard deviation on an examination are $\mu = 74$ and $\sigma = 12$, respectively. Find the scores in standard units of students receiving: (a) 65, (b) 74, (c) 86, (d) 92.

(a) $\quad z = \dfrac{x - \mu}{\sigma} = \dfrac{65 - 74}{12} = -0.75,$ (c) $\quad z = \dfrac{x - \mu}{\sigma} = \dfrac{86 - 74}{12} = 1.0$

(b) $\quad z = \dfrac{x - \mu}{\sigma} = \dfrac{74 - 74}{12} = 0,$ (d) $\quad z = \dfrac{x - \mu}{\sigma} = \dfrac{92 - 74}{12} = 1.5$

6.22. The mean and standard deviation on an examination are $\mu = 74$ and $\sigma = 12$, respectively. Find the grades corresponding to standard scores: (a) -1, (b) 0.5, (c) 1.25, (d) 1.75.

Solving $z = \dfrac{x - \mu}{\sigma}$ for x yields $x = \sigma z + \mu$. Thus

(a) $\quad x = \sigma z + \mu = (12)(-1) + 74 = 62,$ (c) $\quad x = \sigma z + \mu = (12)(1.25) + 74 = 89$

(b) $\quad x = \sigma z + \mu = (12)(0.5) + 74 = 80,$ (d) $\quad x = \sigma z + \mu = (12)(1.75) + 74 = 95$

6.23. Table 6-1 (page 184) uses $\Phi(z)$ to denote the area under the standard normal curve ϕ between 0 and z. Find: (a) $\Phi(1.63)$, (b) $\Phi(0.75)$, (c) $\Phi(1.1)$, (d) $\Phi(4.1)$.

Use Table 6-1 as follows:

(a) To find $\Phi(1.63)$, look down the first column on the left for the row labeled 1.6, and then continue right for the column labeled 3. The entry is 0.448 4. That is, the entry corresponding to row 1.6 and column 3 is 0.448 4. Hence $\Phi(1.63) = 0.448$ 4.

(b) To find $\Phi(0.75)$, look down the first column on the left for the row labeled 0.7, and then continue right for the column labeled 5. The entry is 0.273 4. That is, the entry corresponding to row 0.7 and column 5 is 0.273 4. Hence $\Phi(0.75) = 0.273$ 4.

(c) To find $\Phi(1.1)$, look on the left for the row labeled 1.1. The first entry in this row is 0.364 3 which corresponds to $1.1 = 1.10$. Hence $\Phi(1.1) = 0.364$ 3.

(d) The value of $\Phi(z)$ for any $z \geq 3.9$ is 0.500 0. Thus, $\Phi(4.1) = 0.500$ 0 even though 4.1 is not in the table.

6.24. Let Z be the random variable with standard normal distribution ϕ. Find the value of z if (a) $P(0 \leq Z \leq z) = 0.442$ 9, (b) $P(Z \leq z) = 0.796$ 7, (c) $P(z \leq Z \leq 2) = 0.100$ 0.

(a) Here $z > 0$. Thus, draw a picture of z and $P(0 \leq Z \leq z)$ as in Fig. 6-11(a). Here Table 6-1 can be used directly. The entry 0.442 9 appears to the right of row 1.5 and under column 8. Thus, $z = 1.58$.

(b) Note z must be positive since the probability is greater than 0.5. Thus, draw z and $P(Z \leq z)$ as in Fig. 6-11(b). We have

$$\Phi(z) = P(0 \leq Z \leq z) = P(Z \leq z) - 0.5 = 0.796\ 7 - 0.500\ 0 = 0.296\ 7$$

Since 0.296 7 appears in row 0.8 and column 3, we get $z = 0.83$.

(c) Since $\Phi(2) = 0.477\,2$ exceeds $0.100\,0$, z must lie between 0 and 2. Thus, draw z and $P(z \leq Z \leq 1)$
     as in Fig. 6-11(c). Then

$$\Phi(z) = \Phi(2) - P(z \leq Z \leq 2) = 0.477\,2 - 0.100\,0 = 0.377\,2$$

From Table 6-1, we get $z = 1.16$.

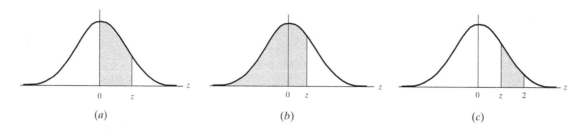

(a)   (b)   (c)

Fig. 6-11

6.25. Let Z be the random variable with standard normal distribution ϕ. Find:
(a) $P(0 \leq Z \leq 1.35)$, (b) $P(-1.21 \leq Z \leq 0)$, (c) $P(Z = 1.5)$.

(a) By definition $\Phi(z)$ is the area under the curve ϕ between 0 and z. Therefore, using Table 6-1,

$$P(0 \leq Z \leq 1.35) = \Phi(1.35) = 0.411\,5$$

(b) By symmetry and Table 6-1,

$$P(-1.21 \leq Z \leq 0) = P(0 \leq Z \leq 1.21) = \Phi(1.21) = 0.386\,9$$

(c) The area under a single point $a = 1.5$ is 0; hence

$$P(Z = 1.5) = 0$$

6.26. Let Z be the random variable with standard normal distribution ϕ. Find:
(a) $P(-1.37 \leq Z \leq 2.01)$,   (b) $P(0.65 \leq Z \leq 1.26)$,   (c) $P(-1.79 \leq Z \leq -0.54)$.

Use the following formula (pictured in Fig. 6-3):

$$P(z_1 \leq Z \leq z_2) = \begin{cases} \Phi(z_2) + \Phi(|z_1|) & \text{if } z_1 \leq 0 \leq z_2 \\ \Phi(z_2) - \Phi(z_1) & \text{if } 0 \leq z_1 \leq z_2 \\ \Phi(|z_1|) - \Phi(|z_2|) & \text{if } z_1 \leq z_2 \leq 0 \end{cases}$$

(a) Here $-1.73 < 0 < 1.26$, which is the first condition in the formula. Hence

$$P(-1.37 \leq Z \leq 2.01) = \Phi(1.26) + \Phi(1.37) = 0.414\,7 + 0.477\,8 = 0.892\,5$$

(b) Here $0 < 0.65 < 1.26$, which is the second condition in the formula. Hence

$$P(0.65 \leq Z \leq 1.26) = \Phi(1.26) - \Phi(0.65) = 0.396\,2 - 0.242\,2 = 0.154\,0$$

(c) Here $-1.79 < -0.54 \leq 0$, which is the third condition in the formula. Hence

$$P(-1.79 \leq Z \leq -0.54) = \Phi(1.79) - \Phi(0.54) = 0.463\,3 - 0.205\,4 = 0.257\,9$$

6.27. Let Z be the random variable with standard normal distribution ϕ. Find the following
one-sided probabilities:
(a) $P(Z \leq -0.22)$,   (b) $P(Z \leq 0.33)$,   (c) $P(Z \geq 0.44)$,   (d) $P(Z \geq -0.55)$

Figure 6-4 shows how to compute one-sided probabilities:

(a) $P(Z \leq -0.22) = 0.5 - \Phi(0.22) = 0.5 - 0.087\,1 = 0.412\,9$

(b) $P(Z \leq 0.33) = 0.5 + \Phi(0.22) = 0.5 + 0.129\,3 = 0.629\,3$

(c) $P(Z \geq 0.44) = 0.5 - \Phi(0.44) = 0.5 - 0.170\,0 = 0.330\,0$

(d) $P(Z \geq -0.55) = 0.5 + \Phi(0.55) = 0.5 + 0.208\,8 = 0.708\,8$

6.28. Suppose that the student IQ scores form a normal distribution with mean $v = 100$ and standard deviation $\sigma = 20$. Find the percentage P of students whose scores fall between:

(a) 80 and 120,   (c) 40 and 160,   (e) over 160,

(b) 60 and 140,   (d) 100 and 120,   (f) less than 80.

All the scores are units of the standard deviation $\sigma = 20$ from the mean $\mu = 100$; hence we can use the 68–95–99.7 rule or Fig. 6-2 instead of Table 6-1 to obtain P as follows:

(a) $P = P(80 \leq \text{IQ} \leq 120) = P(-1 \leq Z \leq 1) = 68\%$

(b) $P = P(60 \leq \text{IQ} \leq 140) = P(-2 \leq Z \leq 2) = 95\%$

(c) $P = P(40 \leq \text{IQ} \leq 160) = P(-3 \leq Z \leq 3) = 99.7\%$

(d) $P = P(100 \leq \text{IQ} \leq 120) = P(0 \leq Z \leq 1) = 68\%/2 = 34\%$

(e) Using (c) and symmetry, we have:

$$P = P(\text{IQ} \geq 160) = P(Z \geq 3) = [1 - 99.7\%]/2 = 0.3\%/2 = 0.15\%$$

(f) $P = P(\text{IQ} \leq 80) = P(Z \leq -1) = 50\% - 34\% = 16\%$

6.29. Suppose the temperature T during May is normally distributed with mean $\mu = 68°$ and standard deviation $\sigma = 6°$. Find the probability p that the temperature during May is

(a) between 70° and 80°,   (b) less than 60°.

First convert the T values into Z values in standard units, using $z = (t - \mu)/\sigma$, draw the appropriate picture, and then use Table 6-1.

(a) We convert as follows:

When $t = 70$, we get $z = (70 - 68)/6 = 0.33$.

When $t = 80$, we get $z = (80 - 68)/6 = 2.00$.

Since $0 < 0.33 < 2.00$, draw Fig. 6-12(a). Then

$$p = P(70 \leq T \leq 80) = P(0.33 \leq Z \leq 2.00)$$
$$= \Phi(2.00) - \Phi(0.33) = 0.477\,2 - 0.129\,3 = 0.347\,9$$

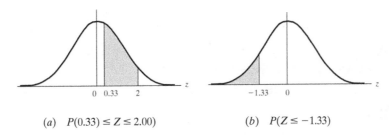

(a) $P(0.33) \leq Z \leq 2.00)$ (b) $P(Z \leq -1.33)$

Fig. 6-12

(b) First we convert as follows:

$$\text{When } t = 60, \text{ we get } z = (60 - 68)/6 = -1.33.$$

This is a one-sided probability with $-1.33 < 0$; hence draw Fig. 6-12(b). Using symmetry and that half the area under the curve is 0.500 0, we obtain

$$P = P(T \le 60) = P(Z \le -1.33) = P(Z \ge 1.33)$$
$$= 0.5 - \Phi(1.33) = 0.500\,0 - 0.408\,2 = 0.091\,8$$

6.30. Suppose the weights W of 800 male students are normally distributed with mean $\mu = 140$ lb and standard deviation $\sigma = 10$ lb. Find the number N of students with weights:

(a) between 138 and 148 lb,   (b) more than 152 lb.

First convert the W values into Z values in standard units, using $z = (w - \mu)\,\sigma$, draw the appropriate figure, and then use Table 6-1 (page 184).

(a) We convert as follows:

$$\text{When } w = 138, \text{ we get } z = (138 - 140)/10 = -0.2.$$
$$\text{When } w = 148, \text{ we get } z = (148 - 140)/10 = 0.8.$$

Since $-0.2 < 0 < 0.8$, draw Fig. 6-13(a). Then

$$P(138 \le W \le 148) = P(-0.2 \le Z \le 0.8)$$
$$= \Phi(0.8) + \Phi(-0.2) = 0.288\,1 + 0.079\,3 = 0.367\,4$$

Thus, $N = 800(0.367\,4) \approx 294$.

(b) We first convert as follows:

$$\text{When } w = 152, \text{ we get } z = (152 - 140)/10 = 1.20$$

This is a one-sided probability with $0 < 1.20$; hence draw Fig. 6-13(b). Using the fact that half the area under the curve is 0.500 0, we obtain

$$P(W \ge 152) = P(Z \ge 1.2) = 0.5 - \Phi(1.2) = 0.500\,0 - 0.384\,9 = 0.115\,1$$

Thus, $N = 800(0.115\,1) \approx 92$.

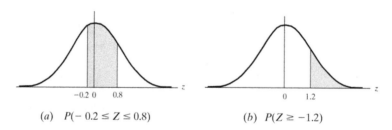

(a) $P(-0.2 \le Z \le 0.8)$ (b) $P(Z \ge -1.2)$

Fig. 6-13

6.31. Let Z be the random variable with standard normal distribution ϕ. Find the value of z if

(a) $P(z \le Z \le 1) = 0.476\,6$,   (b) $P(z \le Z \le 1) = 0.712\,2$.

By Table 6.1, $\Phi(1) = 0.341\,3$, and so $2\Phi(1) = 0.642\,3$.

(a) We have $\Phi(1) < 0.476\,6 < 2\Phi(1)$. Therefore, z is negative and $-z < 1$. Thus, draw z, $-z$, and $P(z \le Z \le 1)$ as in Fig. 6-14(a). Using symmetry, we obtain:

$$\Phi(-z) = 0.476\,6 - 0.341\,3 = 0.125\,3$$

By Table 6.1, $-z = 0.32$. Hence $z = -0.32$.

(b) We have $2\Phi(1) < 0.712\,2$. Therefore, z is negative and $-z > 1$. Thus, draw z, $-z$, and $P(z \le Z \le 1)$ as in Fig. 6-14(b).   Using symmetry, we obtain

$$\Phi(-z) = 0.712\,2 - 0.341\,3 = 0.370\,9$$

By Table 6.1, $-z = 1.13$.   Hence $z = -1.13$.

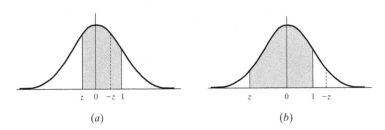

$$(a) \qquad\qquad\qquad\qquad (b)$$

Fig. 6-14

6.32. Use linear interpolation in Table 6-1, which only gives values of $\Phi(z)$ in steps of 0.01 for z, to solve the following:

(a) Find $\Phi(1.233)$ (b)   Find z to the nearest thousandth, if $\Phi(z) = 0.291\,7$.

(a)   The linear interpolation is indicated in Fig. 6-15(a).   We have

$$\frac{x}{17} = \frac{3}{10} \qquad \text{or} \qquad x \approx 5 \qquad \text{and so} \qquad P = 4082 + 5 = 4087$$

Thus, $\Phi(1.233) = 0.408\,7$.

(b)   The linear interpolation is indicated in Fig. 6-15(b) where, by Table 6-1, 2917 lies between 2910 and 2939.   We have

$$\frac{x}{10} = \frac{7}{29} \qquad \text{or} \qquad x \approx 4 \qquad \text{and so} \qquad Q = 0.814$$

That is, $z = 0.814$.

$$10 \left[3 \begin{bmatrix} 1.230 \rightarrow 4082 \\ 1.233 \rightarrow \quad P \\ 1.240 \rightarrow 4099 \end{bmatrix} x \right] 17 \qquad\qquad 29 \left[7 \begin{bmatrix} 2910 \rightarrow 0.810 \\ 2917 \rightarrow \quad Q \\ 2939 \rightarrow 0.820 \end{bmatrix} x \right] 10$$

$$(a) \qquad\qquad\qquad\qquad\qquad (b)$$

Fig. 6-15

NORMAL APPROXIMATION TO THE BINOMIAL DISTRIBUTION

This section of problems uses BP to denote the binomial probability and NP to denote the normal probability.

6.33. A fair coin is tossed 12 times.   Determine the probability P that the number of heads occurring is between 4 and 7 inclusive by using: (a) the binomial distribution, (b) the normal approximation to the binomial distribution.

(a)   Let heads denote a success.   By Theorem 6.1, with $n = 12$ and $p = q = 1/2$,

$$BP(4) = \binom{12}{4}\left(\frac{1}{2}\right)^4\left(\frac{1}{2}\right)^8 = \frac{495}{4096} \qquad BP(6) = \binom{12}{6}\left(\frac{1}{2}\right)^6\left(\frac{1}{2}\right)^6 = \frac{924}{4096}$$

$$BP(5) = \binom{12}{5}\left(\frac{1}{2}\right)^5\left(\frac{1}{2}\right)^7 = \frac{792}{4096} \qquad BP(7) = \binom{12}{7}\left(\frac{1}{2}\right)^7\left(\frac{1}{2}\right)^5 = \frac{792}{4096}$$

Hence $P = \dfrac{495}{4096} + \dfrac{792}{4096} + \dfrac{924}{4096} + \dfrac{792}{4096} = \dfrac{3003}{4096} = 0.733\,2.$

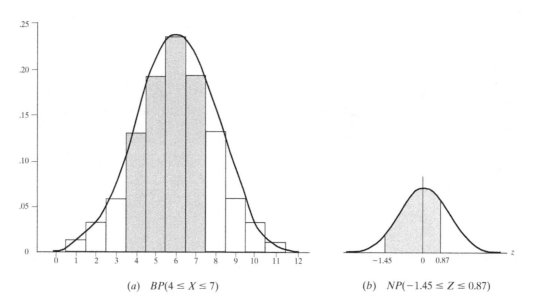

(a) $BP(4 \leq X \leq 7)$ (b) $NP(-1.45 \leq Z \leq 0.87)$

Fig. 6-16

(b)   Here

$$\mu = np = 12(\tfrac{1}{2}) = 6, \qquad \sigma^2 = npq = 12(\tfrac{1}{2})(\tfrac{1}{2}) = 3, \qquad \sigma = \sqrt{3} = 1.73$$

Let X denote the number of heads occurring.   We seek $BP(4 \leq X \leq 7)$ which corresponds to the shaded area in Fig. 6-16(a).   On the other hand, if we assume the data are continuous, in order to apply the binomial approximation, we must find $NP(3.5 \leq X \leq 7.5)$, as indicated in Fig. 6-16(a).   We convert x values to z values in standard units using $Z = (X - \mu)\sigma$.   Thus:

$$3.5 \text{ in standard units} = (3.5 - 6)/1.73 = -1.45$$
$$7.5 \text{ in standard units} = (7.5 - 6)/1.73 = 0.87$$

Then, as indicated by Fig. 6-13(b),

$$P = NP(3.5 \leq X \leq 7.5) = NP(-1.45 \leq Z \leq 0.87)$$
$$= \Phi(0.87) + \Phi(1.45) = 0.308\,7 + 0.426\,5 = 0.734\,3$$

(Note that the relative error $e = |(0.733\,2 - 0.734\,3)/0.733\,2| = 0.001\,5$ is less than 0.2 percent.)

6.34. A fair die is tossed 180 times. Determine the probability P that the face 6 will appear:

(a) between 29 and 32 times inclusive,   (b)   between 31 and 35 times inclusive,

(c) less than 22 times.

This is a binomial experiment $B(n, p)$ with $n = 180$, $p = 1/6$ and $q = 1 - p = 5/6$. Then

$$\mu = np = 180(1/6) = 30, \qquad \sigma^2 = npq = 180(1/6)(5/6) = 25, \qquad \sigma = 5$$

Let X denote the number of times the face 6 appears.

(a)   We seek $BP(29 \leq X \leq 32)$ or, assuming the data are continuous, $NP(28.5 \leq X \leq 32.5)$.   Converting x values into standard units, we get:

$$28.5 \text{ in standard units} = (28.5 - 30)/5 = -0.3$$
$$32.5 \text{ in standard units} = (32.5 - 30)/5 = 0.5$$

Thus, as shown in Fig. 6-17(a),

$$P = NP(28.5 \leq X \leq 32.5) = NP(-0.3 \leq Z \leq 0.5)$$

$$= \Phi(0.5) + \Phi(0.3) = 0.191\,5 + 0.117\,9 = 0.309\,4$$

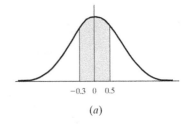 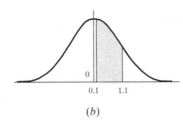 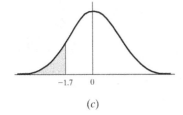

| (a) | (b) | (c) |

Fig. 6-17

(b)   We seek $BP(31 \leq X \leq 35)$ or, assuming the data are continuous, $NP(30.5 \leq X \leq 35.5)$.   Converting x values into standard units, we get:

$$30.5 \text{ in standard units} = (30.5 - 30)/5 = 0.1$$
$$35.5 \text{ in standard units} = (35.5 - 30)/5 = 1.1$$

Thus, as shown in Fig. 6-17(b),

$$P = NP(30.5 \leq X \leq 35.5) = NP(0.1 \leq Z \leq 1.1)$$

$$= \Phi(1.1) - \Phi(0.1) = 0.364\,3 - 0.039\,8 = 0.324\,5$$

(c)   We seek the one-sided probability $P(X < 22)$ or, approximately, $NP(X \leq 21.5)$.   (See remark below and in Section 6.6 on one-sided probabilities.) We have

$$21.5 \text{ in standard units} = (21.5 - 30)/5 = -1.7$$

Therefore, as shown in Fig. 6-17(c), using symmetry and that half the area under the curve is 0.500 0,

$$P = NP(X \leq 21.5) = NP(Z \leq -1.7) = 0.500\,0 - \Phi(1.7) = 0.500\,0 - 0.455\,4 = 0.044\,6$$

Remark: Since the binomial variable is never negative, we should actually replace $BP(X < 22)$ by

$$BP(0 \leq X < 22) \approx NP(-0.5 \leq X \leq 21.5) = NP(-6.1 \leq Z \leq -1.7)$$

$$= NP(Z \leq -1.7) - P(Z \leq -6.1)$$

However, $P(Z \leq -6.1)$ is very small and so it is neglected.

6.35. Among 10,000 random digits, find the probability P that the digit 3 appears: (a) between 975 and 1025 times, (b) at most 950 times.

This is a binomial experiment $B(n, p)$ with $n = 10,000$, $p = 0.1$ and $q = 1 - p = 0.9$. Then

$$\mu = np = 10,000(0.1) = 1000, \qquad \sigma^2 = npq = 10,000(0.1)(0.9) = 900, \qquad \sigma = 30$$

Let X denote the number of times 3 appears.

(a) We seek $BP(975 \leq X \leq 1025)$ or, approximately, $NP(974.5 \leq X \leq 1025.5)$. We have

$$974.5 \text{ in standard units} = (974.5 - 1000)/30 = -0.85$$
$$1025.5 \text{ in standard units} = (1025.5 - 1000)/30 = 0.85$$

Therefore,

$$P = NP(974.5 \leq X \leq 1025.5) = NP(-0.85 \leq Z \leq 0.85)$$
$$= 2\Phi(0.85) = 2(0.302\,6) = 0.605\,2$$

(b) We seek the one-sided probability $BP(X \leq 950)$ or, approximately, $NP(X \leq 950.5)$. (See remark Section 6.6.) We have

$$950.5 \text{ in standard units} = (950.5 - 1000)/30 = -1.65$$

Therefore

$$P = NP(X \leq 950.5) = NP(Z \leq -1.65)$$
$$= 0.500\,0 - \Phi(1.65) = 0.500\,0 - 0.450\,5 = 0.049\,5$$

6.36. Assume that 4 percent of the population over 65 years old has Alzheimer's disease. Suppose a random sample of 3500 people over 65 is taken. Find the probability P that fewer than 150 of them have the disease.

This is a binomial experiment $B(n, p)$ with $n = 3500$, $p = 0.04$, and $q = 1 - p = 0.96$. Then

$$\mu = np = (3500)(0.04) = 140, \qquad \sigma^2 = npq = (3500)(0.04)(0.96) = 134.4, \qquad \sigma = \sqrt{134.4} = 11.6$$

Let X denote the number of people with Alzheimer's disease.
We seek $BP(X < 150)$ or, approximately, $NP(X \leq 149.5)$. We have

$$149.5 \text{ in standard units} = (149.5 - 140)/11.6 = 0.82$$

Therefore

$$P = NP(X \leq 149.5) = NP(Z \leq 0.82) = 0.500\,0 + \Phi(0.82) = 0.500\,0 + 0.293\,9 = 0.793\,9$$

POISSON DISTRIBUTION

6.37. Find: (a) $e^{-1.3}$, (b) $e^{-2.5}$.
Use Table 6-2 (page 192) and the law of exponents.

(a) $e^{-1.3} = (e^{-1})(e^{-0.3}) = (0.368)(0.741) = 0.273$

(b) $e^{-2.5} = (e^{-2})(e^{-0.5}) = (0.135)(0.607) = 0.081\,9$

6.38. For the Poisson distribution $P(k) = f(k; \lambda) = \dfrac{\lambda^k e^{-\lambda}}{k!}$, find:

(a) $f(2; 1)$ (b) $f(3; \frac{1}{2})$,   (c) $f(2; 0.7)$.

Use Table 6-2 to obtain $e^{-\lambda}$.

(a) $f(2; 1) = \dfrac{1^2 e^{-1}}{2!} = \dfrac{e^{-1}}{2} = \dfrac{0.368}{2} = 0.184.$

(b) $f(3, \frac{1}{2}) = \dfrac{(1/2)^3 e^{-0.5}}{3!} = \dfrac{e^{-0.5}}{48} = \dfrac{0.607}{48} = 0.013.$

(c) $f(2; 0.7) = \dfrac{(0.7)^2 e^{-0.7}}{2!} = \dfrac{(0.49)(0.497)}{2} = 0.12.$

6.39. Suppose 300 misprints are distributed randomly throughout a book of 500 pages. Find the probability P that a given page contains: (a) exactly 2 misprints, (b) 2 or more misprints.

We view the number of misprints on one page as the number of successes in a sequence of Bernoulli trials. Here $n = 300$ since there are 300 misprints, and $p = 1/500$, the probability that a misprint appears on a given page. Since p is small, we use the Poisson approximation to the binomial distribution with $\lambda = np = 0.6$.

(a) $P = f(2; 0.6) = \dfrac{(0.6)^2 e^{-0.6}}{2!} = \dfrac{(0.36)(0.549)}{2} = 0.098\,8 \approx 0.1.$

(b) We have

$$P(0 \text{ misprint}) = f(0; 0.6) = \dfrac{(0.6)^0 e^{-0.6}}{0!} = e^{-0.6} = 0.549$$

$$P(1 \text{ misprint}) = f(1; 0.6) = \dfrac{(0.6)^1 e^{-0.6}}{1!} = (0.6)(0.549) = 0.329$$

Then $P = 1 - P(0 \text{ or } 1 \text{ misprint}) = 1 - (0.549 + 0.329) = 0.122.$

6.40. Suppose 2 percent of the items produced by a factory are defective. Find the probability P that there are 3 defective items in a sample of 100 items.

The binomial distribution with $n = 100$ and $p = 0.2$ applies. However, since p is small, we use the Poisson approximation with $\lambda = np = 2$. Thus

$$P = f(3; 2) = \dfrac{2^3 e^{-2}}{3!} = \dfrac{8(0.135)}{6} = 0.180$$

6.41. Show that the Poisson distribution $f(k; \lambda)$ is a probability distribution, that is,

$$\sum_{k=0}^{\infty} f(k; \lambda) = 1$$

By known results of analysis, $e^\lambda = \sum_{k=0}^{\infty} \lambda^k/k!$. Hence

$$\sum_{k=0}^{\infty} f(k; \lambda) = \sum_{k=0}^{\infty} \dfrac{\lambda^k e^{-\lambda}}{k!} = e^{-\lambda} \sum_{k=0}^{\infty} \dfrac{\lambda^k}{k!} = e^{-\lambda} e^\lambda = 1$$

6.42. Prove Theorem 6.5. Let X be a random variable with the Poisson distribution $f(k; \lambda)$. Then (i) $E(X) = \lambda$, (ii) $\mathrm{var}(X) = \lambda$. Hence $\sigma_X = \sqrt{\lambda}$.

(i) Using $f(k; \lambda) = \lambda^k e^{-\lambda}/k!$, we obtain the following where, in the last sum, we drop the term $k = 0$, since its value is 0, and we factor out λ from each term:

$$E(X) = \sum_{k=0}^{\infty} k \cdot f(k; \lambda) = \sum_{k=0}^{\infty} k \frac{\lambda^k e^{-\lambda}}{k!} = \lambda \sum_{k=1}^{\infty} \frac{\lambda^{k-1} e^{-\lambda}}{(k-1)!}$$

Let $s = k - 1$ in the above last sum. As k runs through the values 1 to ∞, s runs through the values 0 to ∞. Using $\Sigma_{s=0}^{\infty} f(s; \lambda) = 1$ by the preceding Problem 6.41, we get

$$E(X) = \lambda \sum_{s=0}^{\infty} \frac{\lambda^s e^{-\lambda}}{s!} = \lambda \sum_{s=0}^{\infty} f(s; \lambda) = \lambda$$

Thus, (i) is proved.

(ii) We compute $E(X^2)$ as follows where, again, in the last sum, we drop the term $k = 0$, since its value is 0, and we factor out λ from each term:

$$E(X^2) = \sum_{k=0}^{\infty} k^2 f(k; \lambda) = \sum_{k=0}^{\infty} k^2 \frac{\lambda^k e^{-\lambda}}{k!} = \lambda \sum_{k=1}^{\infty} k \frac{\lambda^{k-1} e^{-\lambda}}{(k-1)!}$$

Again we let $s = k - 1$ and obtain:

$$E(X^2) = \lambda \sum_{s=0}^{\infty} (s+1) \frac{\lambda^s e^{-\lambda}}{s!} = \lambda \sum_{k=0}^{\infty} (s+1) f(s; \lambda)$$

We break up the last sum into two sums to obtain the following where we use (i) to obtain λ for the first sum and Problem 6.41 to obtain 1 for the second sum:

$$E(X^2) = \lambda \sum_{k=0}^{\infty} s f(s; \lambda) + \lambda \sum_{k=0}^{\infty} f(s; \lambda) = \lambda(\lambda) + \lambda(1) = \lambda^2 + \lambda$$

Hence $\mathrm{var}(X) = E(X^2) - \mu_X^2 = \lambda^2 + \lambda - \lambda^2 = \lambda$

Thus, (ii) is proved.

6.43. Show that if p is small and n is large, then the binomial distribution $B(n, p)$ is approximated by the Poisson distribution $\mathrm{POI}(\lambda)$ where $\lambda = np$. That is, using

$$BP(k) = \binom{n}{k} p^k q^{n-k} \qquad \text{and} \qquad f(k; \lambda) = \frac{\lambda^k e^{-\lambda}}{k!}$$

we get $BP(k) \approx f(k; \lambda)$ where $np = \lambda$.

We have $BP(0) = (1 - p)^n = (1 - \lambda/n)^n$. Taking the natural logarithm of both sides yields:

$$\ln BP(0) = n \ln(1 - \lambda/n)$$

The Taylor expansion of the natural logarithm follows:

$$\ln(1 + x) = x - \frac{x^2}{2} + \frac{x^3}{3} - \cdots$$

Thus $$\ln\left(1 - \frac{\lambda}{n}\right) = -\frac{\lambda}{n} - \frac{\lambda^2}{2n^2} - \frac{\lambda^3}{3n^3} - \cdots$$

Therefore, when n is large,

$$\ln BP(0) = n \ln\left(1 - \frac{\lambda}{n}\right) = -\lambda - \frac{\lambda^2}{2n} - \frac{\lambda^3}{3n^2} - \cdots \approx -\lambda$$

Hence $BP(0) = e^{-\lambda}$.

Furthermore, if p is very small, then $q \approx 1$.   Thus

$$\frac{BP(k)}{BP(k-1)} = \frac{(n-k+1)p}{kq} = \frac{np-(k-1)p}{kq} = \frac{\lambda-(k-1)p}{kq} \approx \frac{\lambda}{k}$$

That is, $BP(k) \approx \lambda BP(k-1)/k$.   Thus, using $BP(0) \approx e^{-\lambda}$, we get

$$BP(1) \approx \lambda e^{-\lambda}, \qquad BP(2) \approx \lambda^2 e^{-\lambda}/2!, \qquad BP(3) \approx \lambda^3 e^{-\lambda}/3!$$

And so on.   That is, by induction, $BP(k) \approx \lambda^k e^{-\lambda}/k! = f(k; \lambda)$.

MISCELLANEOUS DISTRIBUTIONS AND PROBLEMS

6.44. The painted light bulbs produced by a factory are 50 percent red, 30 percent green, and 20 percent blue.   In a sample of 5 bulbs, find the probability P that 2 are red, 1 is green, and 3 are blue.

This is a multinomial distribution.   By Theorem 6.6,

$$P = \frac{5!}{2!\,1!\,2!}(0.5)^2(0.3)(0.2)^2 = 0.9$$

6.45. A committee of 4 is selected at random from a class with 12 students of whom 7 are men.   Find the probability P that the committee contains: (a) exactly 2 men, (b) at least 2 men.

This is a hypergeometric distribution with $N = 12$, $M = 7$, $n = 5$.   There are $N - M = 5$ women.

(a)   There are $C(12, 4)$ ways to choose the 4-person committee, $C(7, 2)$ ways to choose the 2 men, and $C(5, 2)$ ways to choose the 2 women.   Thus (Theorem 6.7)

$$P = P(2) = \frac{C(7, 2)\,C(5, 2)}{C(12, 4)} = \frac{(21)(10)}{495} = 0.424 = 42.4\%$$

(b)   Here $P = P(2) + P(3) + P(4)$.   Hence

$$P = \frac{C(7, 2)\,C(5, 2) + C(7, 3)\,C(5, 1) + C(7, 4)}{C(12, 4)} = 0.848 = 84.8\%$$

6.46. Suppose the probability that team A wins each game in a tournament is 60 percent. A plays until it loses.

(a)   Find the expected number E of games that A plays.

(b)   Find the probability P that A plays in at least 4 games.

(c)   Find the probability P that A wins the tournament if the tournament has 64 teams.   (Thus, a team winning 6 times wins the tournament.)

This is a geometric distribution with $p = 0.4$ and $q = 0.6$.   (A plays until A loses.)

(a)   By Theorem 6.8, $E = 1/p = 1/0.4 = 2.5$.

(b)   The only way A plays at least 4 games is if A wins the first 3 games.   Thus (Theorem 6.8(iv))

$$P = P(k > 3) = q^3 = (0.6)^3 = 0.216 = 21.6\%$$

(c)   Here A must win all 6 games; hence $P = (0.6)^6 = 0.046\,7 = 4.67\%$

6.47, Let X be a random variable with the following geometric distribution:

k	1	2	3	4	5	$\cdots$
$P(k)$	p	qp	q^2p	q^3p	q^4p	$\cdots$

Prove Theorem 6.8(i): $E(X) = 1/p$.

Here all sums are from 1 to ∞. We have

$$E(X) = \sum kq^{k-1}p = p\left(\sum kq^{k-1}\right)$$

Let

$$y = \sum q^k = \frac{1}{1-q}$$

The derivative with respect to q yields

$$\frac{dy}{dq} = \sum kq^{k-1} = \frac{1}{(1-q)^2}$$

Substituting this value for $\sum kq^{k-1}$ in the formula for E yields

$$E = \frac{p}{(1-q)^2} = \frac{p}{p^2} = p$$

(Note that calculus is used to help evaluate the infinite series.)

6.48. Let X be the (uniform) continuous random variable with distribution UNIF(a, b), that is, whose distribution function f is a constant $k = 1/(b-a)$ on the interval $I = [a, b]$ and zero elsewhere. [See Fig. 6-8.] Prove Theorem 6.9: (i) $E(X) = (a+b)/2$. (ii) var$(X) = (b-a)^2/12$, (iii) cumulative distribution $F(x)$ is equal to:

(1) 0 for $x < a$; (2) $(x-a)/(b-a)$ for $a \le x \le b$; (3) 1 for $x > b$. [See Fig. 6-8(b).]

(i) If we view probability as weight or mass, and the mean as the center of gravity, then it is intuitively clear that $\mu = (a+b)/2$. We verify this mathematically using calculus:

$$\mu = E(X) = \int_{-\infty}^{\infty} xf(x)\,dx = \int_{a}^{b} \frac{x}{b-a}\,dx$$

$$= \left[\frac{x^2}{2(b-a)}\right]_{a}^{b} = \frac{b^2}{2(b-a)} - \frac{a^2}{2(b-a)} = \frac{a+b}{2}$$

(ii) We have

$$E(X^2) = \int_{-\infty}^{\infty} x^2 f(x)\,dx = \int_{a}^{b} \frac{x^2}{b-a}\,dx$$

$$= \left[\frac{x^3}{3(b-a)}\right]_{a}^{b} = \frac{b^3}{3(b-a)} - \frac{a^3}{3(b-a)} = \frac{b^2 + ab + a^2}{3}$$

Then

$$\mathrm{var}(X) = E(X^2) - [E(X)]^2 = \frac{b^2 + ab + a^2}{3} - \frac{a^2 + 2ab + b^2}{4} = \frac{(b-a)^2}{12}$$

(iii)   We have three cases:

(1)   For $x < a$:

$$F(x) = \int_{-\infty}^{x} f(t)\, dt = \int_{-\infty}^{x} 0\, dt = 0$$

(2)   For $a \leq x \leq b$:

$$F(x) = \int_{-\infty}^{x} f(t)\, dt = \int_{a}^{x} \frac{1}{b-a}\, dt = \left[\frac{t}{b-a}\right]_{a}^{x} = \frac{x-a}{b-a}$$

(3)   For $x > b$:

Since $F(x)$ is a cumulative distribution function, $F(x) \geq F(b) = 1$.   But

$$F(x) = P(X \leq x) \leq 1$$

Hence $F(x) = 1$.

6.49.   Consider the following normal distribution:

$$f(x) = \frac{1}{\sigma\sqrt{2\pi}} \exp[-1/2(x-\mu)^2/\sigma^2]$$

Show that $f(x)$ is a continuous probability distribution, that is, show that $\int_{-\infty}^{\infty} f(x)\, dx = 1$.

Substituting $t = (x-\mu)/\sigma$ in $\int_{-\infty}^{\infty} f(x)\, dx$, we obtain the integral

$$I = \frac{1}{\sqrt{2\pi}} \int_{-\infty}^{\infty} e^{-t^2/2}\, dt$$

It suffices to show that $I^2 = 1$.   We have

$$I^2 = \frac{1}{2\pi} \int_{-\infty}^{\infty} e^{-t^2/2}\, dt \int_{-\infty}^{\infty} e^{-s^2/2}\, ds = \frac{1}{2\pi} \int_{-\infty}^{\infty} \int_{-\infty}^{\infty} e^{-(s^2-t^2)/2}\, ds\, dt$$

We introduce polar coordinates in the above double integral.   Let $s = r\cos\theta$ and $t = r\sin\theta$.   Then $ds\, dt = r\, dr\, d\theta$, $0 \leq \theta \leq 2\pi$, and $0 \leq r \leq \infty$.   That is,

$$I^2 = \frac{1}{2\pi} \int_{0}^{2\pi} \int_{0}^{\infty} re^{-r^2/2}\, dr\, d\theta$$

But

$$\int_{0}^{\infty} re^{-r^2/2}\, dr = \left[-e^{-r^2/2}\right]_{0}^{\infty} = 1$$

Hence $I^2 = \dfrac{1}{2\pi} \displaystyle\int_{0}^{2\pi} d\theta = 1$ and the theorem is proved.

6.50. Prove Theorem 6.3. Let X be a random variable with the normal distribution

$$f(x) = \frac{1}{\sigma\sqrt{2\pi}} \exp[-1/2(x - \mu)^2/\sigma^2]$$

Then (i) $E(X) = \mu$ and (ii) $\text{var}(X) = \sigma^2$. Hence $\sigma_X = \sigma$.

(i) By definition, $E(X) = \dfrac{1}{\sigma\sqrt{2\pi}} \displaystyle\int_{-\infty}^{\infty} x \exp[-1/2(x - \mu)^2/\sigma^2]\, dx$. Setting $t = (x - \mu)/\sigma$, we obtain

$$E(X) = \frac{1}{\sqrt{2\pi}} \int_{-\infty}^{\infty} (\sigma t + \mu)\, e^{-t^2/2}\, dt = \frac{1}{\sqrt{2\pi}} \int_{-\infty}^{\infty} t e^{-t^2/2}\, dt + \mu \frac{1}{\sqrt{2\pi}} \int_{-\infty}^{\infty} e^{-t^2/2}\, dt$$

But $g(t) = te^{-t^2/2}$ is an odd function, that is, $g(-t) = -g(t)$; hence $\displaystyle\int_{-\infty}^{\infty} te^{-t^2/2}\, dt = 0$. Furthermore,

$\dfrac{1}{\sqrt{2\pi}} \displaystyle\int_{-\infty}^{\infty} e^{-t^2/2}\, dt = 1$, by the preceding problem. Accordingly, $E(X) = \dfrac{\sigma}{\sqrt{2\pi}} \cdot 0 + \mu \cdot 1 = \mu$ as claimed.

(ii) By definition, $E(X^2) = \dfrac{1}{\sigma\sqrt{2\pi}} \displaystyle\int_{-\infty}^{\infty} x^2 \exp[-1/2(x - \mu)^2 \sigma^2]\, dx$. Again setting $t = (x - \mu)/\sigma$, we obtain

$$E(X^2) = \frac{1}{\sqrt{2\pi}} \int_{-\infty}^{\infty} (\sigma t + \mu)^2\, e^{-t^2/2}\, dt$$

$$= \sigma^2 \frac{1}{\sqrt{2\pi}} \int_{-\infty}^{\infty} t^2\, e^{-t^2/2}\, dt + 2\mu\sigma \frac{1}{\sqrt{2\pi}} \int_{-\infty}^{\infty} te^{-t^2/2}\, dt + \mu^2 \frac{1}{\sqrt{2\pi}} \int_{-\infty}^{\infty} e^{-t^2/2}\, dt$$

which reduces as above to $E(X^2) = \sigma^2 \dfrac{1}{\sqrt{2\pi}} \displaystyle\int_{-\infty}^{\infty} t^2 e^{-t^2/2}\, dt + \mu^2$.

We integrate the above integral by parts. Let $u = t$ and $dv = te^{-t^2/2}\, dt$. Then $v = -e^{-t^2/2}$ and $du = dt$. Thus

$$\frac{1}{\sqrt{2\pi}} \int_{-\infty}^{\infty} t^2\, e^{-t^2/2}\, dt = \frac{1}{\sqrt{2\pi}} \left[-te^{-t^2/2} \right]_{-\infty}^{\infty} + \frac{1}{\sqrt{2\pi}} \int_{-\infty}^{\infty} e^{-t^2/2}\, dt = 0 + 1 = 1$$

Consequently, $E(X) = \sigma^2 \cdot 1 + \mu^2 = \sigma^2 + \mu^2$ and

$$\text{var}(X) = E(X^2) - \mu_X^2 = \sigma^2 + \mu^2 - \mu^2 = \sigma^2$$

Thus, the theorem is proved.

Supplementary Problems

BINOMIAL DISTRIBUTION

6.51. Find $P(k)$ for the binomial distribution $B(n, p)$ where:

(a) $n = 5$, $p = 1/3$, $k = 2$; (b) $n = 7$, $p = 1/2$, $k = 3$; (c) $n = 4$, $p = 1/4$, $k = 2$.

6.52. A card is drawn and replaced 3 times from an ordinary 52-card deck. Find the probability that: (a) 2 hearts were drawn, (b) 3 hearts were drawn, (c) at least 1 heart was drawn.

6.53. A box contains 3 red marbles and 2 white marbles. A marble is drawn and replaced 3 times from the box. Find the probability that:
(*a*) 1 red marble was drawn, (*b*) 2 red marbles were drawn, (*c*) at least 1 red marble was drawn.

6.54. The batting average of a baseball player is 0.300. (That is, the probability that he gets a hit is 0.300.) He comes to bat 4 times. Find the probability that he will get: (*a*) exactly 2 hits, (*b*) at least 1 hit.

6.55. The probability that Tom scores on a three-point basketball shot is $p = 0.4$. He shoots $n = 5$ times. Find the probability that he scores: (*a*) exactly 2 times, (*b*) at least once.

6.56. Team A has probability $p = 0.4$ of winning each time it plays. Suppose A plays 4 games. Find the probability that A wins: (*a*) half of the games, (*b*) at least 1 game, (*c*) more than half of the games.

6.57. An unprepared student takes a 5-question true-false quiz and guesses every answer. Find the probability that the student will pass the quiz if at least 4 correct answers is the passing grade.

6.58. A certain type of missile hits its target with probability $p = 1/5$. (*a*) If 3 missiles are fired, find the probability that the target is hit at least once. (*b*) Find the number of missiles that should be fired so that there is at least a 90 percent probability of hitting the target (at least once).

6.59. A card is drawn and replaced in an ordinary 52-card deck. Find the number of times a card must be drawn so that: (*a*) there is an even chance of drawing a heart, (*b*) the probability of drawing a heart is greater than 75 percent.

6.60. A fair die is repeatedly tossed. Find the number of times the die must be tossed so that: (*a*) there is an even chance of tossing a 6, (*b*) the probability of tossing a 6 is greater than 80 percent.

EXPECTED VALUE AND STANDARD DEVIATION

6.61. Team B has probability $p = 0.6$ of winning each time it plays. Let X denote the number of times B wins in 4 games. (*a*) Find the distribution of X. (*b*) Find the mean μ, variance σ^2, and standard deviation σ of X.

6.62. Suppose 2 percent of the bolts produced by a factory are defective. In a shipment of 3600 bolts from the factory, find the expected number E of defective bolts and the standard deviation σ.

6.63. A fair die is tossed 180 times. Find the expected number E of times the face 6 occurs and the standard deviation σ.

6.64. Team A has probability $p = 0.8$ of winning each time it plays. Let X denote the number of times A will win in $n = 100$ games. Find the mean μ, variance σ^2, and standard deviation σ of X.

6.65. Let X be a binomially distributed random variable $B(n, p)$ with $E(X) = 2$ and $\text{var}(X) = 4/3$. Find n and p.

6.66. Consider the binomial distribution $B(n, p)$. Show that

(*a*) $\dfrac{P(k)}{P(k-1)} = \dfrac{(n-k+1)\,p}{kq}$.

(*b*) $P(k-1) < P(k)$ for $k < (n+1)p$ and $P(k-1) > P(k)$ for $k > (n+1)p$.

NORMAL DISTRIBUTION

6.67. Let Z be the standard normal random variable. Find

    (*a*) $P(-0.81 \le Z \le 1.13)$ (*c*) $P(0.53 \le Z \le 2.03)$

    (*b*) $P(-0.23 \le Z \le 1.6)$ (*d*) $P(0.15 \le Z \le 1.50)$

6.68. Let Z be the standard normal random variable. Find

    (*a*) $P(Z \le 0.73)$ (*c*) $P(Z \ge 0.2)$ (*e*) $P(Z = 1.8)$

    (*b*) $P(Z \le 1.8)$ (*d*) $P(Z \ge -1.5)$ (*f*) $P(|Z| \le 0.25)$

6.69. Let X be normally distributed with mean $\mu = 8$ and standard deviation $\sigma = 2$. Find the following without using Table 6-1,

    (*a*) $P(6 \le X \le 10)$ (*c*) $P(4 \le X \le 10)$ (*e*) $P(6 \le X \le 12)$

    (*b*) $P(4 \le X \le 12)$ (*d*) $P(4 \le X \le 6)$ (*f*) $P(8 \le X \le 10)$

6.70. Let X be normally distributed with mean $\mu = 8$ and standard deviation $\sigma = 4$. Find:

    (*a*) $P(5 \le X \le 10)$ (*c*) $P(3 \le X \le 9)$ (*e*) $P(X \ge 15)$

    (*b*) $P(10 \le X \le 15)$ (*d*) $P(3 \le X \le 7)$ (*f*) $P(X \le 5)$

6.71. Suppose the weights of 2000 male students are normally distributed with mean $\mu = 155$ lb and standard deviation $\sigma = 20$ lb. Find the number of students with weights:

    (*a*) not more than 100 lb,   (*c*) between 150 and 175 lb (inclusive),

    (*b*) between 120 and 130 lb (inclusive),   (*d*) greater than or equal to 200 lb.

6.72. Suppose the diameter d of bolts manufactured by a company is normally distributed with mean $\mu = 0.5$ cm and standard deviation $\sigma = 0.4$ cm. A bolt is considered defective if $d \le 0.45$ cm or $d > 0.55$ cm. Find the percentage of defective bolts manufactured by the company.

6.73. Suppose the scores on an examination are normally distributed with mean $\mu = 76$ and standard deviation $\sigma = 15$. The top 15 percent of the students receive A's and the bottom 10 percent receive F's. Find: (*a*) the minimum score to receive an A, (*b*) the minimum score to pass (not to receive an F).

NORMAL APPROXIMATION TO THE BINOMIAL DISTRIBUTION

6.74. A fair coin is tossed 10 times. Find the probability of obtaining between 4 and 7 heads inclusive by using:

    (*a*) the binomial distribution,   (*b*) the normal approximation to the binomial distribution.

6.75. A fair coin is tossed 400 times. Find the probability that the number of heads which occurs differs from 200 by:

    (*a*) more than 10,   (*b*) more than 25 times.

6.76. A fair die is tossed 720 times. Find the probability that the face 6 will occur:
    (*a*) between 100 and 125 times inclusive, (*b*) more than 135 times, (*c*) less than 110 times.

6.77. Among 625 random digits, find the probability that the digit 7 appears:
    (*a*) between 50 and 60 times, (*b*) between 60 and 70 times, (*c*) more than 75 times.

POISSON DISTRIBUTION

6.78. Find: (a) $e^{-1.6}$, (b) $e^{-2.3}$

6.79. For the Poisson distribution $f(k, \lambda) = \lambda^k e^{-\lambda}/k!$, find:

(a) $f(2; 1.5)$, (b) $f(3; 1)$, (c) $f(2; 0.6)$.

6.80. Suppose 220 misprints are distributed randomly throughout a book of 200 pages. Find the probability that a given page contains: (a) no misprints, (b) 1 misprint, (c) 2 misprints, (d) 2 or more misprints.

6.81. Suppose 1 percent of the items made by a machine are defective. In a sample of 100 items, find the probability that the sample contains: (a) no defective item, (b) 1 defective item, (c) 3 or more defective items.

6.82. Suppose 2 percent of the people on the average are left-handed. Find the probability of 3 or more left-handed among 100 people.

6.83. Suppose there is an average of 2 suicides per year per 50,000 population. In a city of 100,000, find the probability that in a given year the number of suicides is: (a) 0, (b) 1, (c) 2, (d) 2 or more.

MISCELLANEOUS DISTRIBUTIONS

6.84. A die is loaded so that the faces occur with the following probabilities:

k	1	2	3	4	5	6
$P(k)$	0.1	0.15	0.15	0.15	0.15	0.3

The die is tossed 6 times. Find the probability that: (a) each face occcurs once, (b) the faces 4, 5, 6 each appear twice.

6.85. A box contains 5 red, 3 white, and 2 blue marbles. A sample of 6 marbles is drawn with replacement, that is, each marble is replaced before the next marble is drawn. Find the probability that:
(a) 3 are red, 2 are white, 1 is blue; (b) 2 are red, 3 are white, 1 is blue; (c) 2 of each color appear.

6.86. A box contains 8 red and 4 white marbles. Find the probability that a sample of size $n = 4$ will contain 2 red and 2 white marbles if the sampling is done: (a) without replacement, (b) with replacement.

6.87. Driving down a main street, the probability is 0.8 that the car meets a green light (go) instead of a red light (stop). (a) Find the expected number E of green lights the car meets before it must stop. (b) If the car "makes" the first 3 lights (they are green), find the expected number F of additional green lights the car meets before it must stop.

6.88. Let X be the continuous uniform random variable UNIF(1, 3). Find $E(X)$, var(X), and cumulative distribution $F(x)$.

6.89. Suppose the life expectancy X (in hours) of a transistor tube is exponential with $\beta = 180$, that is, the following are the distribution $f(x)$ and cumulative distribution $F(x)$ of X:

$$f(x) = (1/180)\, e^{-x/180} \qquad \text{and} \qquad F(x) = 1 - e^{-x/180}$$

Find the probability that the tube will last: (a) less than 36 h, (b) between 36 and 90 h, (c) more than 90 h.

6.90. Let X be the geometric random variable GEO(p). Using the relation $\Sigma_{k=1}^{\infty} k^2 q^k = [q(q+1)]/(1-q)^3$, show that

   (a) $E(X^2) = (2-p)/p^2$,   (b) $\text{var}(X) = (1-p)/p^2$

6.91. Let X be the geometric random variable GEP(p). Prove Theorem 6.8: (iii) Cumulative distribution $F(k) = 1 - q^k$.   (iv) $P(k > r) = q^r$.

6.92. Show that the geometric random variable $X = $ GEO(p) has the "no memory" property, that is,

$$P(k > r + s \,|\, k > s) = P(k > r)$$

Answers to Supplementary Problems

6.51. (a) 80/243; (b) 21/128; (c) 27/128.

6.52. (a) 9/64; (b) 1/64; (c) 37/64.

6.53. (a) 36/1215; (b) 54/125; (c) 117/125.

6.54. (a) 0.254 6; (b) 0.759 9.

6.55. (a) 0.345 6; (b) 0.922.

6.56. (a) 216/625; (b) 544/625; (c) 112/625.

6.57. 6/32 = 18.75%.

6.58. (a) $1 - 64/125 = 61/125$; (b) 11.

6.59. (a) 3; (b) 5.

6.60. (a) 4; (b) 9.

6.61. (a) [0, 1, 2, 3, 4; 16/625, 96/625, 216/625, 216/625, 81/625]; (b) $\mu = 2.4$, $\sigma^2 = 0.96$, $\sigma = 0.98$.

6.62. $E = 72$, $\sigma = 8.4$.

6.63. $E = \mu = 30$, $\sigma = 5$.

6.64. $\mu = 80$, $\sigma^2 = 16$, $\sigma = 4$.

6.65. $n = 6$, $p = 1/3$.

6.67. (a) 0.661 8; (b) 0.536 2; (c) 0.276 9; (d) 0.334 5.

6.68. (a) 0.767 3; (b) 0.964 1; (c) 0.420 7; (d) 0.933 2; (e) 0; (f)0.197 4.

6.69. (a) 68.2%; (b) 95.4%; (c) 81.8%; (d) 13.6%; (e) 81.8%; (f) 34.1%.

6.70. (a) 0.464 9; (b) 0.268 4; (c) 0.493 1; (d) 0.295 7; (e) 0.040 1; (f) 0.226 6.

6.71. (*a*) 6; (*b*) 131; (*c*) 880; (*d*) 24.

6.72. 7.3%.

6.73. (*a*) 92; (*b*) 57.

6.74. (*a*) 0.773 4; (*b*) 0.771 8.

6.75. (*a*) 0.293 8; (*b*) 0.010 8.

6.76. (*a*) 0.688 6; (*b*) 0.001 1.

6.77. (*a*) 0.351 8; (*b*) 0.513 1; (*c*) 0.041 8.

6.78. (*a*) 0.202; (*b*) 0.100.

6.79. (*a*) 0.251; (*b*) 0.061 3; (*c*) 0.988.

6.80. (*a*) 0.333; (*b*) 0.366; (*c*) 0.201; (*d*) 0.301.

6.81. (Here $\lambda = 1$.) (*a*) 0.368; (*b*) 0.368; (*c*) 0.080.

6.82. 0.325.

6.83. (*a*) 0.018 3; (*b*) 0.073 2; (*c*) 0.146 4; (*d*) 0.909.

6.84. (*a*) 0.010 9; (*b*) 0.001 03.

6.85. (*a*) 0.135; (*b*) 0.081 0; (*c*) 0.081 0.

6.86. (*a*) $[(28)(6)]/495 = 0.339 = 33.9\%$; (*b*) $8/27 = 0.296 = 29.6\%$.

6.87. (*a*) $E = 1/0.2 = 5$; (*b*) (No memory) $F = 1/0.2 = 5$.

6.88. (*a*) (Theorem 6.9.) $E(X) = 2$, $\text{var}(X) = 1/3$, $F(x) = (x - 1)/2$.

6.89. (*a*) 0.181; (*b*) 0.212; (*c*) 0.607.

6.90. (*a*) $E(X^2) = \Sigma k^2 pq^{k-1} = (p/q) \Sigma k^2 q^k = (2 - p)p^2$.

6.91. Hint: Use $1 + q + q^2 + \cdots + q^{k-1} = (1 - q^k)/(1 - q)$.

CHAPTER 7

Markov Processes

7.1 INTRODUCTION

This chapter investigates a sequence of repeated trials of an experiment in which the outcome at any step in the sequence depends, at most, on the outcome of the preceding step and not on any other previous outcome. Such a sequence is called a *Markov chain* or *Markov process*.

EXAMPLE 7.1

(a) A box contains 100 light bulbs of which 8 are defective. One light bulb after another is selected from the box and tested to see if it is defective. This is not an example of a Markov process. The outcome of the third trial does depend on the preceding two trials.

(b) Three children, A, B, C, are throwing a ball to each other. A always throws the ball to B, and B always throws the ball to C. However, C is just as likely to throw the ball to B as to A. This is an example of a Markov process. Namely, the child throwing the ball is not influenced by those who previously had the ball.

Elementary properties of vectors and matrices, especially the multiplication of matrices, are required for this chapter. Thus, we begin with a review of vectors and matrices. The entries in our vectors and matrices will be real numbers, and the real numbers will also be called *scalars*.

7.2 VECTORS AND MATRICES

A *vector* $\mathbf{u}$ is a list of n numbers, say, $a_1, a_2, \ldots, a_n$. Such a vector is denoted by

$$\mathbf{u} = [a_1, a_2, \ldots, a_n]$$

The numbers a_i are called the *components* or *entries* of $\mathbf{u}$. If all the $a_i = 0$, then $\mathbf{u}$ is called the *zero vector*. By a *scalar multiple* $k\mathbf{u}$ of $\mathbf{u}$ (where k is a real number), we mean the vector obtained from $\mathbf{u}$ by multiplying each of its components by k, that is,

$$k\mathbf{u} = [ka_1, ka_2, \ldots, ka_n]$$

Two vectors are equal if and only if their corresponding components are equal.

A *matrix* $\mathbf{A}$ is a rectangular array of numbers usually presented in the form

$$\mathbf{A} = \begin{bmatrix} a_{11} & a_{12} & \cdots & a_{1n} \\ a_{21} & a_{22} & \cdots & a_{2n} \\ \cdots\cdots\cdots\cdots\cdots\cdots\cdots \\ a_{m1} & a_{m2} & \cdots & a_{mn} \end{bmatrix}$$

The m horizontal lists of numbers are called the *rows* of $\mathbf{A}$, and the n vertical lists of numbers are its *columns*. Thus, the following are the rows of the matrix $\mathbf{A}$:

$$[a_{11}, a_{12}, \ldots, a_{1n}], [a_{11}, a_{12}, \ldots, a_{1n}], \ldots, [a_{11}, a_{12}, \ldots, a_{1n}]$$

Furthermore, the following are the columns of the matrix $\mathbf{A}$:

$$\begin{bmatrix} a_{11} \\ a_{21} \\ \cdots \\ a_{m1} \end{bmatrix}, \begin{bmatrix} a_{12} \\ a_{22} \\ \cdots \\ a_{m2} \end{bmatrix}, \ldots, \begin{bmatrix} a_{1n} \\ a_{2n} \\ \cdots \\ a_{mn} \end{bmatrix}$$

Observe that the element a_{ij} of $\mathbf{A}$, called the *ij entry*, appears in row i and column j. We frequently denote such a matrix by writing $\mathbf{A} = [a_{ij}]$.

A matrix with m rows and n columns is called an m by n matrix, written $m \times n$. The pair of numbers m and n is called the *size* of the matrix. Two matrices $\mathbf{A}$ and $\mathbf{B}$ are equal, written $\mathbf{A} = \mathbf{B}$, if they have the same size and if corresponding elements are equal. Thus, the equality of two $m \times n$ matrices is equivalent to a system of mn equalities, one for each corresponding pair of elements.

A matrix with only one row may be viewed as a vector and vice versa. A matrix whose entries are all zero is called a *zero* matrix and will usually be denoted by 0.

Square Matrices

A *square matrix* is a matrix with the same number of rows and columns. In particular, a square matrix with n rows and n columns or, in other words, an $n \times n$ matrix, is said to be of *order n* and is called an *n-square* matrix.

The *diagonal* (or *main diagonal*) of an n-square matrix $\mathbf{A} = [a_{ij}]$ consists of the elements

$$a_{11}, \quad a_{22}, \quad \ldots, \quad a_{nn}$$

The n-square matrix with 1's on the diagonal and 0's elsewhere is called the *unit* matrix or *identity* matrix, and will usually be denoted by $\mathbf{I}_n$ or simply $\mathbf{I}$.

Multiplication of Matrices

Now suppose $\mathbf{A}$ and $\mathbf{B}$ are two matrices such that the number of columns of $\mathbf{A}$ is equal to the number of rows of $\mathbf{B}$, say $\mathbf{A}$ is an $m \times p$ matrix and $\mathbf{B}$ is a $p \times n$ matrix. Then the product of $\mathbf{A}$ and $\mathbf{B}$, denoted by $\mathbf{AB}$, is the $m \times n$ matrix $\mathbf{C}$ whose ij entry is obtained by multiplying the elements of row i of $\mathbf{A}$ by the corresponding elements of column j of $\mathbf{B}$ and then adding. That is, if $\mathbf{A} = [a_{ik}]$ and $\mathbf{B} = [b_{kj}]$, then

$$\mathbf{AB} = \begin{bmatrix} a_{11} & \cdots & a_{1p} \\ \cdot & \cdots & \cdot \\ a_{i1} & \cdots & a_{ip} \\ \cdot & \cdots & \cdot \\ a_{m1} & \cdots & a_{mp} \end{bmatrix} \begin{bmatrix} b_{11} & \cdots & b_{1j} & \cdots & b_{1n} \\ \cdot & \cdots & \cdot & \cdots & \cdot \\ \cdot & \cdots & \cdot & \cdots & \cdot \\ \cdot & \cdots & \cdot & \cdots & \cdot \\ b_{p1} & \cdots & b_{pj} & \cdots & b_{pn} \end{bmatrix} = \begin{bmatrix} c_{11} & \cdots & c_{1n} \\ \cdot & \cdots & \cdot \\ \cdot & c_{ij} & \cdot \\ \cdot & \cdots & \cdot \\ c_{m1} & \cdots & c_{mn} \end{bmatrix} = \mathbf{C}$$

where

$$c_{ij} = a_{i1} b_{1j} + a_{i2} b_{2j} + \cdots + a_{ip} b_{pj} = \sum_{k=1}^{p} a_{ik} b_{kj}$$

Namely, the product $\mathbf{AB}$ is the matrix $\mathbf{C} = [c_{ij}]$, where c_{ij} is defined above.

The product $\mathbf{AB}$ is not defined if $\mathbf{A}$ is an $m \times p$ matrix and $\mathbf{B}$ is a $q \times n$ matrix and $p \neq q$. That is, $\mathbf{AB}$ is not defined if the number of columns of $\mathbf{A}$ is not equal to the number of rows of $\mathbf{B}$.

There are special cases of matrix multiplication which are of special interest for us. Suppose $\mathbf{A}$ is an n-square matrix. Then we can form all the *powers* of $\mathbf{A}$, that is,

$$\mathbf{A}^2 = \mathbf{A}\mathbf{A}, \; \mathbf{A}^3 = \mathbf{A}\mathbf{A}^2, \; \mathbf{A}^4 = \mathbf{A}\mathbf{A}^3, \ldots$$

In addition, if $\mathbf{u}$ is a vector with n components, then we can form the product

$$\mathbf{u}\mathbf{A}$$

which is, again, a vector with n components. We call $\mathbf{u} \neq 0$ a *fixed vector* or *fixed point* of $\mathbf{A}$ if $\mathbf{u}$ is "left fixed", that is, not changed, when multiplied by $\mathbf{A}$, that is, if

$$\mathbf{u}\mathbf{A} = \mathbf{u}$$

In this case, for any scalar $k \neq 0$, one can show that

$$(k\mathbf{u})\,\mathbf{A} = k(\mathbf{u}\mathbf{A}) = k\mathbf{u}$$

This yields the following theorem.

Theorem 7.1: If $\mathbf{u}$ is a fixed vector of a matrix $\mathbf{A}$, then every nonzero scalar multiple $k\mathbf{u}$ of $\mathbf{u}$ is also a fixed vector of $\mathbf{A}$.

EXAMPLE 7.2

(a) Let $\mathbf{A} = \begin{bmatrix} 1 & 2 \\ 3 & 4 \end{bmatrix}$. Then

$$\mathbf{A}^2 = \begin{bmatrix} 1 & 2 \\ 3 & 4 \end{bmatrix}\begin{bmatrix} 1 & 2 \\ 3 & 4 \end{bmatrix} = \begin{bmatrix} 1+6 & 2+8 \\ 3+12 & 6+16 \end{bmatrix} = \begin{bmatrix} 7 & 10 \\ 15 & 22 \end{bmatrix}$$

(b) Let $\mathbf{u} = [2, -1]$ and $\mathbf{A} = \begin{bmatrix} 2 & 1 \\ 2 & 3 \end{bmatrix}$. Then

$$\mathbf{u}\mathbf{A} = [2, -1]\begin{bmatrix} 2 & 1 \\ 2 & 3 \end{bmatrix} = [4-2, 2-3] = [2, -1] = \mathbf{u}$$

Thus, $\mathbf{u}$ is a fixed vector of $\mathbf{A}$. Then, as expected from the above theorem, $2\mathbf{u} = [4, -2]$ is also a fixed vector of $\mathbf{A}$, namely,

$$(2\mathbf{u})\,\mathbf{A} = [4, -2]\begin{bmatrix} 2 & 1 \\ 2 & 3 \end{bmatrix} = [8-4, 4-6] = [4, -2] = 2\mathbf{u}$$

7.3 PROBABILITY VECTORS AND STOCHASTIC MATRICES

A vector $\mathbf{q} = [q_1, q_2, \ldots, q_n]$ is called a *probability vector* if its entries are nonnegative and their sum is 1, that is, if:

(i) Each $q_i \geq 0$,   (ii) $q_1 + q_2 + \cdots + q_n = 1$.

Recall that the probability distribution of a sample space S with n points has these two propereties and hence forms a probability vector.

A square matrix $\mathbf{P} = [p_{ij}]$ is called a *stochastic* matrix if each row of $\mathbf{P}$ is a probability vector. Thus, a probability vector may also be viewed as a stochastic matrix.

The following theorem (proved in Problem 7.8) applies. (The proof uses the fact that if $\mathbf{u}$ is a probability vector, then $\mathbf{u}\mathbf{A}$ is also a probability vector.)

Theorem 7.2: Suppose $\mathbf{A}$ and $\mathbf{B}$ are stochastic matrices. Then the product $\mathbf{A}\mathbf{B}$ is also a stochastic matrix. Thus, in particular, all powers $\mathbf{A}^n$ are stochastic matrices.

We now define an important class of stochastic matrices.

Definition: A stochastic matrix $\mathbf{P}$ is said to be *regular* if all the entries of some power $\mathbf{P}^m$ of $\mathbf{P}$ are positive.

EXAMPLE 7.3

(a) The nonzero vector $\mathbf{u} = [3, 1, 0, 5]$ is not a probability vector since the sum of its entries is 9, not 1. However, since the components of $\mathbf{u}$ are nonnegative, there is a unique probability vector $\mathbf{q_v}$ which is a scalar multiple of $\mathbf{u}$. This probability vector $\mathbf{q_v}$ can be obtained by multiplying $\mathbf{u}$ by the reciprocal of the sum of its components. That is, the following is the unique probability vector which is a multiple of $\mathbf{u}$:

$$\mathbf{q_v} = \frac{1}{9}\mathbf{v} = [3/9, 1/9, 0, 5/9]$$

(b) Consider the following two matrices:

$$\mathbf{A} = \begin{bmatrix} 0 & 1 \\ 1/2 & 1/2 \end{bmatrix} \quad \text{and} \quad \mathbf{B} = \begin{bmatrix} 1 & 0 \\ 1/2 & 1/2 \end{bmatrix}$$

Both of them are stochastic matrices. In particular, $\mathbf{A}$ is regular since, as follows, all entries in $\mathbf{A}^2$ are positive:

$$\mathbf{A}^2 = \begin{bmatrix} 0 & 1 \\ 1/2 & 1/2 \end{bmatrix}\begin{bmatrix} 0 & 1 \\ 1/2 & 1/2 \end{bmatrix} = \begin{bmatrix} 1/2 & 1/2 \\ 1/4 & 3/4 \end{bmatrix}$$

On the other hand, one can show that $\mathbf{B}$ is not regular. Specifically

$$\mathbf{B}^2 = \begin{bmatrix} 1 & 0 \\ 3/4 & 1/4 \end{bmatrix} \quad \mathbf{B}^3 = \begin{bmatrix} 1 & 0 \\ 7/8 & 1/8 \end{bmatrix} \quad \mathbf{B}^4 = \begin{bmatrix} 1 & 0 \\ 15/16 & 1/16 \end{bmatrix}$$

and every power $\mathbf{B}^m$ of $\mathbf{B}$ will have 1 and 0 in the first row. Accordingly, $\mathbf{B}$ is not regular.

The fundamental property of regular stochastic matrices is contained in the following theorem whose proof lies beyond the scope of this text.

Theorem 7.3: Let $\mathbf{P}$ be a regular stochastic matrix. Then:

   (i) $\mathbf{P}$ has a unique fixed probability vector $\mathbf{t}$, and the components of $\mathbf{t}$ are all positive.

   (ii) The sequence $\mathbf{P}, \mathbf{P}^2, \mathbf{P}^3, \ldots$ of powers of $\mathbf{P}$ approaches the matrix $\mathbf{T}$ whose rows are each the fixed point $\mathbf{t}$.

   (iii) If $\mathbf{q}$ is any probability vector, then the sequence of vectors

$$\mathbf{q}, \mathbf{qP}, \mathbf{qP}^2, \mathbf{qP}^3, \ldots$$

   approaches the fixed point $\mathbf{t}$.

Note that $\mathbf{P}^n$ approaches $\mathbf{T}$ means that each entry of $\mathbf{P}^n$ approaches the corresponding entry of $\mathbf{T}$, and $\mathbf{qP}^n$ approaches $\mathbf{t}$ means that each component of $\mathbf{qP}^n$ approaches the corresponding component of $\mathbf{t}$.

EXAMPLE 7.4 Consider the following stochastic matrix $\mathbf{P}$ [which is regular since $\mathbf{P}^5$ has only positive entries]:

$$\mathbf{P} = \begin{bmatrix} 0 & 1 & 0 \\ 0 & 0 & 1 \\ 1/2 & 1/2 & 0 \end{bmatrix}$$

Find its unique fixed probability vector $\mathbf{t}$ for $\mathbf{P}$.

Method 1: We seek the probability vector **t** with three components such that **tP** = **t**. The vector **t** can be represented in the form $[x, y, 1 - x - y]$. Accordingly, we form the following matrix equation:

$$[x, y, 1 - x - y] \begin{bmatrix} 0 & 1 & 0 \\ 0 & 0 & 1 \\ 1/2 & 1/2 & 0 \end{bmatrix} = [x, y, 1 - x - y]$$

Multiply the left side of the matrix equation, and then set corresponding components equal to each other. This yields the following system of linear equations:

$$\begin{cases} \frac{1}{2} - \frac{1}{2}x - \frac{1}{2}y = x \\ x + \frac{1}{2} - \frac{1}{2}x - \frac{1}{2}y = y \\ y = 1 - x - y \end{cases} \quad \text{or} \quad \begin{cases} 3x + y = 1 \\ x - 3y = -1 \\ x + 2y = 1 \end{cases} \quad \text{or} \quad \begin{cases} x = \frac{1}{5} \\ y = \frac{4}{5} \end{cases}$$

Thus, **t** = [1/5, 2/5, 2/5].

Method 2: We first seek any fixed vector **u** = $[x, y, z]$ of the matrix **P**. Thus, we form the matrix equation:

$$[x, y, z] \begin{bmatrix} 0 & 1 & 0 \\ 0 & 0 & 1 \\ 1/2 & 1/2 & 0 \end{bmatrix} = [x, y, z] \quad \text{or} \quad \begin{array}{c} \frac{1}{2}z = x \\ x + \frac{1}{2}z = y \\ y = z \end{array}$$

We know that the system has a nonzero solution; hence we can arbitrarily assign a value to one of the unknowns. Set $z = 2$. Then by the first equation $x = 1$ and by the third equation $y = 2$. Thus, **u** = $[1, 2, 2]$ is a fixed point of **P**. But every multiple of **u** is also a fixed point of **P**. Accordingly, multiply **u** by 1/5 to obtain the following unique fixed probability vector of **P**:

$$\mathbf{t} = \tfrac{1}{5}\mathbf{u} = [1/5, 2/5, 2/5]$$

7.4 TRANSITION MATRIX OF A MARKOV PROCESS

A Markov process or chain consists of a sequence of repeated trials of an experiment whose outcomes have the following two properties:

(i)   Each outcome belongs to a finite set $\{a_1, a_2, \ldots, a_n\}$ called the *state space* of the system; if the outcome on the nth trial is a_i, then we say the system is in state a_i at time n or at the nth step.

(ii) The outcome of any trial depends, at most, on the outcome of the preceding trial and not on any other previous outcome.

Accordingly, with each pair of states (a_i, a_j), there is given the probability p_{ij} that a_j occurs immediately after a_i occurs. The probabilities p_{ij} form the following n-square matrix:

$$\mathbf{M} = \begin{bmatrix} p_{11} & p_{12} & \cdots & p_{1n} \\ p_{21} & p_{22} & \cdots & p_{2n} \\ \multicolumn{4}{c}{\cdots\cdots\cdots\cdots\cdots\cdots} \\ p_{n1} & p_{n2} & \cdots & p_{nn} \end{bmatrix}$$

This matrix **M** is called the *transition matrix* of the Markov process.

Observe that with each state a_i there corresponds the ith row $[p_{i1}, p_{i2}, \ldots, p_{in}]$ of the transition matrix **M**. Moreover, if the system is in state a_i, then this row represents the probabilities of all the possible outcomes of the next trial and so it is a probability vector. We state this result formally.

Theorem 7.4: The transition matrix **M** of a Markov process is a stochastic matrix.

EXAMPLE 7.5

(*a*)   A man either takes a bus or drives his car to work each day.   Suppose he never takes the bus 2 days in a row; but if he drives to work, then the next day he is just as likely to drive again as he is to take the bus.

This stochastic process is a Markov chain since the outcome on any day depends only on what happened the preceding day.   The state space is $\{b(\text{bus}),\ d(\text{drive})\}$ and the transition matrix $\mathbf{M}$ follows:

$$\mathbf{M} = \begin{array}{c}\ \\ b \\ d\end{array}\begin{array}{cc}b & d \\ \left[\begin{array}{cc}0 & 1 \\ 1/2 & 1/2\end{array}\right.\end{array}\Big]$$

The first row of the matrix $\mathbf{M}$ corresponds to the fact that the man never takes the bus 2 days in a row, and so he definitely will drive the day after he takes the bus.   The second row of $\mathbf{M}$ corresponds to the fact that the day after he drives he will drive or take the bus with equal probability.

(*b*)   Three children, Ann (*A*), Bill (*B*), and Casey (*C*), are throwing a ball to each other.   Ann always throws the ball to Bill, and Bill always throws the ball to Casey.   However, Casey is just as likely to throw the ball to Bill as to Ann.   The ball throwing is a Markov process with the following transition matrix:

$$\mathbf{M} = \begin{array}{c}\ \\ A \\ B \\ C\end{array}\begin{array}{ccc}A & B & C \\ \left[\begin{array}{ccc}0 & 1 & 0 \\ 0 & 0 & 1 \\ 1/2 & 1/2 & 0\end{array}\right.\end{array}\Big]$$

The first row of the matrix corresponds to the fact that Ann always throws the ball to Bill.   The second row of the matrix corresponds to the fact that Bill always throws the ball to Casey.   The last row of the matrix corresponds to the fact that Casey always throws the ball to Ann or Bill with equal probability (and does not throw the ball to himself).

[Observe that this is the Markov process given in Example 7.1(*b*).]

(*c*)   An elementary school contains 200 boys and 150 girls.   One student is selected after another to take an eye examination.   Let X_n denote the sex of the nth student who takes the examination; hence the following is the state space of the stochastic process:

$$S = \{m(\text{male}),\ f(\text{female})\}$$

This process is not a Markov process.   For example, the probability that the third student is a girl depends not only on the outcome of the first trial but on the outcomes of both the first and second trials.

7.5   STATE DISTRIBUTIONS

Consider a Markov process with transition matrix $\mathbf{M}$.   The kth state distribution of the Markov process is the following probability vector:

$$\mathbf{q_k} = [q_{k1}, q_{k2}, \ldots, q_{kn}]$$

where q_{ki} is the probability that the state a_i occurs at the kth trial of the Markov chain.

Suppose the initial state distribution q_0 (at time $t = 0$) is given.   Then the subsequent state distributions can be obtained by multiplying the preceding state distribution by the transition matrix $\mathbf{M}$.   Namely,

$$q_0\,\mathbf{M} = q_1, \qquad q_1\,\mathbf{M} = q_2, \qquad q_2\,\mathbf{M} = q_3, \ldots$$

Accordingly

$$q_2 = q_1\,\mathbf{M} = (q_0\,\mathbf{M})\,\mathbf{M} = q_0\,\mathbf{M}^2 \qquad q_3 = q_2\,\mathbf{M} = (q_0\,\mathbf{M}^2)\,\mathbf{M} = q_0\,\mathbf{M}^3$$

and so on.   We state this result formally.

Theorem 7.5:   Suppose an initial state distribution q_0 is given.   Then, for $k = 1, 2, \ldots$,

$$q_k = q_{k-1}\,\mathbf{M} = q_0\,\mathbf{M}^k$$

EXAMPLE 7.6 Consider the Markov chain in Example 7.5(*b*) with transition matrix **M**. Suppose Casey is the first person with the ball, that is, suppose $q_0 = [0, 0, 1]$ is the initial probability distribution. Then

$$q_1 = q_0 \mathbf{M} = [0, 0, 1] \begin{bmatrix} 0 & 1 & 0 \\ 0 & 0 & 1 \\ 1/2 & 1/2 & 0 \end{bmatrix} = [1/2, 1/2, 0]$$

$$q_2 = q_1 \mathbf{M} = [1/2, 1/2, 0] \begin{bmatrix} 0 & 1 & 0 \\ 0 & 0 & 1 \\ 1/2 & 1/2 & 0 \end{bmatrix} = [0, 1/2, 1/2]$$

$$q_3 = q_2 \mathbf{M} = [0, 1/2, 1/2] \begin{bmatrix} 0 & 1 & 0 \\ 0 & 0 & 1 \\ 1/2 & 1/2 & 0 \end{bmatrix} = [1/4, 1/4, 1/2]$$

Thus, after 3 throws, the probability that Ann has the ball is 1/4, that Bill has the ball is 1/4, and that Casey has the ball is 1/2.

7.6 REGULAR MARKOV PROCESSES AND STATIONARY STATE DISTRIBUTIONS

A Markov chain is said to be *regular* if its transition matrix **M** is regular. Recall Theorem 7.3: if **M** is regular then **M** has a unique fixed probability vector **t** and, for any probability vector **q**, the sequence

$$\mathbf{q}, \mathbf{qM}, \mathbf{qM}^2, \mathbf{qM}^3, \ldots$$

approaches the unique fixed point **t**. Thus, Theorems 7.3 and 7.5 give us the next basic result.

Theorem 7.6: Suppose the transition matrix **M** of a Markov chain is regular. Then, in the long run, the probability that any state a_j occurs is approximately equal to the component t_j of the unique fixed probability vector **t** of **M**.

Thus, we see that the effect of the initial state distribution in a regular Markov process wears off as the number of steps increases. That is, every sequence of state distributions approaches the fixed probability vector **t** of **M**, which is called the *stationary distribution* of the Markov chain.

EXAMPLE 7.7

(*a*) Consider the Markov process in Example 7.5(*b*) where Ann, Bill, and Casey throw a ball to each other with the following transition matrix:

$$\mathbf{M} = \begin{bmatrix} 0 & 1 & 0 \\ 0 & 0 & 1 \\ 1/2 & 1/2 & 0 \end{bmatrix}$$

By Example 7.4, $\mathbf{t} = [1/5, 2/5, 2/5]$ is the unique fixed probability vector of **M**. Thus, in the long run, Ann will be thrown the ball 20 percent of the time, and Bill and Casey will be thrown the ball 40 percent of the time.

(*b*) Consider the Markov process in Example 7.5(*a*) where a man takes a bus or drives to work with the following transition matrix:

$$\mathbf{M} = \begin{bmatrix} 0 & 1 \\ 1/2 & 1/2 \end{bmatrix}$$

Find the stationary distribution of the Markov process.

We seek a probability vector $\mathbf{t} = [x, 1 - x]$ such that $\mathbf{tM} = \mathbf{t}$. Thus, set

$$[x, 1 - x]\begin{bmatrix} 0 & 1 \\ 1/2 & 1/2 \end{bmatrix} = [x, 1 - x]$$

Multiply the left side of the matrix equation to obtain

$$[\tfrac{1}{2} - \tfrac{1}{2}x, \tfrac{1}{2} + \tfrac{1}{2}x] = [x, 1 - x] \quad \text{or} \quad \begin{cases} \tfrac{1}{2} - \tfrac{1}{2}x = x \\ \tfrac{1}{2} + \tfrac{1}{2}x = 1 - x \end{cases} \quad \text{or } x = \tfrac{1}{3}$$

Thus, $\mathbf{t} = [1/3, 1 - 1/3] = [1/3, 2/3]$ is the unique fixed probability vector of $\mathbf{M}$. Therefore, in the long run, the man will take the bus to work 1/3 of the time, and drive to work the other 2/3 of the time.

Solved Problems

MATRIX MULTIPLICATION

7.1. Let $\mathbf{u} = [1, -2, 4]$ and $\mathbf{A} = \begin{bmatrix} 1 & 3 & -1 \\ 0 & 2 & 5 \\ 4 & 1 & 6 \end{bmatrix}$. Find $\mathbf{uA}$.

The product of the three-component vector $\mathbf{u}$ by the 3×3 matrix $\mathbf{A}$ is again a three-component vector. To obtain the first component of $\mathbf{uA}$, multiply the elements of $\mathbf{u}$ by the corresponding elements of the first column of $\mathbf{A}$ and then add as follows:

$$[1, -2, 4]\begin{bmatrix} 1 & 3 & -1 \\ 0 & 2 & 5 \\ 4 & 1 & 6 \end{bmatrix} = [1(1) - 2(0) + 4(4), \quad , \quad] = [17, \quad , \quad]$$

To obtain the second component of $\mathbf{uA}$, multiply the elements of $\mathbf{u}$ by the corresponding elements of the second column of $\mathbf{A}$ and then add as follows:

$$[1, -2, 4]\begin{bmatrix} 1 & 3 & -1 \\ 0 & 2 & 5 \\ 4 & 1 & 6 \end{bmatrix} = [17, \quad 1(3) - 2(2) + 4(1), \quad] = [17, 3, \quad]$$

To obtain the third component of $\mathbf{uA}$, multiply the elements of $\mathbf{u}$ by the corresponding elements of the third column of $\mathbf{A}$ and then add as follows:

$$[1, -2, 4]\begin{bmatrix} 1 & 3 & -1 \\ 0 & 2 & 5 \\ 4 & 1 & 6 \end{bmatrix} = [17, \quad 3, \quad 1(-1) - 2(5) + 4(6)] = [17, 3, 13]$$

Namely, $$\mathbf{uA} = [17, 3, 13]$$

7.2. Find $\mathbf{AB}$ where $\mathbf{A} = \begin{bmatrix} 1 & 3 \\ 2 & -1 \end{bmatrix}$ and $\mathbf{B} = \begin{bmatrix} 2 & 0 & -4 \\ 5 & -2 & 6 \end{bmatrix}$.

Since $\mathbf{A}$ is 2×2 and $\mathbf{B}$ is 2×3, the product $\mathbf{AB}$ is defined and $\mathbf{AB}$ is a 2×3 matrix. To obtain the first row of the product $\mathbf{AB}$, multiply the first row $(1, 3)$ of $\mathbf{A}$ by the corresponding elements of each of the columns $\begin{bmatrix} 2 \\ 3 \end{bmatrix}, \begin{bmatrix} 0 \\ -2 \end{bmatrix}, \begin{bmatrix} -4 \\ 6 \end{bmatrix}$ of $\mathbf{B}$ and then add

$$\mathbf{AB} = \begin{bmatrix} 2 + 15 & 0 - 6 & -4 + 18 \end{bmatrix} = \begin{bmatrix} 17 & -6 & 14 \end{bmatrix}$$

To obtain the second row of the product $\mathbf{AB}$, multiply the second row $(2, -1)$ of $\mathbf{A}$ by the corresponding elements of each of the columns of $\mathbf{B}$, and then add

$$\mathbf{AB} = \begin{bmatrix} 17 & -6 & 14 \\ 4 - 5 & 0 + 2 & -8 - 6 \end{bmatrix} = \begin{bmatrix} 17 & -6 & 14 \\ -1 & 2 & -14 \end{bmatrix}$$

Remark: **B** is a 2×3 matrix and **A** is a 2×2 matrix, so the number 3 of columns of **B** is not equal to the number 2 of rows of **A**; hence the product **BA** is not defined.

7.3.   Let $\mathbf{A} = \begin{bmatrix} 1 & 2 \\ 3 & 5 \end{bmatrix}$ and $\mathbf{B} = \begin{bmatrix} 4 & 6 \\ 0 & -2 \end{bmatrix}$. Find **AB** and **BA**.

We have

$$\mathbf{AB} = \begin{bmatrix} 4+0 & 6-4 \\ 12+0 & 18-10 \end{bmatrix} = \begin{bmatrix} 4 & -2 \\ 12 & 8 \end{bmatrix}$$

and

$$\mathbf{BA} = \begin{bmatrix} 4+18 & 8+30 \\ 0-6 & 0-10 \end{bmatrix} = \begin{bmatrix} 22 & 38 \\ -6 & -10 \end{bmatrix}$$

Remark: Although the products **AB** and **BA** are defined, they are not equal. In other words, matrix multiplication does not satisfy the commutative law that $\mathbf{AB} = \mathbf{BA}$.

7.4.   Let $\mathbf{A} = \begin{bmatrix} 1 & 3 \\ 2 & 4 \end{bmatrix}$. Find: (*a*) $\mathbf{A}^2$, (*b*) $\mathbf{A}^3$.

(*a*) $\mathbf{A}^2 = \mathbf{AA} = \begin{bmatrix} 1 & 3 \\ 2 & 4 \end{bmatrix}\begin{bmatrix} 1 & 3 \\ 2 & 4 \end{bmatrix}$

$$= \begin{bmatrix} 1(1)+3(2) & 1(3)+3(4) \\ 2(1)+4(2) & 2(3)+4(4) \end{bmatrix} = \begin{bmatrix} 7 & 15 \\ 10 & 22 \end{bmatrix}$$

(*b*) $\mathbf{A}^3 = \mathbf{AA}^2 = \begin{bmatrix} 1 & 3 \\ 2 & 4 \end{bmatrix}\begin{bmatrix} 7 & 15 \\ 10 & 22 \end{bmatrix}$

$$= \begin{bmatrix} 1(7)+3(10) & 1(15)+3(22) \\ 2(7)+4(10) & 2(15)+4(22) \end{bmatrix} = \begin{bmatrix} 37 & 81 \\ 54 & 118 \end{bmatrix}$$

PROBABILITY VECTORS AND STOCHASTIC MATRICES

7.5.   Find a multiple of each vector **v** which is a probability vector $\mathbf{q_v}$:

(*a*) $\mathbf{v} = [2, 1, 2, 0, 3]$,   (*c*) $\mathbf{v} = [2/3, 1, 3/5, 5/6]$,

(*b*) $\mathbf{v} = [1/2, 2/3, 2, 5/6]$,   (*d*) $\mathbf{v} = [0, 0, 0, 0]$

(*a*)   The sum of the components of **v** is 8; hence multiply **v** by 1/8, that is, multiply each component of **v** by 1/8 to obtain the probability vector

$$\mathbf{q_v} = [1/4, 1/8, 1/4, 0, 3/8]$$

(*b*)   First multiply the vector **v** by 6 to eliminate fractions. This yields the vector $\mathbf{v}' = [3, 4, 12, 5]$. The sum of the components of $\mathbf{v}'$ is 24. Then multiply each component of $\mathbf{v}'$ by 1/24 to obtain the probability vector

$$\mathbf{q_v} = [1/8, 1/6, 1/2, 5/24]$$

However, $\mathbf{q_v}$ is also a multiple of **v**.

(*c*)   First multiply the vector **v** by 30 to obtain $\mathbf{v}' = [20, 30, 18, 25]$. The sum of the components of $\mathbf{v}'$ is 93. Then multiply each component of $\mathbf{v}'$ by 1/93 to obtain

$$\mathbf{q_v} = [20/93, 30/93, 18/93, 25/93]$$

(*d*)   Every scalar multiple of the zero vector is the zero vector whose components add up to 0. Thus, no multiple of the zero vector is a probability vector.

7.6. Determine which of the following are stochastic matrices:

$$\mathbf{A} = \begin{bmatrix} 1/3 & 1/3 & 1/3 \\ 1/2 & 0 & 1/2 \end{bmatrix}, \qquad \mathbf{B} = \begin{bmatrix} 3/4 & 1/4 \\ 2/3 & 2/3 \end{bmatrix}, \qquad \mathbf{C} = \begin{bmatrix} 3/2 & -1/2 \\ 1/4 & 3/4 \end{bmatrix}, \qquad \mathbf{D} = \begin{bmatrix} 3/4 & 1/4 \\ 1/2 & 1/2 \end{bmatrix}.$$

   $\mathbf{A}$ is not a stochastic matrix since it is not a square matrix.
   $\mathbf{B}$ is not a stochastic matrix since the sum of the entries in the second row exceeds 1.
   $\mathbf{C}$ is not a stochastic matrix since an entry is negative.
   $\mathbf{D}$ is a stochastic matrix.

7.7. Suppose $\mathbf{A} = [a_{ij}]$ is an n-square stochastic matrix, and $\mathbf{u} = [u_1, u_2, \ldots, u_n]$ is a probability vector. Prove that $\mathbf{u}\mathbf{A}$ is also a probability vector.

   By matrix multiplication:

$$\mathbf{u}\mathbf{A} = [u_1, u_2, \ldots, u_n] \begin{bmatrix} a_{11} & a_{12} & \cdots & a_{1n} \\ a_{21} & a_{22} & \cdots & a_{2n} \\ \cdots\cdots\cdots\cdots\cdots\cdots \\ a_{n1} & a_{n2} & \cdots & a_{nn} \end{bmatrix}$$

$$= [\Sigma_i u_i a_{i1}, \Sigma_i u_i a_{i2}, \ldots, \Sigma_i u_i a_{in}]$$

Since the u_i and the a_{ij} are nonnegative, the components of $\mathbf{u}\mathbf{A}$ are also nonnegative. Thus, it only remains to show that the sum S of the components of $\mathbf{u}\mathbf{A}$ is equal to one. Using the fact that the sum of the entries in any row of $\mathbf{A}$ is equal to one and that the sum of the components of $\mathbf{u}$ is equal to one, that is, using $\Sigma_j a_{ij} = 1$, for any i, and $\Sigma_i u_i = 1$, we get

$$S = \Sigma_i u_i a_{i1} + \Sigma_i u_i a_{i2} + \cdots + \Sigma_i u_i a_{in}$$
$$= u_1 \Sigma_j a_{1j} + u_2 \Sigma_j a_{2j} + \cdots + u_n \Sigma_j a_{nj}$$
$$= u_1(1) + u_2(1) + \cdots + u_n(1)$$
$$= u_1 + u_2 + \cdots + u_n = 1$$

Thus, $\mathbf{u}\mathbf{A}$ is a probability vector.

7.8. Prove Theorem 7.2. Suppose $\mathbf{A}$ and $\mathbf{B}$ are stochastic matrices. Then the product $\mathbf{A}\mathbf{B}$ is also a stochastic matrix. Thus, in particular, all powers $\mathbf{A}^n$ are stochastic matrices.

   Let $\mathbf{s}_i$ denote the ith row of the product matrix $\mathbf{A}\mathbf{B}$. Then $\mathbf{s}_i$ is obtained by multiplying the ith row $\mathbf{r}_i$ of $\mathbf{A}$ by the matrix $\mathbf{B}$, that is,

$$\mathbf{s}_i = \mathbf{r}_i \mathbf{B}$$

Since $\mathbf{r}_i$ is a probability vector and $\mathbf{B}$ is a stochastic matrix, the product $\mathbf{s}_i$ is also a probability vector by the preceding Problem 7.7. Thus, $\mathbf{A}\mathbf{B}$ is a stochastic matrix since each row is a probability vector.

7.9. Let $\mathbf{p} = [p_1, p_2, \ldots, p_n]$ be a probability vector, and let $\mathbf{T}$ be a matrix whose rows are all the same vector $\mathbf{t} = [t_1, t_2, \ldots, t_n]$. Prove that $\mathbf{p}\mathbf{T} = \mathbf{t}$.

   Here we use the fact that $p_1 + p_2 + \cdots + p_n = \Sigma p_i = 1$. We have

$$\mathbf{p}\mathbf{T} = [p_1, p_2, \ldots, p_n] \begin{bmatrix} t_1 & t_2 & \cdots & t_n \\ t_1 & t_2 & \cdots & t_n \\ \cdots\cdots\cdots\cdots\cdots \\ t_1 & t_2 & \cdots & t_n \end{bmatrix}$$

$$= [\Sigma p_i t_1, \Sigma p_i t_2, \ldots, \Sigma p_i t_n]$$
$$= [t_1 \Sigma p_i, t_2 \Sigma p_i, \ldots, t_n \Sigma p_i]$$
$$= [t_1, t_2, \ldots, t_n] = \mathbf{t}$$

REGULAR STOCHASTIC MATRICES AND FIXED PROBABILITY VECTORS

7.10. Find the unique fixed probability vector **t** of the regular stochastic matrix $\mathbf{A} = \begin{bmatrix} 3/4 & 1/4 \\ 1/2 & 1/2 \end{bmatrix}$. Which matrix does $\mathbf{A}^n$ approach as n becomes larger?

We seek a probability vector $\mathbf{t} = [x, 1 - x]$ such that $\mathbf{tA} = \mathbf{t}$. Thus, set

$$[x, 1 - x]\begin{bmatrix} 3/4 & 1/4 \\ 1/2 & 1/2 \end{bmatrix} = [x, 1 - x]$$

Multiply the left side of the above matrix equation and then set corresponding components equal to each other to obtain the following two equations:

$$\tfrac{3}{4}x + \tfrac{1}{2} - \tfrac{1}{2}x = x \quad \text{and} \quad \tfrac{1}{4}x + \tfrac{1}{2} - \tfrac{1}{2}x = 1 - x$$

Solve either equation to obtain $x = 2/3$. Thus, $\mathbf{t} = [2/3, 1/3]$.

The matrix $\mathbf{A}^n$ approaches the matrix **T** whose rows are each the fixed point **t**; hence $\mathbf{A}^n$ approaches

$$\mathbf{T} = \begin{bmatrix} 2/3 & 1/3 \\ 2/3 & 1/3 \end{bmatrix}$$

7.11. Consider the general 2×2 stochastic matrix $\mathbf{M} = \begin{bmatrix} 1 - a & a \\ b & 1 - b \end{bmatrix}$. Prove that the vector $\mathbf{u} = [b, a]$ is a fixed point **M**. (Note that the fixed point **u** of **M** consists of the nondiagonal elements of **M**.)

Matrix multiplication yields

$$\mathbf{uM} = [b, a]\begin{bmatrix} 1 - a & a \\ b & 1 - b \end{bmatrix} = [b - ab + ab, ab + a - ab] = [b, a] = \mathbf{u}$$

Thus, **u** is a fixed point of **M**.

7.12. Use Problem 7.11 to find the unique fixed probability vector **t** of each stochastic matrix:

(a) $\mathbf{A} = \begin{bmatrix} 1/3 & 2/3 \\ 1 & 0 \end{bmatrix}$, (b) $\mathbf{B} = \begin{bmatrix} 1/2 & 1/2 \\ 2/3 & 1/3 \end{bmatrix}$, (c) $\mathbf{C} = \begin{bmatrix} 0.7 & 0.3 \\ 0.8 & 0.2 \end{bmatrix}$.

(a) By Problem 7.11, $\mathbf{u} = [1, 2/3]$ is a fixed point of **A**. Multiply **u** by 3 to obtain the fixed point $[3, 2]$ of **A** which has no fractions. Since the sum of the components of $[3, 2]$ is 5, multiply $[3, 2]$ by 1/5 to obtain the required probability vector $\mathbf{t} = [3/5, 2/5]$.

(b) By Problem 7.11, $\mathbf{u} = [2/3, 1/2]$ is a fixed point of **B**. Multiply **u** by 6 to obtain the fixed point $[4, 3]$ of **B** which has no fractions. Since the sum of the components of $[4, 3]$ is 7, multiply $[4/3]$ by 1/7 to obtain the required probability vector $\mathbf{t} = [4/7, 3/7]$.

(c) By Problem 7.11, $\mathbf{u} = [0.8, 0.3]$ is a fixed point of **C**. Hence $[8, 3]$ and the probability vector $\mathbf{t} = [8/11, 3/11]$ are also fixed points of **C**.

7.13. Find the unique fixed probability vector **t** of the following regular stochastic matrix:

$$\mathbf{P} = \begin{bmatrix} 0 & 1 & 0 \\ 1/6 & 1/2 & 1/3 \\ 0 & 2/3 & 1/3 \end{bmatrix}$$

Which matrix does $\mathbf{P}^n$ approach as n becomes larger?

We first seek any fixed vector $\mathbf{u} = [x, y, z]$ of $\mathbf{P}$.   Thus, set

$$[x, y, z] \begin{bmatrix} 0 & 1 & 0 \\ 1/6 & 1/2 & 1/3 \\ 0 & 2/3 & 1/3 \end{bmatrix} = [x, y, z]$$

Multiply the left side of the above matrix equation and then set corresponding components equal to each other to obtain the following system of three equations:

$$\begin{cases} \frac{1}{6}y = x \\ x + \frac{1}{2}y + \frac{2}{3}z = y \\ \frac{1}{3}y + \frac{1}{3}z = z \end{cases} \quad \text{or} \quad \begin{cases} y = 6x \\ 6x + 3y + 4z = 6y \\ y + z = 3z \end{cases} \quad \text{or} \quad \begin{cases} y = 6x \\ 6x + 4z = 3y \\ y = 2z \end{cases}$$

We know the system has a nonzero solution; hence we can arbitrarily assign a nonzero value to one of the unknowns.   Set $x = 1$.   By the first equation $y = 6$ and by the last equation $z = 3$.   Thus, $\mathbf{u} = [1, 6, 3]$ is a fixed point of $\mathbf{P}$.   Since $1 + 6 + 3 = 10$, the vector

$$\mathbf{t} = [1/10, 6/10, 3/10]$$

is the required unique fixed probability vector of $\mathbf{P}$.

The matrix $\mathbf{P}^n$ approaches the matrix $\mathbf{T}$ whose rows are each the fixed point $\mathbf{t}$; hence $\mathbf{P}^n$ approaches

$$\mathbf{T} = \begin{bmatrix} 1/10 & 6/10 & 3/10 \\ 1/10 & 6/10 & 3/10 \\ 1/10 & 6/10 & 3/10 \end{bmatrix}$$

7.14.   Suppose $\mathbf{P}$ is a stochastic matrix.   [Assume $\mathbf{P}$ is not a 1×1 matrix.]

(a)   Suppose $\mathbf{t} = [1/4, 0, 1/2, 1/4, 0]$ is a fixed point of $\mathbf{P}$.   Explain why $\mathbf{P}$ is not regular.

(b)   Suppose $\mathbf{P}$ has 1 on the diagonal.   Show that $\mathbf{P}$ is not regular.

(a)   Theorem 7.3 tells us that if $\mathbf{P}$ is regular, then $\mathbf{P}$ has a unique probability vector whose components are all positive.   Since $\mathbf{t}$ has zero components, $\mathbf{P}$ is not regular.

(b)   Let $\mathbf{e_k}$ be the vector with 1 in the kth position and 0's elsewhere, that is, $\mathbf{e_k}$ has the following form:

$$\mathbf{e_k} = [0, \ldots, 0, 1, 0, \ldots, 0]$$

Suppose the kth diagonal entry of $\mathbf{P}$ is 1.   Since $\mathbf{P}$ is a stochastic matrix, $\mathbf{e_k}$ must be the kth row of $\mathbf{P}$.   By matrix multiplication, $\mathbf{e_k}$ will be the kth row of all powers of $\mathbf{P}$.   Thus, $\mathbf{P}$ is not regular.

7.15.   Determine which of the following stochastic matrices are regular:

(a) $\mathbf{A} = \begin{bmatrix} 0 & 1 \\ 1 & 0 \end{bmatrix}$ (b) $\mathbf{B} = \begin{bmatrix} 1/2 & 1/4 & 1/4 \\ 0 & 1 & 0 \\ 1/2 & 1/2 & 0 \end{bmatrix}$ (c) $\mathbf{C} = \begin{bmatrix} 0 & 0 & 1 \\ 1/2 & 1/4 & 1/4 \\ 0 & 1 & 0 \end{bmatrix}$

Recall that a stochastic matrix is regular if a power of the matrix has only positive entries.

(a)   We have

$$\mathbf{A}^2 = \begin{bmatrix} 0 & 1 \\ 1 & 0 \end{bmatrix} \begin{bmatrix} 0 & 1 \\ 1 & 0 \end{bmatrix} = \begin{bmatrix} 1 & 0 \\ 0 & 1 \end{bmatrix} = \text{the unit matrix } \mathbf{I}$$

$$\mathbf{A}^3 = \begin{bmatrix} 1 & 0 \\ 0 & 1 \end{bmatrix} \begin{bmatrix} 0 & 1 \\ 1 & 0 \end{bmatrix} = \begin{bmatrix} 0 & 1 \\ 1 & 0 \end{bmatrix} = \mathbf{A}$$

Thus, every even power of $\mathbf{A}$ is the unit matrix $\mathbf{I}$, and every odd power of $\mathbf{A}$ is the matrix $\mathbf{A}$.   Thus, every power of $\mathbf{A}$ has zero entries, and so $\mathbf{A}$ is not regular.

(b) The matrix **B** is not regular since it has a 1 on its diagonal.

(c) Computing $\mathbf{C}^2$ and $\mathbf{C}^3$ yields

$$\mathbf{C}^2 = \begin{bmatrix} 0 & 1 & 0 \\ 1/8 & 5/16 & 9/16 \\ 1/2 & 1/4 & 1/4 \end{bmatrix} \qquad \mathbf{C}^3 = \begin{bmatrix} 1/2 & 1/4 & 1/4 \\ 5/32 & 41/64 & 13/64 \\ 1/8 & 5/16 & 9/16 \end{bmatrix}$$

Since all entries in $\mathbf{C}^3$ are positive, **C** is regular.

MARKOV PROCESSES

7.16. Bob's study habits are as follows. If he studies one night, he is 70 percent sure not to study the next night. On the other hand, if he does not study one night, he is only 60 percent sure not to study the next night as well. Find out how often, in the long run, Bob studies.

This is a Markov process where the states of the system are S (studying) and T (not studying). The transition matrix **M** of the process is as follows:

$$\mathbf{M} = \begin{array}{c} \\ S \\ T \end{array} \begin{array}{cc} S \quad\ T \\ \begin{bmatrix} 0.3 & 0.7 \\ 0.4 & 0.6 \end{bmatrix} \end{array}$$

To discover what happens in the long run, we must find the unique fixed probability vector **t** of **M**. By Problem 7.11, $\mathbf{u} = [0.4, 0.7]$ is a fixed point of **M** and so $\mathbf{t} = [4/11, 7/11]$ is the required fixed probability vector. Thus, in the long run, Bob studies 4/11 of the time.

7.17. A psychologist makes the following assumptions concerning the behavior of mice subjected to a particular feeding schedule. For any particular trial, 80 percent of the mice that went right on a previous experiment will go right on this trial, and 60 percent of those mice that went left on the previous experiment will go right on this trial. Suppose 50 percent of the mice went right on the first trial.

(a) Find the prediction of the psychologist for the next two trials.

(b) When will the process stabilize?

The states of the system are R (right) and L (left), and the transition matrix **M** of the process is as follows:

$$\mathbf{M} = \begin{array}{c} \\ R \\ L \end{array} \begin{array}{cc} R \quad\ L \\ \begin{bmatrix} 0.8 & 0.2 \\ 0.6 & 0.4 \end{bmatrix} \end{array}$$

The probability distribution for the first (initial) trial is $\mathbf{q} = [0.5, 0.5]$.

(a) To predict the probability distribution for the next step (second trial), multiply **q** by **M**. This yields

$$[0.5, 0.5] \begin{bmatrix} 0.8 & 0.2 \\ 0.6 & 0.4 \end{bmatrix} = [0.7, 0.3]$$

Thus, the psychologist predicts 70 percent of the mice will go right and 30 percent will go left on the second trial. To predict the probability distribution for the next step (third trial), multiply the previous distribution by **M**. This yields

$$[0.7, 0.3] \begin{bmatrix} 0.8 & 0.2 \\ 0.6 & 0.4 \end{bmatrix} = [0.74, 0.26]$$

Thus, the psychologist predicts 74 percent of the mice will go right and 26 percent will go left on the third trial.

(b) The process will stabilize when it reaches its fixed probability distribution $\mathbf{t}$. By Problem 7.11, $\mathbf{u} = [0.6, 0.2]$ is a fixed point of $\mathbf{M}$ and so $\mathbf{t} = [3/4, 1/4] = [0.75, 0.25]$. The fourth trial, rounded to two decimal places, gives the state distribution $[0.75, 0.25]$. Thus, the process stabilizes after the third trial.

7.18. Consider a Markov process with initial probability distribution $q_0 = [1/2, 0, 1/2]$ and the following transition matrix:

$$\mathbf{M} = \begin{bmatrix} 0 & 1/2 & 1/2 \\ 1/2 & 1/2 & 0 \\ 0 & 1 & 0 \end{bmatrix}$$

(a) Find the following three probability distributions q_1, q_2, and q_3.

(b) Find the matrix that $\mathbf{M}^n$ approaches as n gets larger.

(a) Multiply q_0 by $\mathbf{M}$ to obtain q_1:

$$q_1 = q_0\,\mathbf{M} = [1/2, 0, 1/2] \begin{bmatrix} 0 & 1/2 & 1/2 \\ 1/2 & 1/2 & 0 \\ 0 & 1 & 0 \end{bmatrix} = [0, 3/4, 1/4]$$

Multiply q_1 by $\mathbf{M}$ to obtain q_2:

$$q_2 = q_1\,\mathbf{M} = [0, 3/4, 1/4] \begin{bmatrix} 0 & 1/2 & 1/2 \\ 1/2 & 1/2 & 0 \\ 0 & 1 & 0 \end{bmatrix} = [3/8, 5/8, 0]$$

Multiply q_2 by $\mathbf{M}$ to obtain q_3:

$$q_3 = q_2\,\mathbf{M} = [3/8, 5/8, 0] \begin{bmatrix} 0 & 1/2 & 1/2 \\ 1/2 & 1/2 & 0 \\ 0 & 1 & 0 \end{bmatrix} = [5/16, 1/2, 3/16]$$

(b) $\mathbf{M}^n$ approaches the matrix $\mathbf{T}$ whose rows are each the unique fixed probability vector $\mathbf{t}$ of $\mathbf{M}$. To find $\mathbf{t}$, first find any fixed vector $\mathbf{u} = [x, y, z]$ of $\mathbf{M}$. Thus

$$[x, y, z] \begin{bmatrix} 0 & 1/2 & 1/2 \\ 1/2 & 1/2 & 0 \\ 0 & 1 & 0 \end{bmatrix} = [x, y, z] \qquad \text{or} \qquad \begin{cases} \frac{1}{2}y = x \\ \frac{1}{2}x + \frac{1}{2}y + z = y \\ \frac{1}{2}x = z \end{cases}$$

Find any nonzero solution of the above system of linear equations. Set $z = 1$. By the third equation $x = 2$, and by the first equation $y = 4$. Thus, $\mathbf{u} = [2, 4, 1]$ is a fixed point of $\mathbf{M}$ and so $\mathbf{t} = [2/7, 4/7, 1/7]$. Accordingly, $\mathbf{M}_n$ approaches the following matrix:

$$\mathbf{T} = \begin{bmatrix} 2/7 & 4/7 & 1/7 \\ 2/7 & 4/7 & 1/7 \\ 2/7 & 4/7 & 1/7 \end{bmatrix}$$

7.19. A salesman S sells in only three cities, A, B, and C. Suppose S never sells in the same city on successive days. If S sells in city A, then the next day S sells in city B. However, if S sells in either B or C, then the next day S is twice as likely to sell in city A as in the other city. Find out how often, in the long run, S sells in each city.

The transition matrix of the Markov process follows:

$$\mathbf{M} = \begin{array}{c} \\ A \\ B \\ C \end{array} \begin{array}{ccc} A & B & C \\ \left[\begin{array}{ccc} 0 & 1 & 0 \\ 2/3 & 0 & 1/3 \\ 2/3 & 1/3 & 0 \end{array}\right] \end{array}$$

The first row $[0, 1, 0]$ comes from the fact that if S sells in city A, then S will always sell in city B the next day. The 2/3 and 1/3 in the second row and third row come from the fact that if S sells in either city B or C, then S is twice as likely to sell in city A the next day than in the other city. (S is never in the same city 2 days in a row.)

We seek the unique fixed probability vector $\mathbf{t}$ of the transition matrix $\mathbf{M}$. To find $\mathbf{t}$, we first find any fixed vector $\mathbf{u} = [x, y, z]$ of $\mathbf{M}$. Thus

$$[x, y, z] \begin{bmatrix} 0 & 1 & 0 \\ 2/3 & 0 & 1/3 \\ 2/3 & 1/3 & 0 \end{bmatrix} = [x, y, z] \qquad \text{or} \qquad \begin{cases} \frac{2}{3}y + \frac{2}{3}x = x \\ x + \frac{1}{3}z = y \\ \frac{1}{3}y = z \end{cases}$$

We find any nonzero solution of the above system of linear equations. Set $z = 1$. By the third equation $y = 3$, and by the first equation $x = 8/3$. Thus, $\mathbf{u} = [8/3, 3, 1]$. Also, $3\mathbf{u} = [8, 9, 3]$ is a fixed point of $\mathbf{M}$. Multiply $3\mathbf{u}$ by $1/(8 + 9 + 3) = 1/20$ to obtain the unique fixed probability vector $\mathbf{t} = [2/5, 9/20, 3/20] = [0.40, 0.45, 0.15]$. Thus, in the long run, S sells 40 percent of the time in city A, 45 percent of the time in B, and 15 percent of the time in C.

7.20. There are 2 white marbles in box A and 3 red marbles in box B. At each step in the process, a marble is selected from each box and the 2 marbles are interchanged. (Thus, box A always has 2 marbles and box B always has 3 marbles.) The system may be described by three states, s_0, s_1, s_2, which denote, respectively, the number of red marbles in box A.

(a) Find the transition matrix $\mathbf{P}$ of the system.

(b) Find the probability that there are 2 red marbles in box A after 3 steps.

(c) Find the probability that, in the long run, there are 2 red marbles in box A.

The three states, s_0, s_1, s_2, may be described as follows:

	s_0	s_1	s_2
Box A	$2W$	$1W, 1R$	$2R$
Box B	$3R$	$1W, 2R$	$2W, 1R$

(a) There are three cases according to the state of the system.

(1) Suppose the system is in state s_0. Then a white marble must be chosen from box A and a red marble from box B, so the system must move to state s_1. Thus, the first row of $\mathbf{P}$ must be $[0, 1, 0]$.

(2) Suppose the system is in state s_1. There are three subcases:

  (i) The system can move to state s_0 if and only if a red marble is selected from box A and a white marble from box B. The probability that this happens is $(1/2)(1/3) = 1/6$.

  (ii) The system can move to state s_2 if and only if a white marble is selected from box A and a red marble from box B. The probability that this happens is $(1/2)(2/3) = 1/3$.

  (iii) By (i) and (ii), the system remains in state s_1 with probability $1 - 1/6 - 1/3 = 1/2$.

Thus, the second row of $\mathbf{P}$ must be $[1/6, 1/2, 1/3]$.

(3) Suppose the system is in state s_2. A red marble must be drawn from box A. If a red marble is selected from box B, probability 1/3, then the system remains in state s_2; but if a white marble is selected from box B, then the system moves to state s_1. The system can never move from s_2 to s_0. Thus, the third row of **P** must be [0, 2/3, 1/3].

Therefore, the required transition matrix is as follows:

$$\mathbf{P} = \begin{array}{c} \\ s_0 \\ s_1 \\ s_2 \end{array}\begin{array}{c} \begin{array}{ccc} s_0 & s_1 & s_2 \end{array} \\ \left[\begin{array}{ccc} 0 & 1 & 0 \\ 1/6 & 1/2 & 1/3 \\ 0 & 2/3 & 1/3 \end{array}\right] \end{array}$$

(b) The system begins in state s_0. Thus, the initial probability distribution is $q_0 = [1, 0, 0]$. Therefore:

$$q_1 = q_0\,\mathbf{P} = [0, 1, 0] \qquad q_2 = q_1\,\mathbf{P} = [1/6, 1/2, 1/3] \qquad q_3 = q_2\,\mathbf{P} = [1/12, 23/36, 5/18]$$

Accordingly, the probability that there are 2 red marbles in box A after three steps is 5/18.

(c) We seek the unique fixed probability vector **t** of the transition matrix **P**. To find **t**, we first find any fixed vector $[x, y, z]$ of **P**. Thus:

$$[x, y, z]\left[\begin{array}{ccc} 0 & 1 & 0 \\ 1/6 & 1/2 & 1/3 \\ 0 & 2/3 & 1/3 \end{array}\right] = [x, y, z] \qquad \text{or} \qquad \begin{cases} \frac{1}{6}y = x \\ x + \frac{1}{2}y + \frac{2}{3}z = y \\ \frac{1}{3}y + \frac{1}{3}z = z \end{cases}$$

We find any nonzero solution of the above system of linear equations. Set, say, $x = 1$. By the first equation $y = 6$, and by the third equation $z = 3$. Thus, $\mathbf{u} = [1, 6, 3]$ is a fixed point of **P**. Multiply **u** by $1/(1 + 6 + 3) = 1/10$ to obtain the unique fixed probability vector $\mathbf{t} = [0.1, 0.6, 0.3]$. Thus, in the long run, 30 percent of the time there will be 2 red marbles in box A.

Remark: Note that the long-run probability distribution is the same as if the 5 marbles were placed in a box, and 2 marbles were selected at random to put in box A.

MISCELLANEOUS PROBLEMS

7.21. The transition probabilities of a Markov process may be described by a diagram, called a *transition diagram* as follows. The states are points (vertices) in the diagram, and a positive probability p_{ij} is denoted by an arrow (edge) from state a_i to the state a_j labelled by p_{ij}. Find the transition matrix **P** of each transition diagram in Fig. 7-1.

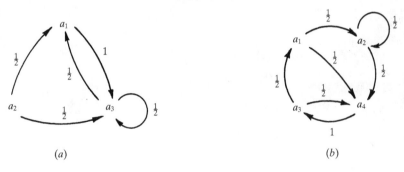

(a)   (b)

Fig. 7-1

(a)   The state space is $S = [a_1, a_2, a_3]$, and hence the transition matrix $\mathbf{P}$ has the following form:

$$\mathbf{P} = \begin{array}{c} \\ a_1 \\ a_2 \\ a_3 \end{array} \begin{array}{ccc} a_1 & a_2 & a_3 \\ \left[\begin{array}{ccc} & & \\ & & \\ & & \end{array} \right] \end{array}$$

Row i of $\mathbf{P}$ is obtained by finding the arrows which emanate from a_i in the diagram; the number attached to the arrow from a_i to a_j is the jth component of row i.   Thus, the following is the required transition matrix:

$$\mathbf{P} = \begin{array}{c} \\ a_1 \\ a_2 \\ a_3 \end{array} \begin{array}{ccc} a_1 & a_2 & a_3 \\ \left[\begin{array}{ccc} 0 & 0 & 1 \\ 1/2 & 0 & 1/2 \\ 1/2 & 0 & 1/2 \end{array} \right] \end{array}$$

(b)   The state space is $S = \{a_1, a_2, a_3, a_4\}$.   The required transition matrix is as follows:

$$\mathbf{P} = \begin{array}{c} \\ a_1 \\ a_2 \\ a_3 \\ a_4 \end{array} \begin{array}{cccc} a_1 & a_2 & a_3 & a_4 \\ \left[\begin{array}{cccc} 0 & 1/2 & 0 & 1/2 \\ 0 & 1/2 & 0 & 1/2 \\ 1/2 & 0 & 0 & 1/2 \\ 0 & 0 & 1 & 0 \end{array} \right] \end{array}$$

7.22.   Suppose the following is the transition matrix of a Markov process:

$$\mathbf{P} = \begin{array}{c} \\ a_1 \\ a_2 \\ a_3 \\ a_4 \end{array} \begin{array}{cccc} a_1 & a_2 & a_3 & a_4 \\ \left[\begin{array}{cccc} 1/2 & 1/2 & 0 & 0 \\ 1/2 & 1/2 & 0 & 0 \\ 1/4 & 1/4 & 1/4 & 1/4 \\ 1/4 & 1/4 & 1/4 & 1/4 \end{array} \right] \end{array}$$

Show that the Markov process is not regular.

Note that once the system enters state a_1 or a_2, then it can never move to state a_3 or a_4, that is, the system remains in the state subspace $\{a_1, a_2\}$.   Thus, every power of $\mathbf{P}$ will have 0 entries in the 3rd and 4th positions of the first and second rows.

$$(1, 3), (1, 4), (2, 3), (2, 4)$$

Supplementary Problems

MATRIX MULTIPLICATION

7.23.   Given $\mathbf{A} = \begin{bmatrix} 1 & -2 & 3 \\ 4 & 1 & -1 \\ 5 & 2 & 3 \end{bmatrix}$.   Find $\mathbf{uA}$ where: (a) $\mathbf{u} = [1, -3, 2]$, (b) $\mathbf{u} = [3, 0, -2]$,

(c) $\mathbf{u} = [4, -1, -1]$.

7.24. Given $\mathbf{A} = \begin{bmatrix} 1 & -1 & 4 \\ 3 & 1 & 5 \end{bmatrix}$ and $\mathbf{B} = \begin{bmatrix} 2 & 1 \\ 6 & -3 \\ 1 & -2 \end{bmatrix}$. Find $\mathbf{AB}$ and $\mathbf{BA}$.

7.25. Given $\mathbf{A} = \begin{bmatrix} 2 & 2 \\ 3 & -1 \end{bmatrix}$. Find $\mathbf{A}^2$ and $\mathbf{A}^3$.

7.26. Given $\mathbf{A} = \begin{bmatrix} 1 & 2 \\ 0 & 1 \end{bmatrix}$. Find (a) $\mathbf{A}^2$, (b) $\mathbf{A}^3$, (c) $\mathbf{A}^n$.

PROBABILITY VECTORS AND STOCHASTIC MATRICES

7.27. Which vectors are probability vectors?
$$\mathbf{u} = [1/4, 1/2, -1/4, 1/2], \qquad \mathbf{v} = [1/2, 0, 1/3, 1/6, 1/6], \qquad \mathbf{w} = [1/12, 1/2, 1/6, 0, 1/4]$$

7.28. Find a scalar multiple of each vector $\mathbf{v}$ which is a probability vector:

(a) $\mathbf{v} = [3, 0, 2, 5, 3]$,   (b) $\mathbf{v} = [2, \frac{1}{2}, 0, \frac{1}{4}, \frac{3}{4}, 1]$,   (c) $\mathbf{v} = [\frac{1}{2}, \frac{1}{6}, 0, \frac{1}{4}]$.

7.29. Which matrices are stochastic?
$$\mathbf{A} = \begin{bmatrix} 0 & 1 & 0 \\ 1/2 & 1/4 & 1/4 \end{bmatrix}, \qquad \mathbf{B} = \begin{bmatrix} 1 & 0 \\ 0 & 1 \end{bmatrix}, \qquad \mathbf{C} = \begin{bmatrix} 0 & 1 \\ 1/2 & 1/4 \end{bmatrix}, \qquad \mathbf{D} = \begin{bmatrix} 1/2 & 1/2 \\ 1/2 & 1/2 \end{bmatrix}$$

REGULAR STOCHASTIC MATRICES AND FIXED PROBABILITY VECTORS

7.30. Find the unique fixed probability vector $\mathbf{t}$ of each matrix:

(a) $\mathbf{A} = \begin{bmatrix} 2/3 & 1/3 \\ 2/5 & 3/5 \end{bmatrix}$,   (b) $\mathbf{B} = \begin{bmatrix} 0.2 & 0.8 \\ 0.5 & 0.5 \end{bmatrix}$,   (c) $\mathbf{C} = \begin{bmatrix} 0.7 & 0.3 \\ 0.6 & 0.4 \end{bmatrix}$

7.31. Find the unique fixed probability vector $\mathbf{t}$ of each matrix:

(a) $\mathbf{A} = \begin{bmatrix} 0 & 1/2 & 1/2 \\ 1/3 & 2/3 & 0 \\ 0 & 1 & 0 \end{bmatrix}$,   (b) $\mathbf{B} = \begin{bmatrix} 0 & 1 & 0 \\ 1/2 & 0 & 1/2 \\ 1/2 & 1/4 & 1/4 \end{bmatrix}$

7.32. Consider the following stochastic matrix:
$$\mathbf{P} = \begin{bmatrix} 0 & 3/4 & 1/4 \\ 1/2 & 1/2 & 0 \\ 0 & 1 & 0 \end{bmatrix}$$

(a) Show that $\mathbf{P}$ is regular.

(b) Find the unique fixed probability vector $\mathbf{t}$ of $\mathbf{P}$.

(c) What matrix does $\mathbf{P}^n$ approach?

(d) What vector does $[1/4, 1/4, 1/2] \mathbf{P}^n$ approach?

7.33. Consider the following stochastic matrix:

$$\mathbf{P} = \begin{bmatrix} 0 & 1/2 & 1/2 & 0 \\ 1/2 & 1/4 & 0 & 1/4 \\ 0 & 0 & 0 & 1 \\ 0 & 1/2 & 0 & 1/2 \end{bmatrix}$$

(a)   Show that $\mathbf{P}$ is regular.

(b)   Find the unique fixed probability vector $\mathbf{t}$ of $\mathbf{P}$.

(c)   What matrix does $\mathbf{P}^n$ approach?

(d)   What vector does $[1/4, 0, 1/2, 1/4]\,\mathbf{P}^n$ approach?

7.34. Consider the following general 3×3 stochastic matrix:

$$\mathbf{P} = \begin{bmatrix} 1-a-b & a & b \\ c & 1-c-d & d \\ e & f & 1-e-f \end{bmatrix}$$

Show that the following vector $\mathbf{v}$ is a fixed point of $\mathbf{P}$:

$$\mathbf{v} = [cf + ce + de, \qquad af + bf + ae, \qquad ad + bd + bc]$$

MARKOV PROCESSES

7.35. John either drives or takes the train to work. If he drives to work, then the next day he takes the train with probability 0.2. On the other hand, if he takes the train to work, then the next day he drives with probability 0.3. Find out how often, in the long run, he drives to work.

7.36. Mary's gambling luck follows a pattern. If she wins a game, the probability of winning the next game is 0.6. However, if she loses a game, the probability of losing the next game is 0.7. There is an even chance that she wins the first game.

(a)   Find the transition matrix $\mathbf{M}$ of the Markov process.

(b)   Find the probability that she wins the second game.

(c)   Find the probability that she wins the third game.

(d)   Find out how often, in the long run, she wins.

7.37. Suppose $q_0 = [1/4, 3/4]$ is the initial state distribution for a Markov process with the following transition matrix:

$$\mathbf{M} = \begin{bmatrix} 1/2 & 1/2 \\ 3/4 & 1/4 \end{bmatrix}$$

(a) Find q_1, q_2, and q_3. (b) Find the vector $\mathbf{v}$ that $q_0\,\mathbf{M}^n$ approaches. (c) Find the matrix that $\mathbf{M}^n$ approaches.

7.38. Suppose $q_0 = [1/2, 1/2, 0]$ is the initial state distribution for a Markov process with the following transition matrix:

$$\mathbf{M} = \begin{bmatrix} 1/2 & 0 & 1/2 \\ 1 & 0 & 0 \\ 1/4 & 1/2 & 1/4 \end{bmatrix}$$

(a) Find q_1, q_2, and q_3. (b) Find the vector $\mathbf{v}$ that $q_0\,\mathbf{M}^n$ approaches. (c) Find the matrix that $\mathbf{M}^n$ approaches.

7.39. Each year Ann trades her car for a new car. If she has a Buick, she trades it in for a Plymouth. If she has a Plymouth, she trades it in for a Ford. However, if she has a Ford, she is just as likely to trade it in for a new Ford as to trade it in for a Buick or for a Plymouth. In 1995 she bought her first car which was a Ford.

(a) Find the probability that she has bought: (i) a 1997 Buick, (ii) a 1998 Plymouth, (iii) a 1998 Ford.

(b) Find out how often, in the long run, she will have a Ford.

MISCELLANEOUS PROBLEMS

7.40. Find the transition matrix corresponding to each transition diagram in Fig. 7-2.

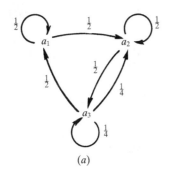

(a)

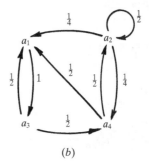

(b)

Fig. 7-2

7.41. Draw a transition diagram for each transition matrix:

$$(a) \quad \mathbf{P} = \begin{array}{c} \\ a_1 \\ a_2 \end{array} \begin{array}{cc} a_1 & a_2 \\ \begin{bmatrix} 1/2 & 1/2 \\ 1/3 & 2/3 \end{bmatrix} \end{array}, \qquad (b) \quad \mathbf{P} = \begin{array}{c} \\ a_1 \\ a_2 \\ a_3 \end{array} \begin{array}{ccc} a_1 & a_2 & a_3 \\ \begin{bmatrix} 0 & 1/2 & 1/2 \\ 1/4 & 1/4 & 1/2 \\ 0 & 1/2 & 1/2 \end{bmatrix} \end{array}$$

7.42. Consider the vector $\mathbf{e_i} = [0, \ldots, 0, 1, 0, \ldots, 0]$ which has 1 in the ith position and 0's elsewhere. Show that, whenever defined, $\mathbf{e_i}\,\mathbf{A} = r_i$, where r_i is the ith row of $\mathbf{A}$.

Answers to Supplementary Problems

7.23. (a) $[-1, -1, 12]$; (b) $[-7, -10, 3]$; (c) $[-5, -11, 10]$.

7.24. $\mathbf{AB} = \begin{bmatrix} 0 & -4 \\ 17 & -10 \end{bmatrix}$; $\qquad \mathbf{BA} = \begin{bmatrix} 5 & -1 & 13 \\ -3 & -9 & 9 \\ -5 & -3 & -6 \end{bmatrix}$.

7.25. $\mathbf{A}^2 = \begin{bmatrix} 10 & 2 \\ 3 & 7 \end{bmatrix}$; $\qquad \mathbf{A}^3 = \begin{bmatrix} 26 & 18 \\ 27 & -1 \end{bmatrix}$.

7.26. $\mathbf{A}^2 = \begin{bmatrix} 1 & 4 \\ 0 & 1 \end{bmatrix}$; $\qquad \mathbf{A}^3 = \begin{bmatrix} 1 & 6 \\ 0 & 1 \end{bmatrix}$; $\qquad \mathbf{A}^n = \begin{bmatrix} 1 & 2n \\ 0 & 1 \end{bmatrix}$.

7.27. Only w.

7.28. (a) $\frac{1}{13}[3, 0, 2, 5, 3]$; (b) $\frac{1}{18}[8, 2, 0, 1, 3, 4]$; (c) $\frac{1}{11}[6, 2, 0, 3]$.

7.29. Only **B** and **D**.

7.30. (a) $[6/11, 5/11]$; (b) $[5/13, 8/13]$; (c) $[2/3, 1/3]$.

7.31. (a) $[2/9, 6/9, 1/9]$; (b) $[5/15, 6/15, 4/15]$.

7.32. (a) $\mathbf{P}^3$ has only positive entries; (b) $\mathbf{t} = [4/13, 8/13, 1/13]$; (c) all rows are $\mathbf{t}$. (d) $\mathbf{t}$.

7.33. (a) $\mathbf{P}^3$ has only positive entries; (b) $\mathbf{t} = [2/11, 4/11, 1/11, 4/11]$; (c) all rows are $\mathbf{t}$. (d) $\mathbf{t}$.

7.35. Transition matrix $\mathbf{M} = \begin{bmatrix} 0.8 & 0.2 \\ 0.3 & 0.7 \end{bmatrix}$. John drives 60 percent of the time.

7.36. (a) $\mathbf{M} = \begin{bmatrix} 0.6 & 0.4 \\ 0.3 & 0.7 \end{bmatrix}$; (b) 45 percent; (c) 43.5 percent; (d) $3/7 \approx 42.9$ percent.

7.37. (a) $q_1 = [11/16, 5/16]$, $q_2 = [37/64, 27/64]$, $q_3 = [155/256, 101/256]$; b $[3/5, 2/5]$; (c) $\begin{bmatrix} 3/5 & 2/5 \\ 3/5 & 2/5 \end{bmatrix}$.

7.38. (a) $q_1 = [3/4, 0, 1/4]$, $q_2 = [7/16, 2/16, 7/16]$, $q_3 = [29/64, 14/64, 21/64]$; (b) $[3/6, 1/6, 2/6]$; (c) all rows are $[3/6, 1/6, 2/6]$.

7.39. (a) (i) 1/9, (ii) 7/27, (iii) 16/27; (b) 50 percent of the time.

7.40. (a) $\begin{bmatrix} 1/2 & 1/2 & 0 \\ 0 & 1/2 & 1/2 \\ 1/2 & 1/4 & 1/4 \end{bmatrix}$; (b) $\begin{bmatrix} 0 & 0 & 1 & 0 \\ 1/4 & 1/2 & 0 & 1/4 \\ 1/2 & 0 & 0 & 1/2 \\ 1/2 & 1/2 & 0 & 0 \end{bmatrix}$.

7.41. See Fig. 7-3.

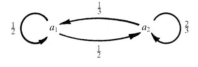

(a)

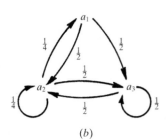

(b)

Fig. 7-3

APPENDIX A

Descriptive Statistics

A.1 INTRODUCTION

Statistics means, on the one hand, lists of numerical data. For example, the weights of the students at a university, or the number of children per family in a city. *Statistics* as a science, on the other hand, is that branch of mathematics which organizes, analyzes, and interprets such raw data.

This appendix will mainly cover topics related to the gathering and description of data, called *descriptive statistics*. It is closely related to probability theory in that the probability model that one develops for the events of a space usually depends on the relative frequencies of such events. The topics of inferential statistics, such as estimation and testing hypothesis, lie beyond the scope of this appendix and text.

The numerical data $x_1, x_2, \ldots$ we consider will either come from a random sample of a larger population or from the larger population itself. We distinguish these two cases using different notation as follows:

n = number of items in a sample	N = number of items in the population
$\bar{x}$ = sample mean	μ = population mean
s^2 = sample variance	σ^2 = population variance
s = sample standard deviation	σ = population standard deviation

Note that Greek letters are used with the population and are called *parameters*, whereas Latin letters are used with the samples and are called *statistics*. First we will give formulas for the data coming from a sample. This will be followed by formulas for the population.

A.2 FREQUENCY TABLES, HISTOGRAMS

One of the first things that one usually does with a large list of numerical data is to collect them into groups (grouped data). A group, sometimes called a *category*, refers to the set of numbers all of which have the same value x_i, or to the set (class) of numbers in a given interval where the midpoint x_i of the interval, called the *class value*, serves as an approximation to the values in the interval. We assume there are k such groups with f_i denoting the number of elements (frequency) in the group with

245

value x_i or class value x_i. Such grouped data yields a table, called a *frequency distribution*, as follows:

Value (or class value)	x_1	x_2	x_3	$\cdots$	x_k
Frequency	f_1	f_2	f_3	$\cdots$	f_k

Thus, the total number of data items is

$$n = f_1 + f_2 + \cdots + f_k = \sum f_i$$

As usual, Σ will denote a summation over all the values of the index, unless otherwise specified.

Our frequency distribution table usually lists, when applicable, the ends of the class intervals, called *class boundaries* or *class limits*. We assume all intervals have the same length called the *class width*. If a data item falls on a class boundary, it is usually assigned to the higher class.

Sometimes the table also lists the *cumulative frequency* function F_s where F_s is defined by

$$F_s = f_1 + f_2 + \cdots + f_s = \sum_{i \leq s} f_i$$

That is, F_s is the sum of the frequencies up to f_s. Thus, $F_k = n$, the number of data items.

The number k of groups that we decide to use to collect our data should not be too small or too large. If it is too small, then we will lose much of the information of the given data; if it is too large, then we will lose the purpose of grouping the data. The rule of thumb is that k should lie between 5 and 12. We illustrate the above with two examples. Note that any such *frequency distribution* can then be pictured as a *histogram* or *frequency polygon*.

EXAMPLE A.1 Suppose an apartment house has $n = 45$ apartments, with the following numbers of tenants:

$$2, \quad 1, \quad 3, \quad 5, \quad 2, \quad 2, \quad 2, \quad 1, \quad 4, \quad 2, \quad 6, \quad 2, \quad 4, \quad 3, \quad 1$$
$$2, \quad 4, \quad 3, \quad 1, \quad 4, \quad 4, \quad 2, \quad 4, \quad 4, \quad 2, \quad 2, \quad 3, \quad 1, \quad 4, \quad 2$$
$$3, \quad 1, \quad 5, \quad 2, \quad 4, \quad 1, \quad 3, \quad 2, \quad 4, \quad 4, \quad 2, \quad 5, \quad 1, \quad 3, \quad 4$$

Observe that the only numbers which appear in the list are 1, 2, 3, 4, 5, and 6. The frequency distribution, including the cumulative frequency distribution, follows:

Number of people	1	2	3	4	5	6
Frequency	8	14	7	12	3	1
Cumulative frequency	8	22	29	41	44	45

The sum of the frequencies is $n = 45$, which is also the last entry in the cumulative frequency row.

Figure A-1 shows the histogram corresponding to the above frequency distribution. The histogram is simply a bar graph where the height of the bar is the frequency of the given number in the list. Similarly, the cumulative frequency distribution could be presented as a histogram; the heights of the bars would be 8, 22, 29, ..., 45.

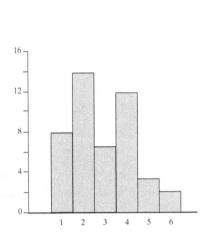

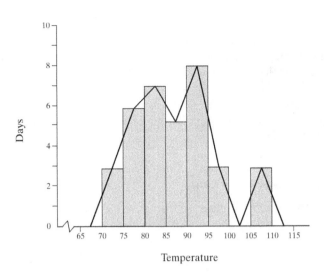

Fig. A-1   Fig. A-2

EXAMPLE A.2 Suppose the 6:00 p.m. temperatures (in degrees Fahrenheit) for a 35-day period are as follows:

$$72.4, \quad 78.2, \quad 86.7, \quad 93.4, \quad 106.1, \quad 107.6, \quad 98.2, \quad 92.0, \quad 81.4, \quad 77.2$$
$$87.9, \quad 82.4, \quad 91.6, \quad 95.0, \quad 92.1, \quad 83.9, \quad 76.4, \quad 78.4, \quad 73.2, \quad 81.4$$
$$86.2, \quad 92.4, \quad 93.6, \quad 84.8, \quad 107.5, \quad 99.2, \quad 94.7, \quad 86.1, \quad 81.0, \quad 77.7$$
$$73.5, \quad 76.0, \quad 80.2, \quad 88.8, \quad 91.3$$

Rather than find the frequency of each individual data item, it is more useful to collect the data in classes as follows (where the temperature 95.0°F is assigned to the higher class 90 to 95 rather than the lower class 85 to 90:

Class boundaries	70–75	75–80	80–85	85–90	90–95	95–100	100–105	105–110
Class value	72.5	77.5	82.5	87.5	92.5	97.5	102.5	107.5
Frequency	3	6	7	5	8	3	0	3
Cumulative frequency	3	9	16	21	29	32	32	35

The class width for this distribution is $w = 5$. The sum of the frequencies is $n = 35$; it is also the last entry in the cumulative frequency row.

Figure A-2 shows the histogram corresponding to the above frequency distribution. It also shows the *frequency polygon* of the data, which is the line graph obtained by connecting the midpoints of the tops of the rectangles in the histogram. Observe that the line graph is extended to the class value 67.5 on the left and to 112.5 on the right. In such a case, the sum of the areas of the rectangles equals the area bounded by the frequency polygon and the x axis.

A.3 MEASURES OF CENTRAL TENDENCY; MEAN AND MEDIAN

There are various ways of giving an overview of data. One way is by graphical descriptions such as the frequency histogram or the frequency polygon discussed above. Another way is to use certain numerical descriptions of the data. Numbers, such as the mean and median, give, in some sense, the central or middle values of the data. The central tendency of our data is discussed in this section.

The next section discusses other numbers, the variance and standard deviation, which measure the dispersion or spread of the data about the mean, and the quartiles, which measure the dispersion or spread of the data about the median.

Many formulas will be designated as (a) or (b) where (a) indicates ungrouped data and (b) indicates grouped data. Unless otherwise stated, we assume that our data come from a random sample of a (larger) population. Separate formulas are given for data which come from the total population itself.

Mean (Arithmetic Mean)

The *arithmetic mean* or simply *mean* of a sample $x_1, x_2, \ldots, x_n$ of n numerical values, denoted by $\bar{x}$ (read: *x*-bar), is the sum of the values divided by the number of values. That is,

Sample mean:
$$\bar{x} = \frac{x_1 + x_2 + \cdots + x_n}{n} = \frac{\Sigma x_i}{n} \qquad (A\text{-}1a)$$

Sample mean:
$$\bar{x} = \frac{f_1 x_1 + f_2 x_2 + \cdots + f_k x_k}{f_1 + f_2 + \cdots + f_k} = \frac{\Sigma f_i x_i}{\Sigma f_i} \qquad (A\text{-}1b)$$

The mean $\bar{x}$ is frequently called the *average value*.

EXAMPLE A.3

(a) Consider the data in Example A.1. Using the frequency distribution, rather than adding up the 45 numbers, we obtain the mean as follows:

$$\bar{x} = \frac{8(1) + 14(2) + 7(3) + 12(4) + 3(5) + 1(6)}{45} = \frac{126}{45} = 2.8$$

In other words, there is an average of 2.8 people living in an apartment.

(b) Consider the data in Example A.2. Using the frequency distribution with class values, rather than the exact 35 numbers, we obtain the mean as follows:

$$\bar{x} = \frac{3(72.5) + 6(77.5) + 7(82.5) + 5(87.5) + 8(92.5) + 3(97.5) + 0(102.5) + 3(107.5)}{35}$$

$$= \frac{3052.5}{35} = 87.2$$

That is, the average 6:00 p.m. temperature is approximately 87.2°F.

Median

Suppose a list $x_1, x_2, \ldots, x_n$ of n data values is sorted in increasing order. The *median* of the data, denoted by

$$\tilde{x} \text{ (read: } x\text{-tilda)}$$

is defined to be the midvalue (if n is odd) or the average of the two middle values (if n is even). That is,

Median:
$$\tilde{x} = \begin{cases} x_{k+1} & \text{when } n \text{ is odd and } n = 2k + 1 \\ \dfrac{x_k + x_{k+1}}{2} & \text{when } n \text{ is even and } n = 2k \end{cases} \qquad (A\text{-}2)$$

Note that $\tilde{x}$ is the average of the $(n/2)$th and $[(n/2) + 1]$th terms when n is even.

Suppose, for example, the following two lists of sorted numbers are given:

$$\text{List } A: \quad 3, 3, 5, 7, 8$$
$$\text{List } B: \quad 1, 2, 5, 5, 7, 8, 8, 9$$

List A has 5 terms; hence the middle term is the third term. Thus, its median $\tilde{x} = 5$. List B has 8 terms; hence there are two middle terms, the fourth term 5 and the fifth term 7. Thus, its median $\tilde{x} = 6$, the average of the two middle terms.

The cumulative frequency distribution can be used to find the median of an arbitrary set of data.

One property of the median $\tilde{x}$ is that there are just as many numbers less than $\tilde{x}$ as there are greater than $\tilde{x}$.

Suppose the data are grouped. The cumulative frequency distribution can be used to find the class with the median. Then the class value is sometimes used as an approximation to the median or, for a better approximation, one can linearly interpolate in the class to find an approximation to the median.

EXAMPLE A.4

(a) Consider the data in Example A.1 which gives the number of tenants in 45 apartments. Here $n = 45$; hence the median $\tilde{x}$ is the twenty-third value. The cumulative frequency row tells us that $\tilde{x} = 3$.

(b) Consider the data in Example A.2 which gives the 6:00 p.m. temperatures for a 35-day period. The median is the eighteenth value, and its exact value can be found by using the original data before they are grouped into classes. Using the grouped data, we can find an approximation to the median in two ways. Note, first, using the cumulative frequency row, that the median is the second value in the group 85–90 which has five values. Thus:

(i) Simply let $\tilde{x} = 87.5$, the class value of the group.

(ii) Linearly interpolate in the class to obtain

$$\tilde{x} = 85 + \tfrac{2}{5}(5) = 87.0$$

Clearly (ii) will usually give a better approximation to the median.

Midrange

The midrange of a sorted sample $x_1, x_2, \ldots, x_n$ is the average of the smallest value x_1 and the largest value x_n. That is,

Midrange: $\qquad\qquad\qquad \text{mid} = \dfrac{x_1 + x_n}{2} \qquad\qquad\qquad (A\text{-}3)$

For the data in Example A.1, $x_1 = 1$ and $x_n = 6$. Thus

$$\text{mid} = \frac{1 + 6}{2} = 3.5$$

For the data in Example A.2, $x_1 = 72.5$ and $x_n = 107.5$. Thus

$$\text{mid} = \frac{72.5 + 107.5}{2} = 90.0$$

(Again we use class values rather than the original data for our formula.)

Additional Measurements

(1) *Weighted Mean (Weighted Arithmetic Mean):* Suppose each value x_i is associated with a nonnegative weighting factor w_i. Then the *weighted mean* is defined as follows:

$$\text{Weighted mean:}\qquad \bar{x} = \frac{\sum w_i x_i}{\sum w_i} = \frac{w_1 x_1 + w_2 x_2 + \cdots + w_k x_k}{w_1 + w_2 + \cdots + w_k} \qquad (A\text{-}4)$$

Here $\sum w_i$ is the total weight. Note that Formula (A-$1b$) is a special case of Formula (A-3) where the weight of x_i is its frequency.

(2) *Grand Mean:* Suppose there are k samples and the ith sample has mean $\bar{x}_i$ and n_i elements. Then the *grand mean*, denoted by $\bar{\bar{x}}$ (read: x-double bar) is defined as follows:

$$\text{Grand mean:}\qquad \bar{\bar{x}} = \frac{\sum n_i \bar{x}_i}{\sum n_i} = \frac{n_1 \bar{x}_1 + n_2 \bar{x}_2 + \cdots + n_k \bar{x}_k}{n_1 + n_2 + \cdots + n_k} \qquad (A\text{-}5)$$

Population Mean

Suppose $x_1, x_2, \ldots, x_N$ are the N numerical values of some population. The formula for the population mean, denoted by the Greek letter μ (read: mu), follows:

$$\text{Population mean:}\qquad \mu = \frac{x_1 + x_2 + \cdots + x_N}{N} = \frac{\sum x_i}{N}$$

$$\text{Population mean:}\qquad \mu = \frac{f_1 x_1 + f_2 x_2 + \cdots + f_k x_k}{f_1 + f_2 + \cdots + f_k} = \frac{\sum f_i x_i}{\sum f_i}$$

(We emphasize that N denotes the number of elements in the population whereas n denotes the number of elements in a sample of the population.)

Remark: Observe that the formula for the population mean μ is the same as the formula for the sample mean $\bar{x}$. On the other hand, there are formulas for the population which are not the same as the corresponding formulas for the sample. For example, the formula for the (population) standard deviation σ (Section A.4) is not the same as the formula for the sample standard deviation s.

A.4   MEASURES OF DISPERSION: VARIANCE AND STANDARD DEVIATION

Consider the following two samples of $n = 7$ numerical values:

List A: 7, 9, 9, 10, 10, 11, 14

List B: 7, 7, 8, 10, 11, 13, 14

Observe that the median (middle value) of each list is $\tilde{x} = 10$. Furthermore, the following shows that both lists have the same mean $\bar{x} = 10$:

$$\text{List } A:\quad \bar{x} = \frac{7 + 9 + 9 + 10 + 10 + 11 + 14}{7} = \frac{70}{7} = 10$$

$$\text{List } B:\quad \bar{x} = \frac{7 + 7 + 8 + 10 + 11 + 13 + 14}{7} = \frac{70}{7} = 10$$

Although both lists have the same first and last elements, the values in list A are clustered more closely about the mean than the values in list B. This section will discuss important ways of measuring such dispersions of data.

Variance and Standard Deviation

Consider a sample of values $x_1, x_2, \ldots, x_n$, and suppose $\bar{x}$ is the mean of a sample. The difference $x_i - \bar{x}$ is called the *deviation* of the data value x_i from the mean $\bar{x}$; it is positive or negative accordingly as x_i is greater or less than $\bar{x}$. The sample variance, denoted by s^2, is defined as the sum of the squares of the deviations divided by $n - 1$. Namely,

Sample variance: $\qquad s^2 = \dfrac{(x_1 - \bar{x})^2 + (x_2 - \bar{x})^2 + \cdots + (x_n - \bar{x})^2}{n - 1} = \dfrac{\Sigma(x_i - \bar{x})^2}{n - 1}$ $\qquad$ (A-6a)

Sample variance: $\qquad s^2 = \dfrac{f_1(x_1 - \bar{x})^2 + f_2(x_2 - \bar{x})^2 + \cdots + f_k(x_k - \bar{x})^2}{f_1 + f_2 + \cdots + f_k - 1} = \dfrac{\Sigma f_i(x_i - \bar{x})^2}{\Sigma f_i - 1}$ $\qquad$ (A-6b)

The nonnegative square root of the sample variance s^2, denoted by s, is called the *sample standard deviation*. That is,

Sample standard deviation: $\qquad\qquad s = \sqrt{s^2}$ $\qquad\qquad$ (A-7)

If the data are organized into classes, then we use the ith class value for x_i in the above Formula (A-6b).

The data in most applications and examples will come from some sample; hence we may simply say variance and standard deviation, omitting the adjective "sample".

Since each squared deviation is nonnegative, so is the variance s^2. Moreover, s^2 is zero precisely when all the data values are all equal (and, therefore, are all equal to the mean $\bar{x}$). Accordingly, if the data are more spread out, then the variance s^2 and the standard deviation s will be larger.

One advantage of the use of the standard deviation s over the variance s^2 is that the standard deviation s will have the same units as the original data.

EXAMPLE A.5 Consider the lists A and B above.

(a) List A has a mean $\bar{x} = 10$. The following are the deviations of the 7 data values:

$$7 - 10 = -3, \quad 9 - 10 = -1, \quad 9 - 10 = -1, \quad 10 - 10 = 0, \quad 10 - 10 = 0, \quad 11 - 10 = 1, \quad 14 - 10 = 4$$

The squares of the deviations are as follows:

$$(-3)^2 = 9, \quad (-1)^2 = 1, \quad (-1)^2 = 1, \quad 0^2 = 0, \quad 0^2 = 0, \quad 1^2 = 1, \quad 4^2 = 16$$

Also, $n - 1 = 7 - 1 = 6$. Therefore, the sample variance s^2 and standard deviation s are derived as follows:

$$s^2 = \frac{9 + 1 + 1 + 0 + 0 + 1 + 16}{6} = \frac{28}{6} = 4.67$$

and $\qquad\qquad\qquad\qquad s = \sqrt{4.67} \approx 2.16$

(b) List B also has a mean $\bar{x} = 10$. The deviations of the data and their squares follow:

$$(-3)^2 = 9, \ (-3)^2 = 9, \ (-2)^2 = 4, \ 0^2 = 0, \ 1^2 = 1, \ 3^2 = 9, \ 4^2 = 16$$

Again, $n - 1 = 6$. Accordingly, the sample variance s^2 and standard deviation s are derived as follows:

$$s^2 = \frac{9 + 9 + 4 + 0 + 1 + 9 + 16}{6} = \frac{48}{6} = 8$$

and $\qquad\qquad\qquad\qquad s = \sqrt{8} \approx 2.83$

Note that list B, which exhibits more dispersion than A, has a larger variance and standard deviation than list A.

Alternate Formulas for Sample Variance

Alternate formulas for the sample variance, that is, which are equivalent to Formulas (A-6a) and (A-6b) are as follows:

Sample variance: $$s^2 = \frac{\Sigma x_i^2 - (\Sigma x_i)^2/n}{n-1}$$ (A-8a)

Sample variance: $$s^2 = \frac{\Sigma f_i x_i^2 - (\Sigma f_i x_i)^2/\Sigma f_i}{\Sigma f_i - 1}$$ (A-8b)

Again, if the data are organized into classes, then we use the class values as approximations to the original values in the above Formula (A-8b).

Although Formulas (A-8a) and (A-8b) may look more complicated than Formulas (A-6a) and (A-6b), they are usually more convenient to use. In particular, these formulas only use one subtraction in the numerator, and they can be used without first calculating the sample mean $\bar{x}$.

EXAMPLE A.6   Consider the following $n = 9$ data values:

$$3, \quad 5, \quad 8, \quad 9, \quad 10, \quad 12, \quad 13, \quad 15, \quad 20$$

Find: (a) mean $\bar{x}$, (b) variance s^2 and standard deviation s.

First construct the following table where the two numbers on the right, 95 and 1217, denote the sums Σx_i and Σx_i^2, respectively:

x_i	3	5	8	9	10	12	13	15	20	95
x_i^2	9	25	64	81	100	144	169	225	400	1217

(It is currently common practice and notationally convenient to write numbers and their sum horizontally rather than vertically.)

(a)   By Formula (A-1a), where $n = 9$,

$$\bar{x} = (\Sigma x_i)/n = 95/9 = 10.56$$

(b)   Here we use Formula (A-8a) with $n = 9$ and $n - 1 = 8$:

$$s^2 = \frac{1217 - (95)^2/9}{8} = \frac{1217 - 1002.78}{8} \approx 26.78$$

Then $$s = \sqrt{26.78} = 5.17$$

Note that if we used Formula (A-6a), we would need to subtract $x = 10.56$ from each x_i before squaring.

EXAMPLE A.7   Consider the data in Example A.1 which gives the number of tenants in 45 apartments. The sample mean $\bar{x} = 2.8$ was obtained in Example A.3. Find the sample variance s^2 and the sample standard deviation s.

First extend the frequency distribution table of the data as follows (where SUM refers to Σf_i, $\Sigma f_i x_i$, and $\Sigma f_i x_i^2$):

Number of people x_i	1	2	3	4	5	6	SUM
Frequency f_i	8	14	7	12	3	1	45
$f_i x_i$	8	28	21	48	15	6	126
x_i^2	1	4	9	16	25	36	
$f_i x_i^2$	8	56	63	192	75	36	430

Then $$s^2 = \frac{430 - (126)^2/45}{44} \approx 1.75 \quad \text{and} \quad s \approx 1.32$$

Note that $n = 45$ and $n - 1 = 44$.

Measures of Position: Quartiles and Five-Number Summary

Consider a set of n data values $x_1, x_2, \ldots, x_n$ which are arranged in increasing order. Recall that the median $M = \tilde{x}$ of the data values has been defined as a number for which, at most, half of the values are less than M and, at most, half of the numbers are greater than M. Here "half" means $n/2$ when n is even and $(n - 1)/2$ when n is odd. Specifically

$$\text{Median } M = \begin{cases} \dfrac{x_k + x_{k+1}}{2} & \text{when } n \text{ is even and } n = 2k \\ x_{k+1} & \text{when } n \text{ is odd and } n = 2k + 1 \end{cases}$$

The first, second, and third quartiles, Q_1, Q_2, Q_3, are defined as follows:

Q_1 = median of the first half of the values

$Q_2 = M$ = median of all the values

Q_3 = median of the second half of the values

The *5-number summary* of the data is the following quintuple:

$$[L, Q_1, M, Q_3, H]$$

where $L = x_1$ is the lowest value, Q_1, $M = Q_2$, Q_3, are the quartiles, and $H = x_n$ is the highest value.

The *range* of the above data is the distance between the lowest and highest value, and the *interquartile range* (IQR) is the distance between the first and third quartiles; namely,

$$\text{range} = H - L \quad \text{and} \quad \text{IQR} = Q_3 - Q_1$$

Observe that

Range Interval: $[L, H]$ contains 100 percent of the data values.

IQR Interval: $[Q_1, Q_3]$ contains about 50 percent of the data values.

Also, observe that the 5-number summary $[L, Q_1, M, Q3, H]$ or, equivalently, the 4 intervals,

$$[L, Q_1], \qquad [Q_1, M], \qquad [M, Q_3], \qquad [Q_3, H]$$

divide the data into 4 sets where each set contains about 25 percent of the data values.

EXAMPLE A.8 Consider the following two lists of $n = 7$ numerical values:

List A: 7, 9, 9, 10, 10, 11, 14

List B: 7, 7, 8, 10, 11, 13, 14

The median of both lists is the fourth value $M = 10$. Find the quartiles Q_1 and Q_3, the 5-number summary $[L, Q_1, M, Q3, H]$, and the range and interquartile range (IQR) of each list. Compare the range and IQR of both lists.

(a)   The median $M = 10$ of list A divides the set into the first half $\{7, 9, 9\}$ and the second half $\{10, 11, 14\}$. Hence $Q_1 = 9$ and $Q_3 = 11$. Also, $L = 7$ is the lowest value and $H = 14$ is the highest value. Thus, the 5-number summary of list A follows:

$$[L, Q_1, M, Q3, H] = [7, 9, 10, 11, 14]$$

Furthermore

$$\text{range} = H - L = 14 - 7 = 7 \quad \text{and} \quad \text{IQR} = Q_3 - Q_1 = 11 - 9 = 2$$

(b)   The median $M = 10$ of list B divides the set into the first half $\{7, 7, 8\}$ and the second half $\{11, 13, 14\}$.   Hence $Q_1 = 7$ and $Q_3 = 13$.   Also, $L = 7$ is the lowest value and $H = 14$ is the highest value.   Thus, the 5-number summary of list B is as follows:

$$[L, Q_1, M, Q3, H] = [7, 7, 10, 13, 14]$$

Furthermore   range $= 14 - 7 = 7$ and   IQR $= 13 - 7 = 6$

Although list B exhibits more dispersion than list A, the ranges of both lists are the same.   However, the IQR $= 6$ of list B is much larger than the IQR $= 2$ of list A.   Generally speaking, the IQR usually gives a more accurate description of the dispersion of a list than the range since the range may be strongly influenced by a single small or large value.

EXAMPLE A.9   Consider the following list of $n = 30$ numerical values:

4   5   5   7   8   8   9   10   10   11   11   11   12   12   12

13   13   14   14   14   15   16   16   18   18   19   19   20   22   25

Find the median M, the quartiles Q_1 and Q_3, the 5-number summary $[L, Q_1, M, Q3, H]$, and the range and interquartile range (IQR) of the data.

Here $n = 30$ is even, so the median M is the average of the fifteenth and sixteenth values.   Thus

$$M = \frac{12 + 13}{2} = 12.5$$

The first quartile Q_1 is the mean of the first half (first 15) numbers, so $Q_1 = 10$, the eighth number of the first half sublist.   The third quartile Q_3 is the mean of the second half (second 15) numbers, so $Q_3 = 16$, the eighth number of the second half sublist.   Here, $L = 4$ and $H = 25$, so the 5-number summary follows:

$$[L, Q_1, M, Q3, H] = [4, 10, 12.5, 16, 25]$$

Furthermore: range $= H - L = 25 - 4 = 21$ and   IQR $= Q_3 - Q_1 = 16 - 10 = 6$

A.5   BIVARIATE DATA, SCATTERPLOTS, CORRELATION COEFFICIENTS

Quite often in statistics it is desired to determine the relationship, if any, between two variables, such as between age and weight, weight and height, years of education and salary, amount of daily exercise and cholesterol level, and so on.   Letting x and y denote the two variables, the data will consist of a list of pairs of numerical values

$$(x_1, y_1), (x_2, y_2), (x_3, y_3), \ldots, (x_n, y_n)$$

where the first values correspond to the variable x and the second values correspond to y.

As with a single variable, we can describe such *bivariate data* both graphically and numerically.   Our primary concern is to determine whether there is a mathematical relationship, such as a linear relationship, between the data.

It should be kept in mind that a statistical relationship between two variables does not necessarily imply there is a *causal* relationship between them.   For example, a strong relationship between weight and height does not imply that one variable causes the other.   On the other hand, eating more does usually increase the weight of a person but it does not usually mean there will be an increase in the height of the person.

Scatterplots

Consider a list of pairs of numerical values representing variables x and y.   The *scatterplot* of the data is simply a picture of the pairs of values as points in a coordinate plane $\mathbf{R}^2$.   The picture sometimes indicates a relationship between the points as illustrated in the following examples.

EXAMPLE A.10

(a) Consider the following data where x denotes the ages of 6 children and y denotes the corresponding number of correct answers in a 10-question test:

x	5	6	6	7	7	8
y	6	6	7	8	9	9

The scatterplot of the data appears in Fig. A-3(a).   The picture of the points indicates, roughly speaking, that the number of correct answers increases as the age increases.   We then say that x and y have a *positive correlation*.

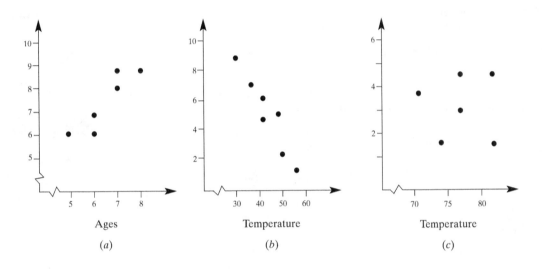

Ages

(a)

Temperature

(b)

Temperature

(c)

Fig. A-3

(b) Consider the following data where x denotes the average daily temperature, in degrees Fahrenheit, and y denotes the corresponding daily natural gas consumption, in cubic feet:

x	50	45	40	38	32	40	55
y	2.5	5.0	6.2	7.4	8.3	4.7	1.8

The scatterplot of the data appears in Fig. A-3(b).   The picture of the points indicates, roughly speaking, that the gas consumption decreases as the temperature increases.   We then say that x and y have a *negative correlation*.

(c) Consider the following data where x denotes the average daily temperature, in degrees Fahrenheit, over a 6-day period and y denotes the corresponding number of defective traffic lights:

x	72	78	75	74	78	76
y	4	5	5	2	2	3

The scatterplot of the data appears in Fig. A-3(c).   The picture of the points indicates that there is no apparent relationship between x and y.

Correlation Coefficient

Scatterplots indicate graphically whether there is a linear relationship between two variables x and y. A numeric indicator of such a linear relationship is the *sample correlation coefficient r* of x and y, which is defined as follows:

Sample correlation coefficient: $$r = \frac{\Sigma (x_i - \bar{x})(y_i - \bar{y})}{\sqrt{\Sigma (x_i - \bar{x})^2 \Sigma (y_i - \bar{y})^2}} \qquad (A\text{-}9)$$

We assume the denominator in Formula (A-9) is not zero. It can be shown that the correlation coefficient r has the following properties:

(1) $-1 \leq r \leq 1$.

(2) $r > 0$ if y tends to increase as x increases and $r < 0$ if y tends to decrease as x increases.

(3) The stronger the linear relationship between x and y, the closer r is to -1 or 1; the weaker the linear relationship between x and y, the closer r is to 0.

An alternate formula for computing r is given below; we then illustrate the above properties of r with examples.

Another numerical measurement of bivariate data with variables x and y is the *sample covariance* which is denoted and defined as follows:

Sample covariance of x and y: $$s_{xy} = \frac{\Sigma (x_i - \bar{x})(y_i - \bar{y})}{n - 1} \qquad (A\text{-}10)$$

Formula (A-9) can now be written in the more compact form as

Sample correlation coefficient: $$r = \frac{s_{xy}}{s_x s_y}$$

where s_x and s_y are the sample standard deviations of x and y, respectively, and s_{xy} is the sample covariance of x and y defined above.

An alternate formula for computing the correlation coefficient r follows:

$$r = \frac{\Sigma x_i y_i - (\Sigma x_i)(\Sigma y_i)/n}{\sqrt{\Sigma x_i^2 - (\Sigma x_i)^2/n} \; \sqrt{\Sigma y_i^2 - (\Sigma y_i)^2/n}} \qquad (A\text{-}11)$$

This formula is very convenient to use after forming a table with the values of x_i, y_i, x_i^2, y_i^2, $x_i y_i$, and their sums, as illustrated below.

EXAMPLE A.11 Find the correlation coefficient r for each data set in Example A.10.

(a) Construct the following table which gives the x, y, x^2, y^2, and xy values, and the last column gives the corresponding sums:

x	5	6	6	7	7	8	39
y	6	6	7	8	9	9	45
x^2	25	36	36	49	49	64	259
y^2	36	36	49	64	81	81	347
xy	30	36	42	56	63	72	299

Now use Formula (A-11) and the number of points is $n = 6$ to obtain

$$r = \frac{299 - (39)(45)/6}{\sqrt{259 - (39)^2/6} \; \sqrt{347 - (45)^2/6}} = \frac{6.50}{\sqrt{5.50} \; \sqrt{9.50}} = 0.899$$

Here r is close to 1, which is expected since the scatterplot in Fig. A-3(a) indicates a strong positive linear relationship between x and y.

(b) Construct the following table which gives the x, y, x^2, y^2 and xy values, and the last column gives the corresponding sums:

x	50	45	40	38	32	40	55	300
y	2.5	5.0	6.2	7.4	8.3	4.7	1.8	35.9
x^2	2,500	2,025	1,600	1,444	1,024	1,600	3,025	13,218
y^2	6.25	25.00	38.44	54.76	68.89	22.09	3.24	218.67
xy	125.0	225.0	248.0	281.2	265.6	188.0	99.0	1,431.8

Formula (A-11), with $n = 7$, yields

$$r = \frac{1431.8 - (300)(35.9)/7}{\sqrt{13,218 - (300)^2/7}\ \sqrt{218.67 - (35.9)^2/7}} = \frac{-106.77}{\sqrt{360.86}\ \sqrt{34.554}} = -0.9562$$

Here r is close to -1, and the scatterplot in Fig. A-3(b) indicates a strong negative linear relationship between x and y.

(c) Construct the following table which gives the x, y, x^2, y^2, and xy values, and the last column gives the corresponding sums:

x	72	78	75	74	78	76	453
y	4	5	5	2	2	3	21
x^2	5,184	6,084	5,625	5,476	6,084	5,776	34,229
y^2	16	25	25	4	4	9	83
xy	288	390	375	148	156	228	1,585

Formula (A-11), with $n = 6$, yields

$$r = \frac{1585 - (453)(21)/6}{\sqrt{34,229 - (453)^2/6}\ \sqrt{83 - (21)^2/6}} = \frac{-0.500}{\sqrt{27.50}\ \sqrt{9.5}} = -0.031$$

Here r is close to 0, which is expected since the scatterplot in Fig. A-3(c) indicates no linear relationship between x and y.

A.6 METHODS OF LEAST SQUARES, REGRESSION LINE, CURVE FITTING

Suppose a scatterplot of the data points (x_i, y_i) indicates a linear relationship between variables x and y or, alternately, suppose the correlation coefficient r of x and y is close to 1 or -1. Then the next step is to find a line L that, in some sense, fits the data. The line L we choose is called the *least-squares line*. This section discusses this line, and then we discuss more general types of curve fitting.

Least-Squares Line

Consider a given set of data points $P_i(x_i, y_i)$ and any (nonvertical) linear equation L. Let y_i^* denote the y value of the point on L corresponding to x_i. Furthermore, let $d_i = y_i - y_i^*$, the difference between the actual value of y and the value of y on the curve or, in other words, the vertical (directed) distance between the point P_i and the line L, as shown in Fig. A-4. The sum

$$\Sigma d_i^2 = d_1^2 + d_2^2 + \cdots + d_n^2$$

is called the *squares error* between the line L and the data points.

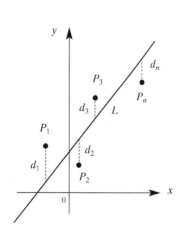

Fig. A-4

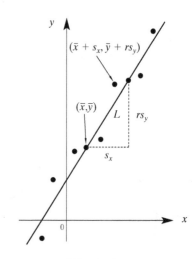

Fig. A-5

The *least-squares line* or the *line of best fit* or the *regression line of y on x* is, by definition, the line L whose squares error is as small as possible. It can be shown that such a line L exists and is unique. Let a denote the y intercept of the line L and let b denote its slope, that is, suppose the following is the equation of the line L:

$$y = a + bx$$

Then a and b can be obtained from the following two equations, called the *normal equations*, in the two unknowns a and b where n is the number of points:

$$\text{Normal equations:} \begin{cases} na + (\Sigma x_i)b = \Sigma y_i \\ (\Sigma x_i)a + (\Sigma x_i^2)b = \Sigma x_i y_i \end{cases} \tag{A-12}$$

In particular, the slope b and y intercept a can also be obtained from the following formula (where r is the correlation coefficient):

$$b = \frac{rs_y}{s_x} \quad \text{and} \quad a = \bar{y} - b\bar{x} \tag{A-13}$$

Formula *(A-13)* is usually used instead of Formula *(A-12)* when one needs, or has already found, the means $\bar{x}$ and $\bar{y}$, the standard deviations s_x and s_y, and the correlation r of the given data points.

Graphing the line L of best fit requires at least two points on L. The second equation in Formula *(A-13)* tells us that $(\bar{x}, \bar{y})$ lies on the regression line L since

$$\bar{y} = (\bar{y} - b\bar{x}) + b\bar{x} = a + b\bar{x}$$

also, the first equation in Formula *(A-13)* then tells us that the point $(x + s_x, y + rs_y)$ is also on L. These points are also pictured in Fig. A-5.

Remark: Recall that the above line L which minimizes the squares of the vertical distances from the given points P_i to L is called the *regression line* of y on x; it is usually used when one views y as a function of x. A line L' also exists which minimizes the squares of the horizontal distances of the points P_i from L'; it is called the *regression line of x on y*. Given any two variables, the data usually indicate that one of them depends upon the other; we then let x denote the independent variable and let y denote the dependent variable. For example, suppose the variables are age and height. We normally assume height is a function of age, so we would let x denote age and y denote height. Accordingly, unless otherwise stated, our least-squares lines will be regression lines of y on x.

EXAMPLE A.12 Find the line L of best fit for the first two scatterplots in Fig. A-3.

(a) By the table in Example A.11(a),

$$\Sigma x_i = 39 \qquad \Sigma y_i = 45 \qquad \Sigma x_i^2 = 259 \qquad \Sigma x_i y_i = 299$$

Also, there are $n = 6$ points. Substitution in the normal equations in Formula $(A\text{-}12)$ yields the following system:

$$6a + 39b = 45$$
$$39a + 259b = 299$$

The solution of the system follows:

$$a = -\frac{2}{11} = -0.18 \qquad b = \frac{13}{11} = 1.18$$

Thus, the following is the line L of best fit.

$$y = -0.18 + 1.18x$$

To graph L, we need only plot two points on L and then draw the line through these points. Setting $x = 5$ and $x = 8$, we obtain the two points:

$$A(5, 5.7) \qquad \text{and} \qquad B(8, 9.3)$$

and then we draw L, as shown in Fig. A-6(a).

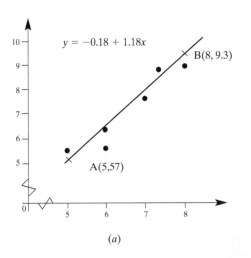

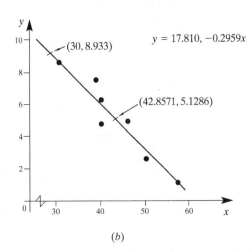

(a)

(b)

Fig. A-6

(b) Here we use Formula $(A\text{-}13)$ rather than Formula $(A\text{-}12)$. By Example A.11(b), with $n = 7$, we obtain

$$r = -0.9562 \qquad \bar{x} = 300/7 = 42.86 \qquad \bar{y} = 35.9/7 = 5.129$$

Using Formulas $(A\text{-}8)$ and $(A\text{-}9)$, we obtain

$$s_x = \sqrt{\frac{13{,}218 - (300)^2/7}{6}} = 7.7552 \qquad \text{and} \qquad s_y = \sqrt{\frac{218.67 - (35.9)^2/7}{6}} = 2.3998$$

Substituting these values in Formula $(A\text{-}12)$, we get

$$b = \frac{(-0.956\ 2)(2.399\ 8)}{7.755\ 2} = -0.295\ 9 \qquad \text{and} \qquad a = 5.128\ 6 - (-0.295\ 9)(42.857) = 17.810$$

Thus, the line L of best fit follows:

$$y = 17.810 - 0.295\ 9x$$

The graph of L, obtained by plotting $(30, 8.933)$ and $(42.857\ 1, 5.128\ 6)$ (approximately) and drawing the line through these points, is shown in Fig. A-6(b).

Curve Fitting

Sometimes the scatterplot does not indicate a linear relationship between the variables x and y, but one may visualize some other standard (well-known) curve, $y = f(x)$, which may approximate the data, called an *approximate curve*. Several such standard curves, where letters other than x and y denote constants, follow:

(1) Parabolic curve: $y = a_0 + a_1 x + a_2 x^2$

(2) Cubic curve: $y = a_0 + a_1 x + a_2 x^2 + a_3 x^3$

(3) Hyperbolic curve: $y = \dfrac{1}{a + bx}$ or $\dfrac{1}{y} = a + bx$

(4) Exponential curve: $y = ab^x$ or $\log y = a_0 + a_1 x$

(5) Geometric curve: $y = ax^b$ or $\log y = \log a + b \log x$

(6) Modified exponential curve: $y = ab^x + c$

(7) Modified geometrical curve: $y = ax^b + c$

Pictures of some of these standard curves appear in Fig. A-7.

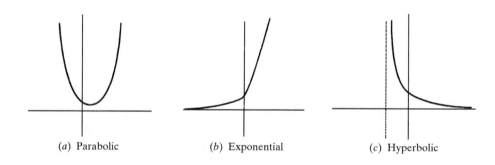

(a) Parabolic   (b) Exponential   (c) Hyperbolic

Fig. A-7

Generally speaking, it is not easy to decide which curve to use for a given set of data points. On the other hand, it is usually easier to determine a linear relationship by looking at the scatterplot or by using the correlation coefficient. Thus, it is standard procedure to find the scatterplot of transformed data. Specifically:

(a) If $\log y$ versus x indicates a linear relationship, use the exponential curve (type 4).

(b) If $1/y$ versus x indicates a linear relationship, use the hyperbolic curve (type 3).

(c) If $\log y$ versus $\log x$ indicates a linear relationship, use the geometric curve (type 5).

Once one decides upon the type of curve to be used, then that particular curve is the one that minimizes the squares error. We state this formally:

Definition: Consider a collection of curves and a given set of data points. The *best-fitting* or *least-squares* curve C in the collection is the curve which minimizes the sum

$$\Sigma d_i^2 = d_1^2 + d_2^2 + \cdots + d_n^2$$

(where d_i denotes the vertical distance from a data point $P_i(x_i, y_i)$ to the curve C).

Just as there are formulas to compute the constants a and b in the regression line L for a set of data points, so there are formulas to compute the constants in the best-fitting curve C in any of the above types (collections) of curves. The derivation of such formulas usually involves calculus.

EXAMPLE A.13 Consider the following data which indicates exponential growth:

x	1	2	3	4	5	6
y	6	18	55	160	485	1460

Find the least-squares exponential curve C for the data, and plot the data points and C on the plane $\mathbf{R}^2$.

The curve C has the form $y = ab^x$ where a and b are unknowns. The logarithm (to base 10) of $y = ab^x$ yields

$$\log y = \log a + x \log b = a' + b'x$$

where $a' = \log a$ and $b' = \log b$. Thus, we seek the least-squares line L for the following data:

x	1	2	3	4	5	6
$\log y$	0.778 2	1.255 3	1.740 4	2.204 1	2.685 7	3.164 4

Using the normal equations in Formula $(A\text{-}12)$ for L, we get

$$a' = 0.302\,8 \qquad b' = 0.476\,7$$

The antiderivatives of a' and b' yield, approximately,

$$a = 2.0 \qquad b = 3.0$$

Thus, $y = 2(3^x)$ is the required exponential curve C. The data points and C are plotted in Fig. A-8.

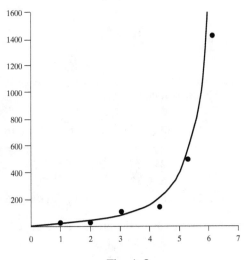

Fig. A-8

Solved Problems

FREQUENCY DISTRIBUTION, MEAN AND MEDIAN

A.1. Consider the following frequency distribution which gives the number f of students who got x correct answers on a 20-question exam:

x (correct answers)	9	10	12	13	14	15	16	17	18	19	20
f (number of students)	1	2	1	2	7	2	1	7	2	6	4

(a) Display the data in a histogram and a frequency polygon.

(b) Find the mean $\bar{x}$, median M, and midrange of the data.

(a) The histogram appears in Fig. A-9. The frequency polygon also appears in Fig. A-9; it is obtained from the histogram by connecting the midpoints of the tops of the rectangle in the histogram.

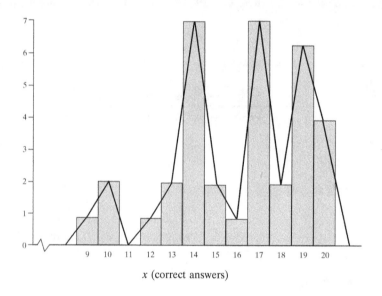

x (correct answers)

Fig. A-9

(b) First we extend our frequency table to include the cumulative distribution function cf, the products $f_i x_i$, and the sums Σf_i and $\Sigma x_i f_i$ as follows:

x	9	10	12	13	14	15	16	17	18	19	20	
f	1	2	1	2	7	2	1	7	2	6	4	35
cf	1	3	4	6	13	15	16	23	25	31	35	
fx	9	20	12	26	98	30	16	119	36	154	80	560

Here we use Formula $(A\text{-}1b)$ which gives the mean $\bar{x}$ for grouped data:

$$\bar{x} = \frac{\Sigma f_i x_i}{\Sigma f_i} = \frac{560}{35} = 16$$

There are $n = 35$ scores, so the mean M is the eighteenth score. The row cf in the table tells us that 16 is the sixteenth score, and 17 is the seventeenth to twenty-third scores. Hence the mean

$$M = 17, \text{ the eighteenth score}$$

The midrange is the average of the first score 9 and the last score 20; hence:

$$\text{mid} = \frac{9 + 20}{2} = 14.5$$

A.2. Consider the following $n = 20$ data items:

$$3 \quad 5 \quad 3 \quad 4 \quad 4 \quad 7 \quad 6 \quad 5 \quad 2 \quad 4$$
$$2 \quad 5 \quad 5 \quad 6 \quad 4 \quad 3 \quad 5 \quad 4 \quad 5 \quad 5$$

(a) Construct the frequency distribution f and cumulative distribution cf of the data, and display the data in a histogram.

(b) Find the mean $\bar{x}$, median M, and midrange of the data.

(a) Construct the following frequency distribution table which also includes the products $f_i x_i$ and the sums Σf_i and $\Sigma x_i f_i$:

x	2	3	4	5	6	7	
f	2	3	5	7	2	1	20
cf	2	5	10	17	19	20	
fx	4	9	20	35	12	7	87

Note that the first line of the table consists of the range of numbers, from 2 to 7. The second line (frequency) can be obtained by either counting the number of times each number occurs or by going through the list one number after another and keeping a *tally count*, a running account as each number occurs. The histogram is shown in Fig. A-10(a).

(b) Here we use Formula (A-1b) which gives the mean $\bar{x}$ for grouped data:

$$\bar{x} = \frac{\Sigma f_i x_i}{\Sigma f_i} = \frac{87}{20} = 4.7$$

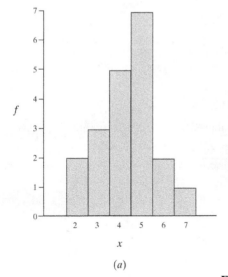

(a)

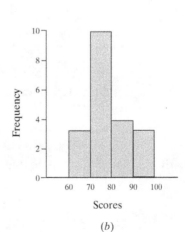

(b)

Fig. A-10

There are $n = 20$ numbers, so the mean M is the average of the tenth and eleventh numbers. The row cf in the table tells us that 4 is the tenth number and 5 is the eleventh number. Hence

$$M = \frac{4 + 5}{2} = 4.5$$

The midrange is the average of the first number 2 and the last number 7; hence

$$\text{mid} = \frac{2 + 7}{2} = 4.5$$

A.3. Consider the following $n = 20$ scores on a statistic exam:

$$74 \quad 80 \quad 65 \quad 85 \quad 95 \quad 72 \quad 76 \quad 72 \quad 93 \quad 84$$
$$75 \quad 75 \quad 60 \quad 74 \quad 75 \quad 63 \quad 78 \quad 87 \quad 90 \quad 70$$

(a) Construct the frequency distribution f table where the data are grouped into four classes:

$$60\text{–}70, \ 70\text{–}80, \ 80\text{–}90, \ 90\text{–}100$$

The table should include the class values x_i and the cumulative distribution cf of the data. (Recall that if a number falls on a class boundary, it is assigned to the higher class.) Also, display the data in a histogram.

(b) Find the mean $\bar{x}$, median M, and midrange of the data.

(a) Construct the following frequency distribution table which also includes the products $f_i x_i$ and the sums Σf_i and $\Sigma x_i f_i$:

Class	60–70	70–80	80–90	90–100	
x_i	65	75	85	95	
f	3	10	4	3	20
cf	3	13	17	20	
fx_i	195	750	340	285	1570

The histogram is shown in Fig. A-10(b).

(b) Using the class values x_i, Formula (A-1b) yields

$$\bar{x} = \frac{\Sigma f_i x_i}{\Sigma f_i} = \frac{1570}{20} = 78.5$$

There are $n = 20$ numbers, so the mean M is the average of the tenth and eleventh class scores which we approximate using their class values. The row cf in the table tells us that 75 is the approximation of the tenth and eleventh scores. Thus

$$M = 75$$

The midrange is the average of the first class value 65 and the last class value 95; hence

$$\text{mid} = \frac{65 + 95}{2} = 80$$

A.4. The yearly rainfall, measured to the nearest tenth of a centimeter, for a 30-year period follows:

$$42.3 \quad 35.7 \quad 47.5 \quad 31.2 \quad 28.3 \quad 37.0 \quad 41.3 \quad 32.4 \quad 41.3 \quad 29.3$$
$$34.3 \quad 35.2 \quad 43.0 \quad 36.3 \quad 35.7 \quad 41.5 \quad 43.2 \quad 30.7 \quad 38.4 \quad 46.5$$
$$43.2 \quad 31.7 \quad 36.8 \quad 43.6 \quad 45.2 \quad 32.8 \quad 30.7 \quad 36.2 \quad 34.7 \quad 35.3$$

(a) Construct the frequency distribution f table where the data are grouped into 10 classes:

$$28\text{–}30,\ 30\text{–}32,\ 32\text{–}34,\ \ldots,\ 46\text{–}48$$

The table should include the class values (cv) x_i and the cumulative distribution (cf) of the data.

(b) Find the mean $\bar{x}$, median M, and midrange of the data.

(a) Construct the following frequency distribution table which also includes the products $f_i x_i$ and the sums Σf_i and $\Sigma x_i f_i$:

Class	28–30	30–32	32–34	34–36	36–38	38–40	40–42	42–44	44–46	46–48	
cv x_i	29	31	33	35	37	39	41	43	45	47	
f	2	4	2	6	4	1	3	5	1	2	30
cf	2	6	8	14	18	19	22	27	28	30	
$f x_i$	58	124	66	210	148	39	123	215	45	94	1122

(b) Using the class values x_i, Formula (A-1b) yields

$$\bar{x} = \frac{\Sigma f_i x_i}{\Sigma f_i} = \frac{1122}{30} = 37.4$$

There are $n = 30$ numbers, so the mean M is the average of the fifteenth and sixteenth class values. The row cf in the table tells us that 37 is the fifteenth and sixteenth class value. Thus

$$M = 37$$

The midrange is the average of the first class value 29 and the last class value 47; hence

$$\text{mid} = \frac{29 + 47}{2} = 39$$

MEASURES OF DISPERSION: VARIANCE, STANDARD DEVIATION, IQR

A.5. Consider the following $n = 10$ data values:

$$1, 2, 2, 3, 4, 5, 7, 8, 9, 9$$

(a) Find the sample mean $\bar{x}$.
(b) Find the variance s^2 and standard deviation s.
(c) Find the median M, 5-number summary $[L, Q_1, M, Q_2, H]$, range, and interquartile range (IQR) of the data.

(a) The mean $\bar{x}$ is the "average" of the numbers, the sum of the values divided by the number $n = 10$ of values:

$$\bar{x} = \frac{1 + 2 + 2 + 3 + 4 + 5 + 7 + 8 + 9 + 9}{10} = \frac{50}{10} = 5$$

(b) **Method 1:** Here we use Formula (A-6a). We have

$$s^2 = \frac{\Sigma (x_i - \bar{x})^2}{n - 1} = \frac{16 + 9 + 9 + 4 + 1 + 0 + 4 + 9 + 16 + 16}{9} = \frac{84}{9} = 9.33 \quad \text{and} \quad s = \sqrt{9.33} = 3.05$$

Method 2: Here we use Formula (A-8a). First construct the following table where the two numbers on the right, 50 and 334, denote the sums Σx_i and Σx_i^2, respectively:

x	1	2	2	3	4	5	7	8	9	9	50
x_i^2	1	4	4	9	16	25	49	64	81	81	334

We have

$$s^2 = \frac{\Sigma x_i^2 - (\Sigma x_i)^2/n}{n-1} = \frac{334 - (50)^2/10}{9} = \frac{84}{9} = 9.33 \quad \text{and} \quad s = \sqrt{9.33} = 3.05$$

(c) Here $n = 10$ is even; hence the median M is the average of the fifth and sixth values. Thus

$$M = \frac{4+5}{2} = 4.5$$

The mean $M = 4.5$ divides the 10 items into two halves, $A = \{1, 2, 2, 3, 4\}$ and $B = \{5, 7, 8, 9, 9\}$, each with 5 numbers. The first quartile Q_1 is the median (middle element) of the first half A, so $Q_1 = 2$; the third quartile Q_3 is the median (middle number) of the second half B, so $Q_3 = 8$. Here $L = 1$ is the lowest number and $H = 9$ is the highest number. Thus, the 5-number summary of the data follows:

$$[L, Q_1, M, Q_2, H] = [1, 2, 4, 5, 8, 9]$$

Furthermore: range $= H - L = 9 - 1 = 8$ and   IQR $= Q_3 - Q_1 = 8 - 2 = 6$

A.6. The ages of $n = 30$ children living in an apartment complex are as follows:

$$2 \;\; 3 \;\; 3 \;\; 1 \;\; 2 \;\; 2 \;\; 3 \;\; 4 \;\; 4 \;\; 3 \;\; 2 \;\; 2 \;\; 6 \;\; 2 \;\; 4$$
$$1 \;\; 2 \;\; 6 \;\; 4 \;\; 2 \;\; 2 \;\; 3 \;\; 7 \;\; 1 \;\; 2 \;\; 3 \;\; 2 \;\; 4 \;\; 2 \;\; 6$$

(a) Find the frequency distribution of the data.

(b) Find the sample mean $\bar{x}$, variance s^2, and standard deviation s for the data.

(c) Find the median M, the 5-number summary $[L, Q_1, M, Q_2, H]$, the range, and the IQR (interquartile range) of the data.

(a) Construct the following frequency table which also includes the cumulative distribution cf function; products $f_i x_i$, x_i^2, $f_i x_i^2$; and the sums Σf_i, $\Sigma f_i x_i$, and $\Sigma f_i x_i^2$:

x	1	2	3	4	5	6	7	
f	3	12	6	5	0	3	1	30
cf	3	15	21	26	26	29	30	
fx	3	24	18	20	0	18	7	90
x^2	1	4	9	16	25	36	49	
fx^2	3	48	54	80	0	108	49	342

(b) We have

$$\bar{x} = \frac{\Sigma f_i x_i}{\Sigma f_i} = \frac{90}{30} = 3$$

Also $s^2 = \dfrac{\Sigma f_i x_i^2 - (\Sigma f_i x_i)^2/n}{n-1} = \dfrac{342 - (90)^2/30}{29} = \dfrac{72}{29} = 2.48$ and $s = \sqrt{2.48} = 1.58$

(c) Here $n = 30$ is even; hence the median M is the average of the fifteenth and sixteenth ages. The row cf in the table tells us that 2 is the fifteenth age and 3 is the sixteenth age. Thus

$$M = \frac{2+3}{2} = 2.5$$

The mean $M = 2.5$ divides the 30 items into two halves, each with 15 ages. The first quartile Q_1 is the median of the first 15 ages, so Q_1 is the eighth age; the third quartile Q_3 is the median of the last 15 ages, so Q_3 is the twenty-third age. Using the cf row in the table, we obtain

$$Q_1 = 2 \quad \text{and} \quad Q_3 = 4$$

Furthermore, $L = 1$ is the lowest number and $H = 7$ is the highest number. Thus, the 5-number summary of the data follows:

$$[L, Q_1, M, Q_2, H] = [1, 2, 2.5, 4, 7]$$

Furthermore: range $= H - L = 7 - 1 = 6$ and   IQR $= Q_3 - Q_1 = 4 - 2 = 2$

A.7. Consider the following list of $n = 18$ data values:

$$2, \ 7, \ 4, \ 1, \ 6, \ 4, \ 8, \ 15, \ 12, \ 7, \ 3, \ 16, \ 1, \ 2, \ 11, \ 5, \ 15, \ 4$$

(a) Find the median M.

(b) Find the quartiles Q_1 and Q_3, the 5-number summary $[L, Q_1, M, Q_2, H]$, the range, and the IQR (interquartile range) of the data.

(a) First arrange the data in numerical order:

$$1, \ 1, \ 2, \ 2, \ 3, \ 4, \ 4, \ 4, \ 5, \ 6, \ 7, \ 7, \ 8, \ 11, \ 12, \ 15, \ 15, \ 16$$

There are $n = 18$ values, so the median M is the average of the ninth and tenth values. Thus

$$M = \frac{5+6}{2} = 5.5$$

(b) Q_1 is the median of the nine values, from 1 to 5, less than M. Thus, $Q_1 = 3$, the fifth value. Q_3 is the median of the nine values, from 6 to 16, greater than M. Thus, $Q_3 = 11$, the fifth value. Also, $L = 1$ is the lowest number and $H = 16$ is the highest number. Thus, the 5-number summary of the data follows:

$$[L, Q_1, M, Q_2, H] = [1, 3, 5.5, 11, 16]$$

Furthermore: range $= H - L = 16 - 1 = 15$ and   IQR $= Q_3 - Q_1 = 11 - 3 = 8$

A.8. Consider the following frequency distribution:

x	1	2	3	4	5	6
f	2	4	6	8	3	2

(a) Find the sample mean $\bar{x}$, variance s^2, and standard deviation s for the data.

(b) Find the median M, the quartiles Q_1 and Q_3, the 5-number summary $[L, Q_1, M, Q_2, H]$, the range, and the IQR (interquartile range) of the data.

(a) Extend the frequency table to include the cumulative distribution cf function; products $f_i x_i$, x_i^2, $f_i x_i^2$; and the sums Σf_i, $\Sigma f_i x_i$, and $\Sigma f_i x_i^2$ as follows:

x	1	2	3	4	5	6	
f	2	4	6	8	3	2	25
cf	2	6	12	20	23	25	
fx	2	8	18	32	15	12	87
x^2	1	4	9	16	25	36	
fx^2	2	16	54	128	75	72	347

Therefore
$$\bar{x} = \frac{\Sigma f_i x_i}{\Sigma f_i} = \frac{87}{25} = 3.48$$

Also $\quad s^2 = \dfrac{\Sigma f_i x_i^2 - (\Sigma f_i x_i)^2/n}{n-1} = \dfrac{347 - (87)^2/25}{24} = \dfrac{44.24}{24} = 1.84 \quad$ and $\quad s = \sqrt{1.84} = 1.36$

(b) Here $n = 25$ is odd; hence the median M is the thirteenth number. The row cf in the table tells us that $M = 4$. The mean $M = 4$ divides the 25 numbers into two halves, each with 12 numbers. The first quartile Q_1 is the median of the first 12 number, so Q_1 is the average of the sixth number 2 and the seventh number 3. Thus, $Q_1 = 2.5$. The third quartile Q_3 is the median of the last 12 numbers, the fourteenth to twenty-fifth numbers, so Q_3 is the average of the nineteenth number 4 and twentieth number 4. Thus, $Q_3 = 4$. Furthermore, $L = 1$ is the lowest number and $H = 6$ is the highest number. Thus, the 5-number summary of the data is as follows:

$$[L, Q_1, M, Q_2, H] = [1, 2.5, 4, 4, 6]$$

Furthermore: $\quad$ range $= H - L = 7 - 1 = 6 \quad$ and $\quad$ IQR $= Q_3 - Q_1 = 4 - 2 = 2$

MISCELLANEOUS PROBLEMS INVOLVING ONE VARIABLE

A.9. An English class for foreign students consists of 20 French students, 25 Italian students, and 15 Spanish students. On an exam, the French students average 78, the Italian students 75, and the Spanish students 76. Find the mean grade for the class.

Here we use Formula (A-5) for the grand mean (the weighted mean of the means) with

$$n_1 = 20 \qquad n_2 = 25 \qquad n_3 = 15 \qquad x_1 = 78 \qquad x_2 = 75 \qquad x_3 = 76$$

This yields
$$\bar{\bar{x}} = \frac{\Sigma n_i \bar{x}_i}{\Sigma n_i} = \frac{20(78) + 25(75) + 15(76)}{20 + 25 + 15} = \frac{4575}{60} = 76.25$$

That is, 76.25 is the mean grade for the class.

A.10. A history class contains 10 freshmen, 15 sophomores, 10 juniors, and 5 seniors. On an exam, the freshmen average 72, the sophomores 76, the juniors 78, and the seniors 80. Find the mean grade for the class.

Here we use Formula (A-5) for the grand mean with

$$n_1 = 10, \quad n_2 = 15, \quad n_3 = 10, \quad n_4 = 5, \quad x_1 = 72, \quad x_2 = 76, \quad x_3 = 78, \quad x_4 = 80$$

Therefore:
$$\bar{\bar{x}} = \frac{\Sigma n_i \bar{x}_i}{\Sigma n_i} = \frac{10(72) + 15(76) + 10(78) + 5(80)}{10 + 15 + 10 + 5} = \frac{3040}{40} = 76$$

That is, 76 is the mean grade for the class.

BIVARIATE DATA

A.11. Consider data sets whose scatterplots appear in Fig. A-11. Estimate the correlation coefficient r for each data set if the choice is one of -1.5, -0.9, 0.0, 0.9, 1.5.

     The correlation coefficient r must lie in the interval $[-1, 1]$. Moreover, r is close to 1 if the data are approximately linear with positive slope, r is close to -1 if the data are approximately linear with negative slope, and r is close to 0 if there is no relationship between the points. Accordingly:

(a) r is close to 1 since there appears to be a strong linear relationship between the points with positive slope; hence $r \approx 0.9$.

(b) $r \approx 0.0$ since there appears to be no relationship between the points.

(c) r is close to -1 since there appears to be a strong linear relationship between the points but with negative slope; hence $r \approx -0.9$.

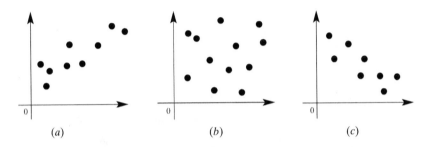

Fig. A-11

A.12. Consider the following list of data values:

x	4	2	10	5	8
y	8	12	4	10	2

(a) Plot the data in a scatterplot.

(b) Compute the correlation coefficient r.

(c) For the x and y values, find the means $\bar{x}$ and $\bar{y}$, and standard deviations s_x and s_y.

(d) Find L, the least-squares line $y = a + bx$.

(e) Graph L on the scatterplot in part (a).

(a)   The scatterplot (with L) is shown in Fig. A-12(a).

(b)   Construct the following table which contains the x, y, x^2, y^2, and xy values and where the last column gives the corresponding sums:

x	4	2	10	5	8	29
y	8	12	4	10	2	36
x^2	16	4	100	25	64	209
y^2	64	144	16	100	4	328
xy	32	24	40	50	16	162

Now use Formula $(A\text{-}11)$ and the number of points $n = 5$ to obtain

$$r = \frac{162 - [(29)(36)]/5}{\sqrt{209 - (29)^2/5}\,\sqrt{328 - (36)^2/5}} = \frac{-46.8}{\sqrt{40.8}\,\sqrt{68.8}} = -0.883\,3$$

(The fact that r is close to -1 is expected since the scatterplot indicates a strong linear relationship with negative slope.)

(c)   Use the above table and Formula (*A-1a*) to obtain

$$\bar{x} = \Sigma x_i/n = 29/5 = 5.8 \qquad \text{and} \qquad \bar{y} = \Sigma y_i/n = 36/5 = 7.2$$

Also, by Formulas (*A-8a*) and (*A-7*),

$$s_x = \sqrt{\frac{209 - (29)^2/5}{4}} = 3.194 \qquad \text{and} \qquad s_y = \sqrt{\frac{328 - (36)^2/5}{4}} = 4.147$$

(d)   Substitute r, s_x, s_y into Formula (*A-13*) to obtain the slope b of the least-squares line L:

$$b = \frac{rs_y}{s_x} = \frac{(-0.883\,3)(4.147)}{3.194} = -1.147$$

Now substitute $\bar{x}$, $\bar{y}$, and b into Formula (*A-13*) to determine the y intercept a of L:

$$a = \bar{y} - b\bar{x} = 7.2 - (-1.147)(5.8) = 13.85$$

Hence L is

$$y = 13.85 - 1.147x$$

Alternately, we can find a and b using the normal equations in Formula (*A-12*) with $n = 5$:

$$\begin{array}{ll} na + \Sigma xb = \Sigma y & \qquad 5a + 29b = 36 \\ \Sigma xa + \Sigma x^2 b = \Sigma xy & \text{or} \quad 29a + 209b = 162 \end{array}$$

(These equations would be used if we did not also want r, $\bar{x}$ and $\bar{y}$, and s_x and s_y.)

(e)   To graph L, we find two points on L and draw the line through them.   One of the two points is

$$(\bar{x}, \bar{y}) = (5.8, 7.2)$$

(which always lies on any least-squares line).   Another point is $(10, 2.4)$, which is obtained by substituting $x = 10$ in the regression equation L and solving for y.   The line L appears in the scatterplot in Fig. A-12(*a*).

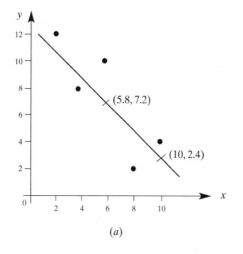

(a)

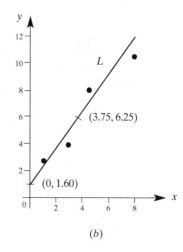

(b)

Fig. A-12

A.13. Repeat Problem A.12 for the following data:

x	1	3	4	7
y	3	4	8	10

(a) The scatterplot (with L) is shown in Fig. A-12(b).

(b) Construct the following table which contains the x, y, x^2, y^2, and xy values and where the last column give the corresponding sums:

x	1	3	4	7	15
y	3	4	8	10	25
x^2	1	9	16	49	75
y^2	9	16	64	100	289
xy	3	12	32	70	117

Now use Formula (A-11) and the number of points $n = 4$ to obtain

$$r = \frac{117 - [(15)(25)]/4}{\sqrt{75 - (15)^2/4}\sqrt{189 - (25)^2/4}} = \frac{23.25}{\sqrt{18.75}\sqrt{32.75}} = 0.938\,2$$

(The fact that r is close to $+1$ is expected since the scatterplot indicates a strong linear relationship with positive slope.)

(c) Use the above table and Formula (A-$1a$) to obtain

$$\bar{x} = \Sigma x_i/n = 15/4 = 3.75 \qquad \text{and} \qquad \bar{y} = \Sigma y_i/n = 25/4 = 6.25$$

Also, by Formulas (A-$8a$) and (A-7):

$$s_x = \sqrt{\frac{75 - (15)^2/4}{3}} = 2.5 \qquad \text{and} \qquad s_y = \sqrt{\frac{189 - (25)^2/4}{3}} = 3.304$$

(d) Substitute r, s_x, s_y into Formula (A-13) to obtain the slope b of the least-squares line L:

$$b = \frac{rs_y}{s_x} = \frac{(0.967\,5)(4.03)}{2.5} = 1.24$$

Now substitute $\bar{x}$, $\bar{y}$, and b into Formula (A-13) to determine the y intercept a of L:

$$a = \bar{y} - b\bar{x} = 6.25 - (1.24)(3.75) = 1.60$$

Hence L is

$$y = 1.60 + 1.24x$$

Alternately, we can find a and b using the normal equations in Formula (A-12) with $n = 4$:

$$\begin{array}{ll} na + \Sigma xb = \Sigma y & 4a + 15b = 25 \\ \Sigma xa + \Sigma x^2 b = \Sigma xy & \text{or} \quad 15a + 75b = 117 \end{array}$$

(These equations would be used if we did not also want r, $\bar{x}$ and $\bar{y}$, and s_x and s_y.)

(e) To graph L, we find two points on L and draw the line through them. One point is $(\bar{x}, \bar{y}) = (3.75, 6.25)$. Another point is $(0, 1.60)$, the y intercept. The line L appears on the scatterplot in Fig. A-12(b).

A.14. The definition of the *sample covariance* s_{xy} of variables x and y follows:

$$s_{xy} = \frac{\Sigma(x_i - \bar{x})(y_i - \bar{y})}{n - 1}$$

Find s_{xy} for the data in: (*a*) Problem A.12, (*b*) Problem A.13.

(*a*) The above formula for s_{xy} yields

$$\begin{aligned}
s_{xy} &= [(4 - 5.8)(8 - 7.2) + (2 - 5.8)(12 - 7.2) + (10 - 5.8)(4 - 7.2) + (5 - 5.8)(10 - 7.2) \\
&\quad + (8 - 5.8)(2 - 7.2)] \\
&= [-1.44 - 18.24 - 13.44 - 2.24 - 11.44]/4 = -46.8/4 = -11.7
\end{aligned}$$

We note that the variances s_x and s_y are always nonnegative but the covariance s_{xy} can be negative, which indicates that y tends to decrease as x increases.

(*b*) The above formula for s_{xy} yields

$$\begin{aligned}
s_{xy} &= [(1 - 3.75)(3 - 6.25) + (3 - 3.75)(4 - 6.75)(4 - 3.75)(6 - 8.25) + (7 - 3.75)(10 - 6.25)]/3 \\
&= [8.937\,5 + 2.062\,5 - 0.562\,5 + 12.187\,5]/3 = 22.625/3 = 7.542
\end{aligned}$$

The covariance here is positive which indicates that y tends to increase as x increases.

A.15. Let W denote the number of American women graduating with a doctoral degree in mathematics in a given year. Suppose that, for certain years, W has the following values:

Year	1985	1990	1995	2000
W	28	36	40	45

We assume that the increase, year by year, is approximately linear and that it will increase linearly in the near future. Estimate W for the years 2005, 2008, and 2010.

Our estimation will use a least-squares line L. For notational and computational convenience we let the year 1980 be a base for our x values. Hence we set

$$x = \text{year} - 1980 \quad \text{and} \quad y = \text{number } W \text{ of women getting doctoral degrees}$$

Thus, we seek the line $y = a + bx$ of best fit for the data where the unknowns a and b will be determined by the following *normal equations (A-12)*:

$$na + (\Sigma x)b = \Sigma y \qquad (\Sigma x)a + (\Sigma x^2)b = \Sigma xy$$

[We do not use Formula (*A-13*) for a and b since we do not need the correlation coefficient r nor do we need the values s_x, s_y, $\bar{x}$, and $\bar{y}$.]

The sums in the above system are obtained by constructing the following table which contains the x, y, x^2, and xy values and where the last column gives the corresponding sums:

x	5	10	15	20	50
y	28	36	40	45	149
x^2	25	100	225	400	750
xy	140	360	600	900	2000

Substitution in the above normal equations, with $n = 4$, yields

$$\begin{array}{lcl}
4a + 50b = 149 & & 4a + 50b = 149 \\
& \text{or} & \\
50a + 750b = 2000 & & a + 15b = 40
\end{array}$$

The solution of the system is $a = 23.5$ and $b = 1.1$. Thus, the following is our least-square line L:

$$y = 23.5 + 1.1x \qquad\qquad (A\text{-}14)$$

The (x, y) points and the line L are plotted in Fig. A-13(*a*).

Substitute 25 (2005), 28 (2008), and 30 (2010) for x in Formula (A-14) to obtain 51, 54.3, and 56.5, respectively. Thus, one would expect that, approximately, $W = 51$, $W = 54$, and $W = 57$ women will receive doctoral degrees in the years 2005, 2008, and 2010, respectively.

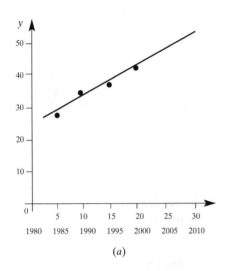

(a)

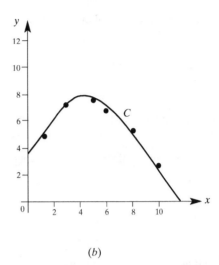

(b)

Fig. A-13

A.16. Find the least-square parabola C for the following data:

x	1	3	5	6	9	10
y	5	7	8	7	5	3

Plot C and the data points in the plane $\mathbf{R}^2$.

The parabola C has the form $y = a + bx + cx^2$ where the unknowns a, b, c are obtained from the following normal equations [which are analogous to the normal equations for the least-square line L in Formula (A-12)]:

$$na + (\Sigma x)b + (\Sigma x^2)c = \Sigma y$$
$$(\Sigma x)a + (\Sigma x^2)b + (\Sigma x^3)c = \Sigma xy$$
$$(\Sigma x^2)a + (\Sigma x^3)b + (\Sigma x^4)c = \Sigma x^2 y$$

The sums in the above system are obtained by constructing the following table which contains the x, y, x^2, x^3, x^4, xy, and $x^2 y$ values and where the last column gives the corresponding sums:

x	1	3	5	6	9	10	34
y	5	7	8	7	5	3	35
x^2	1	9	25	36	81	100	252
x^3	1	27	125	216	729	1,000	2,098
x^4	1	81	625	1,296	6,561	10,000	18,564
xy	5	21	40	42	45	30	183
$x^2 y$	5	63	200	252	405	300	1,225

Substitution in the above normal equations, with $n = 6$, yields

$$6a + 34b + 252c = 35, \qquad 34a + 252b + 2098c = 183, \qquad 252a + 2098b + 18{,}564c = 1225$$

The solution of the system yields

$$a = \frac{12,845}{3,687} = 3.48 \qquad b = \frac{4,179}{2,458} = 1.70 \qquad c = -\frac{1,279}{7,374} = -0.173$$

Thus, the required parabola C follows:

$$y = 3.48 + 1.70x - 0.173x^2$$

The given data points and C are plotted in Fig. A-13(b).

A.17. Derive the normal equations Formula (A-12) for the least-squares line L for n data points $P_i(x_i, y_i)$.

We want to minimize the following least-square error:

$$D = \Sigma d_i^2 = \Sigma [y_i - (a + bx_i)]^2 = \Sigma [a + bx_i - y_i]^2$$

where D may be viewed as a function of a and b. The minimum may be obtained by setting the partial derivatives D_a and D_b, equal to zero. The partial derivatives follow:

$$D_a = \Sigma 2(a + bx_i - y_i) \qquad \text{and} \qquad D_b = \Sigma 2(a + bx_i - y_i)x_i$$

Setting $D_a = 0$ and $D_b = 0$, we obtain the following required equations:

$$na + (\Sigma x_i)b = \Sigma y_i \qquad (\Sigma x_i)a + (\Sigma x_i^2)b = \Sigma x_i y_i$$

Supplementary Problems

FREQUENCY DISTRIBUTIONS, MEAN AND MEDIAN

A.18. The frequency distribution of the weekly wages, in dollars, of a group of unskilled workers follows:

Weekly wages	140–160	160–180	180–200	200–220	220–240	240–260	260–280
Workers	18	24	32	20	8	6	2

(a) Display the data in a histogram and a frequency polygon.

(b) Find the mean $\bar{x}$, median M, and midrange of the data.

A.19. The amounts of 45 personal loans from a loan company follow:

$700	$450	$725	$1125	$675	$1650	$750	$400	$1050
$500	$750	$850	$1250	$725	$475	$925	$1050	$925
$850	$625	$900	$1750	$700	$825	$550	$925	$850
$475	$750	$550	$725	$575	$575	$1450	$700	$450
$700	$1650	$925	$500	$675	$1300	$1125	$775	$850

(a) Group the data into classes with class width $w = \$200$ and beginning with $400, and construct the frequency and cumulative frequency distribution for the grouped data.

(b) Display the frequency distribution in a histogram.

(c) Find the mean $\bar{x}$, median M, and midrange of the data.

A.20. The daily number of station wagons rented by an automobile rental agency during a 30-day period follows:

$$7 \quad 10 \quad 6 \quad 7 \quad 9 \quad 4 \quad 7 \quad 9 \quad 9 \quad 8 \quad 5 \quad 5 \quad 7 \quad 8 \quad 4$$
$$6 \quad 9 \quad 7 \quad 12 \quad 7 \quad 9 \quad 10 \quad 4 \quad 7 \quad 5 \quad 9 \quad 8 \quad 9 \quad 5 \quad 7$$

(a) Construct the frequency and cumulative frequency distribution for the data.

(b) Find the mean $\bar{x}$, median M, and midrange of the data.

A.21. The following denotes the number of people living in each of 35 apartments:

$$1 \quad 1 \quad 1 \quad 1 \quad 1 \quad 1 \quad 1 \quad 2 \quad 2 \quad 2 \quad 2 \quad 2 \quad 2 \quad 2 \quad 2 \quad 2 \quad 2 \quad 2 \quad 2 \quad 2$$
$$3 \quad 3 \quad 3 \quad 3 \quad 4 \quad 4 \quad 4 \quad 4 \quad 4 \quad 5 \quad 5 \quad 5 \quad 6 \quad 6 \quad 7$$

(a) Construct the frequency and cumulative frequency distribution for the data.

(b) Find the mean $\bar{x}$, median M, and midrange of the data.

A.22. The students in a mathematics class are divided into four groups:

(a) much greater than the median,   (c) little below the median,

(b) little above the median,   (d) much below the median.

On which group should the teacher concentrate in order to increase the median of the class?   Mean of the class?

MEASURES OF DISPERSION: VARIANCE, STANDARD DEVIATION, IQR

A.23. The prices of 1 lb of coffee in 7 stores follow:

$$\$5.58, \quad \$6.18, \quad \$5.84, \quad \$5.75, \quad \$5.67, \quad \$5.95, \quad \$5.62$$

(a) Find the mean $\bar{x}$, variance s^2, and standard deviation s.

(b) Find the median M, 5-number summary, and IQR of the data.

A.24. For a given week, the following were the average daily temperatures:

$$35°F, \quad 33°F, \quad 30°F, \quad 36°F, \quad 40°F, \quad 37°F, \quad 38°F$$

(a) Find the mean $\bar{x}$, variance s^2, and standard deviation s.

(b) Find the median M, 5-number summary, and IQR of the data.

A.25. During a given month, the 10 salespeople in an automobile dealership sold the following number of automobiles:

$$13, \quad 17, \quad 10, \quad 18, \quad 17, \quad 9, \quad 17, \quad 13, \quad 15, \quad 14$$

(a) Find the mean $\bar{x}$, variance s^2, and standard deviation s.

(b) Find the median M, 5-number summary, and IQR of the data.

A.26. The ages of students at a college dormitory are recorded, producing the following frequency distribution:

Age x	17	18	19	20	21
Frequency f	5	20	17	6	2

(a) Find the sample mean $\bar{x}$ and standard deviation s.

(b) Find the median M, 5-number summary, and IQR of the data.

A.27. The following distribution gives the number of hours of overtime during 1 month for employees of a company:

Overtime, hours	0	1	2	3	4	5	6	7	8	9	10
Employees	10	2	4	2	6	4	2	4	6	2	8

(a) Find the sample mean $\bar{x}$ and standard deviation s.

(b) Find the median M, 5-number summary, and IQR.

A.28. The following are 40 test scores:

$$52 \quad 55 \quad 58 \quad 58 \quad 60 \quad 61 \quad 64 \quad 66 \quad 66 \quad 68 \quad 72 \quad 75 \quad 75 \quad 75 \quad 76 \quad 76 \quad 77 \quad 77 \quad 78 \quad 78$$
$$80 \quad 80 \quad 81 \quad 82 \quad 82 \quad 84 \quad 85 \quad 85 \quad 85 \quad 86 \quad 88 \quad 90 \quad 92 \quad 95 \quad 95 \quad 95 \quad 100 \quad 100 \quad 100 \quad 100$$

(a) Group the data into 5 classes with class width $w = 10$ beginning with 50 and construct the frequency and cumulative frequency distribution for the grouped data.

(b) Find the sample mean $\bar{x}$ and standard deviation s of the grouped data.

(c) Find the median M, 5-number summary, and IQR of the original data.

(d) Find the median M, 5-number summary, and IQR of the grouped data.

A.29. The following distribution gives the number of visits for medical care by 80 patients during a 1-year period:

Number of visits x	0	1	2	3	4	6	8
Number of patients f	14	21	8	15	7	10	5

(a) Find the sample mean $\bar{x}$ and standard deviation s.

(b) Find the median M, 5-number summary, and IQR.

MISCELLANEOUS PROBLEMS INVOLVING ONE VARIABLE

A.30. The students at a small school are divided into 4 groups: A, B, C, D. The number n of students in each group and the mean score $\bar{x}$ of each group follow:

$$A: n = 80, \bar{x} = 78; \qquad B: n = 60, \bar{x} = 74; \qquad C: n = 85, \bar{x} = 77; \qquad D: n = 75, \bar{x} = 80$$

Find the mean of the school.

A.31. The *mode* of a list of numerical data is the value which occurs most often and more than once. Find the mode of the data in Problems: (a) A.20, (b) A.21, (c) A.26, (d) A.27.

BIVARIATE DATA

A.32. Consider the following list of data values:

x	3	1	6	3	4
y	7	2	14	8	10

(a) Draw a scatterplot of the data.

(b) Compute the correlation coefficient r. [*Hint*: First find Σx_i, Σy_i, Σx_i^2, Σy_i^2, $\Sigma x_i y_i$ and then use Formula (*A-11*).]

(c)   For the x and y values, find the means $\bar{x}$ and $\bar{y}$ and standard deviations s_x and s_y.

(d)   Find L, the least-squares line $y = a + bx$.

(e)   Graph L on the scatterplot in part (a).

A.33.   Repeat Problem A.32 for the following list of data values:

x	1	2	4	6
y	5	4	3	1

A.34.   Find the covariance s_{xy} of the variables x and y in: (a) Problem A.32, (b) Problem A.33. (See Problem A.14 for the definition of s_{xy}.)

A.35.   Suppose 7 people in a company are interviewed, yielding the following data where x is the number of years of service and y is the number of people who reviewed the work of the person:

x	2	3	3	5	6	6	8
y	15	14	13	11	10	9	7

(a)   Draw a scatterplot of the data.

(b)   Find L, the least-squares line $y = a + bx$.

(c)   Graph L on the scatterplot in part (a).

(d)   Predict the number y of people who reviewed the work of another person if the number of years worked by the person is: (i) $x = 1$, (ii) $x = 7$, (iii) $x = 9$.

A.36.   Consider the following bivariate data:

x	0.2	0.4	0.9	1.2	3.0
y	3.0	1.0	0.5	0.4	0.2

(a)   Find the correlation coefficient r.   [*Hint*: First find Σx_i, Σy_i, Σx_i^2, Σy_i^2, $\Sigma x_i y_i$ and then use Formula $(A\text{-}11)$.]

(b)   Plot x against y in a scatterplot.

(c)   Find the least-squares line L for the data and graph L on the scatterplot in (b).

(d)   Find the least-squares hyperbolic curve C which has the form $y = 1/(a + bx)$ or $1/y = a + bx$ and plot C on the scatterplot in (b).   [*Hint*: Find the least-squares line for the data points $(x_i, 1/y_i)$.]

(e)   Which curve, L or C, best fits the data?

A.37.   The following table lists average male weight, in pounds, and height, in inches, for certain ages which range from 1 to 21:

Age	1	3	6	10	13	16	21
Weight	20	30	45	60	95	140	155
Height	28	36	44	50	60	66	70

Find the correlation coefficient r for: (a) age and weight, (b) age and height, (c) weight and height.

A.38.   Let $x =$ age, $y =$ height in Problem A.37. (a) Plot x against y in a scatterplot. (b) Find the line L of best fit. (c) Graph L on the scatterplot in part (a).

A.39. Let x = weight, y = height in Problem A.37.   (a) Plot x against y in a scatterplot.   (b) Find the line L of best fit.   (c) Graph L on the scatterplot in part (a).

A.40. Find the least-squares exponential curve $y = ab^x$ for the following data:

x	1	2	3	4	5	6
y	6	12	24	50	95	190

Answers to Supplementary Problems

A.18.   (a) See Fig. A-14(a); (b) $\bar{x}$ = \$190.36, M = \$190, mid = \$210.

A.19.   (a)   The frequency distribution (where the wage is divided by \$100 for notational convenience) follows:

Amount/100	4–6	6–8	8–10	10–12	12–14	14–16	16–18
Number of loans	11	14	10	4	2	1	3

   (b)   The histogram is shown in Fig. A-14(b).

   (c) $\bar{x}$ = \$842.22, M = \$700, mid = \$1100.

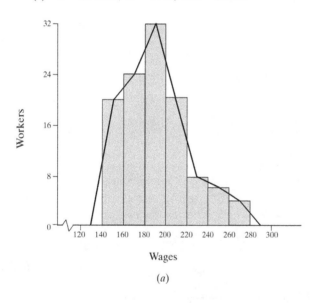

Wages

(a)

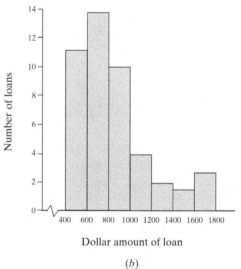

Dollar amount of loan

(b)

Fig. A-14

A.20.   (a)   The distributions follow:

Daily number of wagons	4	5	6	7	8	9	10	11	12
Frequency	3	4	2	8	3	7	2	0	1
Cumulative frequency	3	7	9	17	20	27	29	29	30

   (b) $\bar{x}$ = 7.3, M = 7, mid = 8.

A.21. (a) The frequency and cumulative frequency distributions follow:

Number of people	1	2	3	4	5	6	7
Frequency	7	13	4	5	3	2	1
Cumulative frequency	7	20	24	29	32	34	35

(b) $\bar{x} = 2.8$, $M = 2$, mid $= 4$.

A.22. Group (c) to increase the median; likely (b) and (c) to increase the mean.

A.23. (a) $\bar{x} = \$5.80$, $s^2 = 0.021\,8$, $s \approx 0.15$; (b) $M = \$5.75$, [5.58, 5.62, 5.75, 5.95, 6.18], IQR $= \$0.33$.

A.24. (a) $\bar{x} = 35.67$, $s^2 = 2.37$, $s \approx 1.54$; (b) $M = 36.5$, [30, 33, 36.5, 38, 40], IQR $= 5$.

A.25. (a) $\bar{x} = 14.3$, $s^2 = 9.57$, $s \approx 3.1$; (b) $M = 14.5$, [9, 13, 14.5, 17, 18], IQR $= 4$.

A.26. (a) $\bar{x} = 18.6$, $s^2 = 0.939$, $s \approx 0.97$; (b) $M = 18.5$, [17, 18, 18.5, 19, 21], IQR $= 1$.

A.27. (a) $\bar{x} = 4.92$, $s^2 = 12.97h^2$, $s \approx 3.60h$; (b) $M = 5$, [0, 2, 5, 8, 10], IQR $= 6$.

A.28. (a) The distributions with class values follow:

Scores	50–60	60–70	70–80	80–90	90–100
Frequency	4	6	10	11	9
Cumulative frequency	4	10	20	31	40
Class value	55	65	75	85	95

Remark: The scores 100 are put in the 90–100 group since there are no scores higher than 100. If there were scores higher than 100, then the scores 100 would be put the next higher 100–110 group.

(b) $\bar{x} = 78.8$, $s^2 = 222.93$, $s \approx 14.9$.
(c) $M = 79$, [52, 69, 79, 87, 100], IQR $= 18$.
(d) $M = 80$, [50, 70, 80, 85, 100], IQR $= 15$.

A.29. (a) $\bar{x} = 2.625$, $s^2 = 5.43$, $s \approx 2.3$; (b) $M = 2$, [0, 1, 2, 4, 8], IQR $= 3$.

A.30. $\bar{\bar{x}} = 77.42$.

A.31. (a) 7; (b) 2; (c) 18; (d) 0.

A.32. (a) See Fig. A-15(a).

(b) $\Sigma x_i = 17$, $\Sigma y_i = 41$, $\Sigma x_i^2 = 71$, $\Sigma y_i^2 = 413$, $\Sigma x_i y_i = 171$, $r = \dfrac{31.6}{\sqrt{13.2}\,\sqrt{76.8}} = 0.99$.

(c) $\bar{x} = 3.5$, $\bar{y} = 8.2$, $s_x^2 = 3.30$, $s_x = 1.82$, $s_y^2 = 19.2$, $s_y = 4.38$.

(d) $y = -0.13 + 2.38x$.

A.33. (a) See Fig. A-15(b).

(b) $\Sigma x_i = 13$, $\Sigma y_i = 13$, $\Sigma x_i^2 = 57$, $\Sigma y_i^2 = 51$, $\Sigma x_i y_i = 31$, $r = -\dfrac{11.25}{\sqrt{14.75}\,\sqrt{8.75}} = -0.99$.

(c) $\bar{x} = 3.25$, $\bar{y} = 3.25$, $s_x^2 = 4.92$, $s_x = 2.22$, $s_y^2 = 2.92$, $s_y = 1.71$.

(d) $y = 5.75 - 0.77x$.

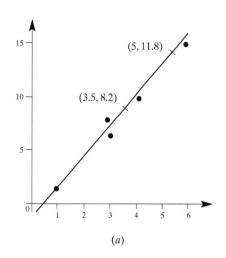 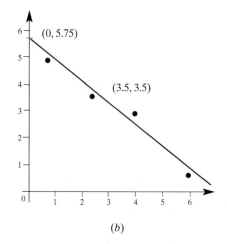

(a)   (b)

Fig. A-15

A.34. (a) $s_{xy} = (15.5 + 0.60 + 0.10 + 0.90 + 14.5)/4 = 7.90$.

   (b) $s_{xy} = (-3.937\,5 - 0.937\,5 - 0.187\,5 - 6.187\,5)/3 = -3.75$.

A.35. (a) and (c) See Fig. A-16(a); (b) $y = 18.5 - 1.5x$; (d) (i) 17, (ii) 8, (iii) 5.

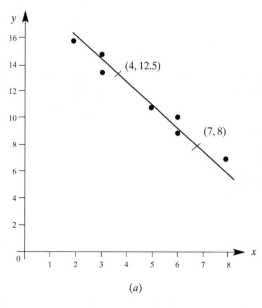 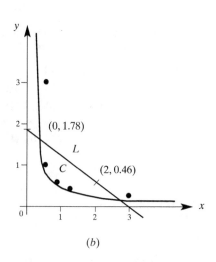

(a)   (b)

Fig. A-16

A.36. (a) $\Sigma x_i = 5.7$, $\Sigma y_i = 5.1$, $\Sigma x_i^2 = 11.45$, $\Sigma y_i^2 = 10.45$, $\Sigma x_i y_i = 2.53$, $r = -\dfrac{3.284}{\sqrt{4.952}\,\sqrt{5.248}} = -0.644$.

   (b) See Fig. A-16(b).

   (c) $y = 1.78 - 0.66x$ and Fig. A-16(b).

   (d) $y = 1/1.6x$ and Fig. A-16(b).

   (e) C is a better fit.

A.37.   (*a*) $r = 0.98$; (*b*) $r = 0.98$; (*c*) $r = 0.97$.

A.38.   (*a*) and (*c*) See Fig. A-17(*a*); (*b*) $y = 29.22 + 2.13x$.

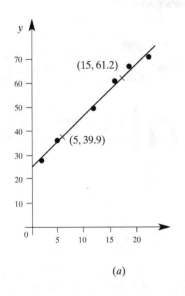

(*a*)

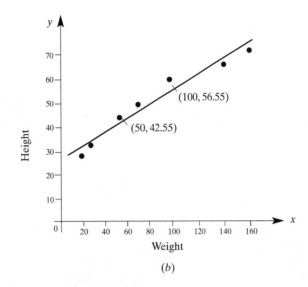

(*b*)

Fig. A-17

A.39.   (*a*) and (*c*) See Fig. A-17(*b*); (*b*) $y = 28.55 + 0.28x$.

A.40. $y = 3(2^x)$.

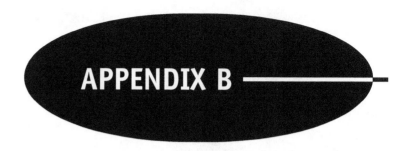

APPENDIX B

Chi-Square Distribution

B.1 INTRODUCTION

One fundamental question in probability and statistical analysis is whether or not a pattern of observed data fits a given distribution such as a uniform, binomial, or normal distribution or some prior distribution. Clearly, the data would not fit the distribution exactly, so we would want to have some criteria of "goodness of fit". The chi-square distribution, denoted by χ^2 and defined below, gives such criteria. (Here χ is the Greek letter chi.)

The chi-square distribution is also used to decide whether or not certain variables are independent. For example, a pollster might want to know whether or not, say, the sex, ethnic background, or salary range of a person is a factor in his or her vote in an election or for some type of legislation.

The formal definition of the χ^2 distribution follows.

Definition: Let $Z_1, Z_2, \ldots, Z_k$ be k independent standard normal distributions. Then

$$\chi^2 = Z_1^2 + Z_2^2 + \cdots + Z_k^2$$

is called the *chi-square distribution* with *k degrees of freedom*.

The number k of degrees of freedom, which can be any positive integer including 1, is frequently denoted by "df". Thus, there is a χ^2 distribution for each k. Figure B-1 pictures the distribution for $k = 1, 4, 6, 8$. The distribution is not symmetric and is skewed to the right. However, for large k the distribution is close to the normal distribution.

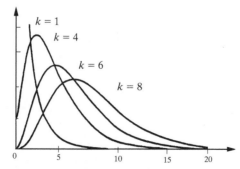

Fig. B-1. Chi-square distribution for *k* degrees of freedom.

282

B.2   GOODNESS OF FIT, NULL HYPOTHESIS, CRITICAL VALUES

Suppose a collection of data, say, given by a frequency distribution with n categories, is obtained from a sample size exceeding 30.   Moreover, suppose we want to decide by some test whether or not the data fit some specific distribution.   Let H_0 denote the assumption that it does, that is:

H_0: Hypothesis that the data fits a given distribution.

Here H_0 is called the *null hypothesis*.

Letting "obs" denote observed data and letting "exp" denote expected data (obtained from the given distribution), the *chi-square value* or *chi-square statistic* for the given data measures the weighted squares of the differences, that is,

$$\chi^2 = \sum \frac{(\text{obs} - \text{exp})^2}{\text{exp}}$$

Assuming that the expected values are not too small (usually, not less than 5), then the above random variable has (approximately) the chi-square distribution with

$$\text{df} = n - 1$$

degrees of freedom.   The formula $\text{df} = n - 1$ comes from the fact that a given frequency distribution with n categories involves probabilities where $n - 1$ of the probabilities determines the nth probability.   (See remark below.)

Clearly, the smaller the χ^2 value, the better the fit.   However, if χ^2 is too "large", that is, if χ^2 exceeds some given *critical value c*, we say that the fit is poor, and we reject H_0.   The critical value c is determined by preassigning a *significance level* α where:

$$\alpha = \text{probability that } \chi^2 \text{ exceeds critical value } c = P(\chi^2 \geq c)$$

Frequently used choices of α are $\alpha = 0.10$, $\alpha = 0.05$, and $\alpha = 0.005$.

Table B-1 gives critical values for some commonly used significance levels.   The significance level α represents the shaded area in the graph appearing in the table.   We emphasize that if the χ^2 value exceeds the critical value c (falls in the shaded area), then we say that we reject the null hypothesis H_0 at the α significance level.

The following remarks are in order:

Remark 1:   The observed data come from a sample from a larger population, so the chi-square values form a discrete random variable.   This random variable closely approximates the continuous χ^2 distribution when the sample size exceeds 30.

Remark 2:   The χ^2 distribution also assumes that each individual expected value is not too small; one rule-of-thumb (noted above) is that no expected value is less than 5.

Remark 3:   The formula $\text{df} = n - 1$ assumes that the size is the only statistic of the sample that is used.   If additional statistics of the sample are used, such as the mean $\bar{x}$ or standard deviation s, then the degrees of freedom df will be smaller.   (See Examples B.4 and B.6.)

Table B-1   Chi-Square Distribution
(α = Probability That χ^2 Exceeds Critical Value c)

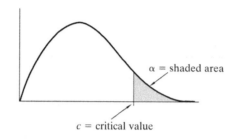

α = shaded area

c = critical value

df	α				
	0.10	0.05	0.025	0.010	0.005
1	2.71	3.84	5.02	6.63	7.88
2	4.61	5.99	7.38	9.21	10.60
3	6.25	7.81	9.35	11.34	12.84
4	7.78	9.49	11.14	13.28	14.86
5	9.24	11.07	12.83	15.09	16.75
6	10.64	12.59	14.45	16.81	18.55
7	12.02	14.07	16.01	18.48	20.28
8	13.36	15.51	17.54	20.09	21.96
9	14.68	16.92	19.02	21.67	23.59
10	16.99	18.31	20.48	23.21	25.19
11	17.28	19.68	21.92	24.72	26.76
12	18.55	21.03	23.34	26.22	28.30
13	19.81	22.36	24.74	27.69	29.82
14	21.06	23.68	26.12	29.14	31.32
15	22.31	25.00	27.49	30.58	32.80
16	23.54	26.30	28.85	32.00	34.27
17	24.77	27.59	30.19	33.41	35.72
18	25.99	28.87	31.53	34.81	37.16
19	27.20	30.14	32.85	36.19	38.58
20	28.41	31.41	34.17	37.57	40.00
25	34.38	37.65	40.65	44.31	46.93
30	40.26	43.77	46.98	50.89	53.67
40	51.81	55.76	59.34	63.69	66.77
50	63.17	67.51	71.42	76.15	79.49
100	118.50	124.30	129.60	135.80	140.20

B.3 GOODNESS OF FIT FOR UNIFORM AND PRIOR DISTRIBUTIONS

This section gives applications of the χ^2 distribution to goodness-of-fit problems involving a uniform distribution and a prior distribution.

EXAMPLE B.1 Uniform Distribution A company introduces a new product in 4 locations, A, B, C, D. The number of items sold during a weekend follow:

Location	A	B	C	D
Number of items sold	80	65	70	85

Let H_0 be the (null) hypothesis that location does not make a difference. Apply the chi-square test at the $\alpha = 0.10$ significance level (90 percent reliability) to accept or reject the null hypothesis H_0.

The total number of items sold was 300. Assuming the null hypothesis H_0 of a uniform distribution, the expected sales at each location would be 75. The χ^2 value for the data follows:

$$\chi^2 = \sum \frac{(\text{obs} - \text{exp})^2}{\text{exp}}$$

$$= \frac{(80 - 75)^2}{75} + \frac{(65 - 75)^2}{75} + \frac{(70 - 75)^2}{75} + \frac{(85 - 75)^2}{75} = 3.33$$

There are df $= 4 - 1 = 3$ degrees of freedom. This is derived from the fact that, assuming the number of items sold is 300, the sales at 3 of the locations determine the sales at the fourth location. Table B-1 shows that the critical χ^2 value for df $= 3$ at the $\alpha = 0.10$ significance level is $c = 6.25$, which is pictured in Fig. B-2. Since $3.33 < 6.25$, we accept the null hypothesis H_0 of a uniform distribution, that is, that the evidence indicates that location does not make a difference.

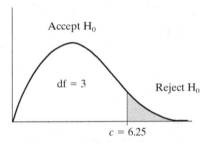

Fig. B-2

EXAMPLE B.2 Prior Distribution The following table lists the percentage of grades of a professor in a certain course for previous years and the number of such grades for 100 of a professor's students for the current year:

Grade	A	B	C	D	F
Previous years	10%	30%	40%	15%	5%
Current years	15	23	32	22	8

Consider the following null hypothesis:

H_0: Current students are typical compared to previous students.

Use a chi-square test at the $\alpha = 0.05$ significance level to accept or reject the null hypothesis H_0.

There are 100 current student grades, so the number of students also gives the percentage.   Thus the χ^2 value of the data follows:

$$\chi^2 = \sum \frac{(\text{obs} - \text{exp})^2}{\text{exp}}$$

$$= \frac{(15 - 10)^2}{10} + \frac{(23 - 30)^2}{30} + \frac{(32 - 40)^2}{40} + \frac{(22 - 15)^2}{15} + \frac{(8 - 5)^2}{5} = 10.8$$

There are df $= 5 - 1 = 4$ degrees of freedom.   This is derived from the fact that there are 100 students so that any 4 of the entries in the distribution table tell us the fifth entry.   Table B-1 shows that the critical χ^2 value for df $= 4$ at the $\alpha = 0.05$ significance level is $c = 9.45$, which is pictured in Fig. B-3.   Since $10.8 > 9.45$, we reject the null hypothesis H_0 that the current students are typical of previous students.

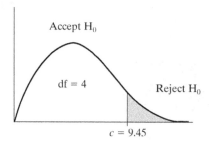

Fig. B-3

B.4   GOODNESS OF FIT FOR BINOMIAL DISTRIBUTION

This section gives applications of the χ^2 distribution to goodness-of-fit problems involving the binomial distribution.

EXAMPLE B.3   Binomial Distribution with Probability p Given   There are 4 special tourist sights A, B, C, D in a city.   A poll of 600 tourists indicated the following number of sights visited by each tourist:

Number of sights	0	1	2	3	4
Number of tourists	130	240	170	52	8

Let H_0 be the null hypothesis that the distribution is binomial with $p = 0.70$.   Test the hypothesis at the $\alpha = 0.10$ significance level.

The binomial distribution with $n = 4$ and $p = 0.7$ follows:

$$P(0) = (0.7)^4 = 0.240, \qquad P(2) = 6(0.3)^2 (0.7)^2 = 0.265, \qquad P(4) = (0.3)^4 = 0.008$$
$$P(1) = 4(0.3)(0.7)^3 = 0.412, \qquad P(3) = 4(0.3)^3 (0.7) = 0.076$$

Multiplying the probabilities by the number of 600 tourists gives the following expected data:

Number of sights	0	1	2	3	4
Expected number of tourists	144	247	159	45	5

The χ^2 value of the data follows:

$$\chi^2 = \sum \frac{(\text{obs} - \text{exp})^2}{\text{exp}} = \frac{(130 - 144)^2}{144} + \frac{(140 - 247)^2}{247}$$

$$+ \frac{(170 - 159)^2}{159} + \frac{(52 - 45)^2}{45} + \frac{(8 - 5)^2}{5} = 5.21$$

There are df $= 5 - 1 = 4$ degrees of freedom. This is derived from the fact that the 5 numbers in the table are related by the equation that their sum is 600. Thus, any 4 of the numbers determine the fifth number.

Table B-1 tells us that $c = 7.78$ is the critical value for df $= 4$ and $\alpha = 0.10$, and this relationship is pictured in Fig. B-4. Since $5.21 < 7.78$, we accept the null hypothesis H_0 that the distribution is binomial with $p = 0.70$.

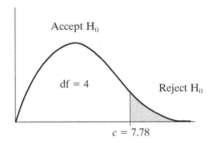

Fig. B-4

Remark: Suppose only 200 tourists were polled instead of 600. Although the sample size does satisfy the condition that it exceeds 30, the expected number of tourists visiting all 4 sights would only be 2, which is less than 5. Thus, with a sample of 200, we would not use the chi-square test to test the hypothesis that the distribution is binomial with $p = 0.7$.

EXAMPLE B.4 Binimial Distribution Using the Sample to Estimate p A factory makes light bulbs and ships them in packets of 4. Suppose 5000 packets are tested and the number of defective bulbs in each packet is recorded yielding the following distribution:

Number of defective bulbs	0	1	2	3	4
Number of packets	1975	2170	740	110	5

Let H_0 be the null hypothesis that the distribution of defective bulbs is binomial. Test the hypothesis at the $\alpha = 0.05$ significance level.

Here $n = 4$ but p is not given. Thus, we use the sample proportion $\hat{p}$ (read: p hat) of defective bulbs as an estimate of p. The number d of defective bulbs in all the packets follows:

$$d = 0(1975) + 1(2170) + 2(740) + 3(110) + 4(5) = 4000$$

The total number b of bulbs is $4(5,000) = 20,000$. Thus, we set

$$p = \hat{p} = \frac{d}{b} = \frac{4,000}{20,000} = 0.2$$

The binomial distribution with $n = 4$, $p = 0.2$ and $q = 1 - 0.2 = 0.8$ follows:

$P(0) = (0.8)^4 = 0.409\,6,$ $P(2) = 6(0.2)^2\,(0.8)^2 = 0.153\,6,$ $P(4) = (0.2)^4 = 0.001\,6$

$P(1) = 4(0.2)(0.8)^3 = 0.409\,6,$ $P(3) = 4(0.2)^3\,(0.8) = 0.025\,6,$

Multiplying the probabilities by 5000, the number of packets, yields the following expected distribution:

Number of defective bulbs	0	1	2	3	4
Expected number of packets	2048	2048	768	128	8

The χ^2 value of the data follows:

$$\chi^2 = \sum \frac{(\text{obs} - \text{exp})^2}{\text{exp}} = \frac{(1975 - 2048)^2}{2048} + \frac{(2170 - 2048)^2}{2048}$$

$$+ \frac{(740 - 768)^2}{768} + \frac{(110 - 128)^2}{128} + \frac{(5 - 8)^2}{8} = 14.5$$

Finding the number of degrees of freedom in this example is different than in the previous example. Here there are two statistics taken from the sample: (a) the size of the sample (5000 packets) and (b) the proportion $\hat{p}$ of defective bulbs (or equivalently, 4000 defective bulbs). Thus, the five entries in the frequency table are related by the following two equations:

$$x_0 + x_1 + x_2 + x_3 + x_4 = 5000 \qquad 0x_0 + 1x_1 + 2x_2 + 3x_3 + 4x_4 = 20{,}000$$

where x_k denotes the number of packets with k defective bulbs. Accordingly, there are only df = 5 − 2 = 3 degrees of freedom; that is, any three of the data entries in the table will yield the remaining two using the two equations.

Table B-1 tells us that $c = 7.81$ is the critical value for df = 3 and $\alpha = 0.05$, as pictured in Fig. B-5. Since 14.5 > 7.81, we reject the null hypothesis H_0 that the distribution is binomial.

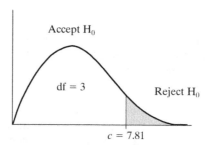

Fig. B-5

B.5 GOODNESS OF FIT FOR NORMAL DISTRIBUTION

This section gives applications of the χ^2 distribution to goodness-of-fit problems involving the normal distribution.

EXAMPLE B.5 Normal Distribution with Given μ and σ Suppose the commuting time T, in minutes, of 300 students at a college has the following distribution:

Time	<20	20–30	30–40	40–50	>50
Number of students	13	75	120	66	26

Consider the following null hypothesis:

H_0: Distribution is normal with mean $\mu = 35$ and standard deviation $\sigma = 10$.

Test the null hypothesis at the $\alpha = 0.10$ significance level.

Using the formula $z = (T - \mu)/\sigma$, we derive the following z values corresponding to the above T values:

T value	20	30	40	50
z value	−1.5	−0.5	0.5	1.5

Figure B-6 shows the normal curve with the T values, the corresponding z values, and the probability distribution for these values obtained from Table 6-1 (page 184) of the standard normal distribution.

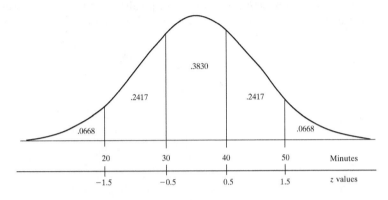

Fig. B-6

Multiplying each probability in Fig. B-6 by 300 gives the following expected numbers of students for the given time periods:

Time	<20	20–30	30–40	40–50	>50
Expected number of students	20	72.5	115	72.5	20

The χ^2 value of the data follows:

$$\chi^2 = \sum \frac{(\text{obs} - \text{exp})^2}{\text{exp}} = \frac{(13 - 20)^2}{20} + \frac{(75 - 72.5)^2}{72.5} + \frac{(120 - 115)^2}{115}$$

$$+ \frac{(66 - 72.5)^2}{72.5} + \frac{(26 - 20)^2}{20} = 5.14$$

There are df = 5 − 1 = 4 degrees of freedom. This is derived from the fact that the five numbers in the table are related by the equation that their sum is 300. Thus, any four of the numbers determine the fifth number.

Table B-1 tells us that $c = 7.78$ is the critical value for df = 4 and $\alpha = 0.10$. Since 5.14 < 7.78, we accept the null hypothesis H_0 that the distribution of commuting times are normal with $\mu = 35$ and $\sigma = 10$.

EXAMPLE B.6 Normal Distribution Using Sample for μ and σ Suppose the heights h, in inches, of a sample of 500 male students at a college have mean $\bar{x} = 66$, standard deviation $s = 4$, and the following distribution:

Height	<58	58–62	62–66	66–70	70–74	>74
Number of students	7	72	162	176	65	18

Consider the following null hypothesis:

H_0: Distribution is normal with $\mu = \bar{x} = 66$ and $\sigma = s = 4$.

We emphasize that, unlike the previous example, the mean μ and standard deviation σ of the normal distribution is not given but is estimated from the sample.   Test the null hypothesis H_0 at the: (a) $\alpha = 0.10$ significance level, (b) $\alpha = 0.05$ significance level.

Using the formula $z = (h - \mu)/\sigma$, we derive the following z values corresponding to the above h values:

h value	58	62	66	70	74
z value	-2	-1	0	1	2

Figure B-7 shows the normal curve with the h values, the corresponding z values, and the probability distribution for these values obtained from Table 6-1 of the standard normal distribution.

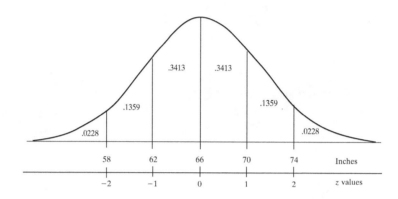

Fig. B-7

Multiplying each probability in Fig. B-7 by 500 gives the following expected numbers of students for the given height ranges:

Height	<58	58–62	62–66	66–70	70–74	>74
Expected number of students	11	68	171	171	68	11

The χ^2 value of the data follows:

$$\chi^2 = \sum \frac{(\text{obs} - \text{exp})^2}{\text{exp}} = \frac{(7 - 11)^2}{11} + \frac{(72 - 68)^2}{68} + \frac{(162 - 171)^2}{171}$$

$$+ \frac{(176 - 171)^2}{171} + \frac{(65 - 68)^2}{68} + \frac{(18 - 11)^2}{11} = 6.89$$

Finding the number of degrees of freedom in this example is different than in the previous example.   Here there are three statistics taken from the sample: the size, the mean, and the standard deviation.   Each statistic yields an equation relating the six numbers in the frequency table: the sum is 500, the mean is 66, and the standard deviation is 4—and the three equations are independent.   Thus, any three of the six data entries in the table will yield the remaining three data entries using the three equations.   Accordingly, in this example, there are only df $= 6 - 3 = 3$ degrees of freedom, not 5 as in the previous example.

(a)   Table B-1 tells us that $c = 6.25$ is the critical value for df $= 3$ and $\alpha = 0.10$.   Since $6.89 > 6.25$, we reject the null hypothesis H_0 that the distribution of heights is normal at the $\alpha = 10\%$ significance level.

(b)   Table B-1 tells us that $c = 7.81$ is the critical value for df $= 3$ and $\alpha = 0.05$.   Since $6.89 < 7.81$, we accept the null hypothesis H_0 that the distribution of heights is normal at the $\alpha = 5\%$ significance level.

B.6 CHI-SQUARE TEST FOR INDEPENDENCE

This section gives applications of the χ^2 distribution to problems involving the independence of various attributes.   For example, one may want to test whether or not there is a "gender gap" (or "age gap") in an election, that is, whether the vote for a given candidate or for some piece of legislation does or does not depend on the gender (or age) of the voter.

Since the chi-square test is not accurate for small values, we will assume, as before, that our sample exceeds 30 and that no expected frequency is less than 5.

EXAMPLE B.7   An engineering college has 4 programs:

(i)   electrical,   (ii)   chemical,   (iii)   mechanical,   (iv)   civil

Suppose 500 students, of which 300 are male and 200 are female, are distributed in the 4 programs as follows:

	Electrical	Chemical	Mechanical	Civil	Total
Male	100	80	70	50	300
Female	50	50	50	50	200
Total	150	130	120	100	500

The 300 and 200 in the last column and the 150, 130, 120, 100 in the last row are called *marginal totals*, and the 500 is called the *grand total*.

Let H_0 be the null hypothesis that the program choice is independent of gender.   Test the null hypothesis at the: (*a*) $\alpha = 0.10$ significance level, (*b*) $\alpha = 0.05$ significance level.

First we want to find the expected eight entries in the table assuming independence.   Note $300/500 = 60$ percent of the students are male and $150/500 = 30$ percent of the students are studying electrical engineering.   Thus, the expected number of male students taking electrical engineering is obtained by multiplying the product of the probabilities by the total number of students, yielding

$$(60\%)(30\%)(500) = 90$$

Equivalently, the expected number can be obtained by multiplying the two marginal totals and dividing by the grand total, that is,

$$\text{Expected entry} = \frac{(\text{row total})(\text{column total})}{\text{grand total}} = \frac{(300)(150)}{500} = 90$$

This formula is derived from the fact that

$$(60\%)(30\%)(500) = \frac{300}{500} \cdot \frac{150}{500} \cdot 500 = \frac{(300)(150)}{500} = 90$$

The other seven expected numbers are obtained similarly.

Furthermore, rather than forming a new table with the expected values, we add each expected value after the corresponding observed value in the above table, say, as follows:

	Electrical	Chemical	Mechanical	Civil	Total
Male	100/90	80/78	70/72	50/60	300
Female	50/60	50/52	50/48	50/40	200
Total	150	130	120	100	500

Some texts place the expected value below or diagonally down from the observed value.   The χ^2 value of the data is easily obtained from the table as follows:

$$\chi^2 = \sum \frac{(\text{obs} - \text{exp})^2}{\text{exp}}$$

$$= \frac{(100 - 90)^2}{90} + \frac{(80 - 78)^2}{78} + \frac{(70 - 72)^2}{72} + \frac{(50 - 60)^2}{60}$$

$$+ \frac{(50 - 60)^2}{60} + \frac{(50 - 52)^2}{52} + \frac{(50 - 48)^2}{48} + \frac{(50 - 40)^2}{40} = 7.21$$

There are df $= (2 - 1)(4 - 1) = 3$ degrees of freedom.   This is derived from the fact that the marginal values are given, and so:

(i)   Any one value in a column will determine the other value.

(ii)   Any three columns will determine the fourth column.

Thus, for example, given the first three entries in the first row will give us the other five entries in the table.

(a)   Table B-1 tells us that $c = 6.25$ is the critical value for df $= 3$ and $\alpha = 0.10$.   Since $7.21 > 6.25$, we reject the null hypothesis H_0 at the $\alpha = 0.10$ significance level that the program choice at the college is independent of gender.

(b)   Table B-1 tells us that $c = 7.81$ is the critical value for df $= 3$ and $\alpha = 0.05$.   Here $7.21 < 7.81$.   Thus, at the $\alpha = 0.05$ significance level, we accept the null hypothesis H_0 that the program choice at the college is independent of gender.

Remark: The above calculation for the degrees of freedom df is true in general.   That is, suppose an attribute A has r categories and another attribute B has c categories yielding a table with r rows and c columns.   Then

$$df = (r - 1)(c - 1)$$

gives the number of degrees of freedom.   This comes from the fact that:

(i)   Any $r - 1$ entries in a column determine the rth entry in the column.

(ii)   Any $c - 1$ columns determine the cth column.

EXAMPLE B.8   A town asks its voters whether or not it should build a new park where the vote could be:

(i)   yes,   (ii)   no,   (iii)   abstain.

A poll of 1000 of the voters yields the following data where voters were divided into three age categories, 18–30, 30–50, 50–70:

	Yes	No	Abstain	Total
18–30	170	60	20	250
31–50	255	140	55	450
51–70	175	100	25	300
Total	600	300	100	1000

Let H_0 be the null hypothesis that the vote is independent of age.   (That is, there is no age gap in the vote.) Test the null hypothesis at the $\alpha = 0.10$ significance level.

First we find the expected nine entries in the table where we assume independence.   Specifically, we use the formula:

$$\text{Expected entry} = \frac{\text{(row total)(column total)}}{\text{grand total}}$$

We add the nine expected values to the above table as follows:

	Yes	No	Abstain	Total
18–30	170/150	60/75	20/25	250
31–50	255/270	140/135	55/45	450
51–70	175/180	100/90	25/30	300
Total	600	300	100	1000

The χ^2 value of the data follows:

$$\chi^2 = \sum \frac{(\text{obs} - \text{exp})^2}{\text{exp}} = \frac{(170 - 150)^2}{150} + \frac{(60 - 75)^2}{75} + \frac{(20 - 25)^2}{25}$$

$$+ \frac{(255 - 270)^2}{270} + \frac{(140 - 135)^2}{135} + \frac{(55 - 45)^2}{45} + \frac{(175 - 180)^2}{180}$$

$$+ \frac{(100 - 90)^2}{90} + \frac{(25 - 30)^2}{30} = 5.32$$

The number of degrees of freedom, as noted by the above remark, is obtained by

$$df = (r - 1)(c - 1) = (3 - 1)(3 - 1) = 4$$

This is derived from the fact that the marginal values are given and so:

(i)   Any two values in a column will determine the third value.

(ii) Any two columns will determine the third column.

Thus, for example, given the four entries in the upper left corner of the table, we can obtain the other five entries.

Table B-1 tells us that $c = 7.78$ is the critical value for df = 4 and $\alpha = 0.10$.   Since $5.32 < 7.78$, we accept the null hypothesis H_0 that the vote for the park is independent of age.

B.7 CHI-SQUARE TEST FOR HOMOGENEITY

Two populations are said to be *homogeneous* with respect to some grouping criteria if they have the same percentage distribution.   This section gives applications of the χ^2 distribution to problems involving *homogeneity*, that is, whether different populations are homogeneous.

The χ^2 test of homogeneity in this section uses the same type of data table which was used in the χ^2 test for independence in the last section.   We note, however, that the χ^2 test of independence involves a single population whereas the χ^2 test for homogeneity involves two different populations.

Again we note that the χ^2 test is not accurate for small values; hence, as before, we assume that our sample exceeds 30 and that no expected frequency is less than 5.

EXAMPLE B.9 A sociologist decides to study the distribution of adults (18 years and above) in two cities, New York and Boston, where the distribution has three categories, under 30 years, 30–60 years, over 60 years. She takes a sample of 150 from New York and a sample of 100 from Boston and obtains the following data:

	<30	30–60	>60	Total
New York	51	77	22	150
Boston	29	63	8	100
Total	80	140	30	250

Let H_0 be the null hypothesis that the adult age distribution in New York and Boston is homogeneous. Test the null hypothesis at the $\alpha = 0.05$ significance level.

The following is the main idea behind our testing procedure:

If the cities are homogeneous, then the data from the combined population will give a better estimate of the age percentages than the data from either individual city.

Thus, we estimate

$$\text{Percentage under 30} \approx \frac{\text{total number under 30 in two cities}}{\text{total number sampled}} = \frac{\text{column total}}{\text{grand total}} = \frac{80}{250} = 32\%$$

Accordingly, with a sample of 150 from New York and 100 from Boston, we would expect the following number of adults under 30 in each sample:

$$\text{New York: } 32\%(150) = 48 \qquad \text{Boston: } 32\%(100) = 32$$

Observe that we are multiplying each row total by the corresponding percentage (column total/grand total). Accordingly, we can again obtain these results using the following formula:

$$\text{Expected entry} = \frac{(\text{row total})(\text{column total})}{\text{grand total}}$$

Thus, we could have proceeded as follows:

$$\text{New York: } \frac{(150)(80)}{250} = 48 \qquad \text{Boston: } \frac{(100)(80)}{250} = 32$$

The other four expected entries are obtained similarly.

We add the six expected values to our original table as follows:

	<30	30–60	>60	Total
New York	51/48	77/84	22/18	150
Boston	29/32	63/56	8/12	100
Total	80	140	30	250

The χ^2 value of the data follows:

$$\chi^2 = \sum \frac{(\text{obs} - \text{exp})^2}{\text{exp}} = \frac{(51 - 48)^2}{48} + \frac{(77 - 84)^2}{84} + \frac{(22 - 18)^2}{18}$$

$$+ \frac{(29 - 32)^2}{32} + \frac{(63 - 56)^2}{56} + \frac{(8 - 12)^2}{12}$$

$$= \frac{9}{48} + \frac{49}{84} + \frac{16}{18} + \frac{9}{32} + \frac{49}{56} + \frac{16}{12} = 4.15$$

There are df = $(2 - 1)(3 - 1) = 2$ degrees of freedom. This is derived from the fact that the marginal values are given, and so:

(i) Any one value in a column will determine the other value.

(ii) Any two columns will determine the third column.

Thus, for example, given the first two entries in the first row, we can obtain the other four entries in the table.

Table B-1 tells us that $c = 4.61$ is the critical value for df = 2 and $\alpha = 0.10$. Since $4.15 < 4.62$, we accept the null hypothesis H_0 at the $\alpha = 0.10$ significance level, that is, that the age distribution of adults in New York and Boston is homogeneous.

Remark: The above calculation for the degrees of freedom df is true in general, that is, if the data table has r rows and c columns. Then the following formula gives the number of degrees of freedom:

$$\text{df} = (r - 1)(c - 1)$$

As noted above, this formula comes from the fact that any $r - 1$ entries in a column determine the rth entry in the column, and any $c - 1$ columns determine the cth column.

Solved Problems

GOODNESS OF FIT

B.1.   A die is tossed 60 times yielding the following distribution:

Face value	1	2	3	4	5	6
Frequency	7	11	10	14	6	12

Let H_0 be the null hypothesis that the die is fair.

(a) Find the χ^2 value.   (b) Find the degrees of freedom df.

(c) Test H_0 at the $\alpha = 0.10$ significance level.

(a)   The die was tossed 60 times and there are 6 possible face values.   Therefore, assuming the die is fair, the expected number of times each face occurs is 10.   Thus, the χ^2 value for the data follows:

$$\chi^2 = \sum \frac{(\text{obs} - \text{exp})^2}{\text{exp}} = \frac{(7 - 10)^2}{10} + \frac{(11 - 10)^2}{10} + \frac{(10 - 10)^2}{10}$$

$$+ \frac{(14 - 10)^2}{10} + \frac{(6 - 10)^2}{10} + \frac{(12 - 10)^2}{10}$$

$$= \frac{9}{10} + \frac{1}{10} + \frac{0}{10} + \frac{16}{10} + \frac{16}{10} + \frac{4}{10} = 4.6$$

(b) There are df $= 6 - 1 = 5$ degrees of freedom. This is derived from the fact that the die was tossed 60 times, so the number of times 5 of the faces occur determines the number of times the sixth face occurs.

(c) Table B-1 shows that the critical χ^2 value for df $= 5$ at the $\alpha = 0.10$ significance level is $c = 9.24$. Since $4.6 < 9.24$, we accept the null hypothesis H_0 that the die is fair.

B.2. Suppose the following table gives the percentage of the number of persons per household in the United States for a given year.

Number of persons	1	2	3	4	5 or more
Percentage of households	20	30	18	15	17

Suppose a survey of 1000 households in Philadelphia for the year yielded the following data:

Number of persons	1	2	3	4	5 or more
Number of households	270	210	200	100	220

Let H_0 be the (null) hypothesis that the distribution of people in households in Philadelphia is the same as the national distribution.

(a) Find the χ^2 value.   (c) Test H_0 at the $\alpha = 0.10$ significance level.

(b) Find the degrees of freedom df.   (d) Test H_0 at the $\alpha = 0.05$ significance level.

(a) Since there are 1000 households, we divide each data value by 1000 to obtain the following percentages for Philadelphia:

Number of persons	1	2	3	4	5 or more
Percentage of households	27	21	20	10	22

Thus, the χ^2 value for the data follows:

$$\chi^2 = \sum \frac{(\text{obs} - \text{exp})^2}{\text{exp}} = \frac{(27 - 20)^2}{20} + \frac{(21 - 30)^2}{30} + \frac{(20 - 18)^2}{18} + \frac{(10 - 15)^2}{15} + \frac{(22 - 17)^2}{17}$$

$$= \frac{49}{20} + \frac{81}{30} + \frac{4}{18} + \frac{25}{15} + \frac{25}{17} = 8.50$$

(b) There are df $= 5 - 1 = 4$ degrees of freedom. This is derived from the fact that four of the five percentages determines the fifth percentage.

(c) Table B-1 shows that the critical χ^2 value for df $= 4$ at the $\alpha = 0.10$ significance level is $c = 7.78$. Since $8.50 > 7.78$, we reject (at the $\alpha = 0.10$ significance level) the null hypothesis H_0 that the Philadelphia distribution is similar to the national distribution.

(d) Table B-1 shows that the critical χ^2 value for df $= 4$ at the $\alpha = 0.05$ significance level is $c = 9.49$. Since $8.50 < 9.49$, we accept (at the $\alpha = 0.05$ significance level) the null hypothesis H_0 that the Philadelphia distribution is similar to the national distribution.

B.3. A poll is taken of 160 families in New York with 4 children yielding the following family sex distribution (where B denotes boys and G denotes girls):

Sex distribution	4B	3B, 1G	2B, 2G	1B, 3G	4G
Frequency	9	46	54	38	13

Let H_0 be the null hypothesis that the New York distribution is binomial with $p = 1/2$.

(a) Find the expected distribution.   (c) Find the degrees of freedom df.

(b) Find the χ^2 value.   (d) Test H_0 at the $\alpha = 0.10$ significance level.

(a) The binomial distribution with $n = 4$ and $p = 0.5$ follows:

x	0	1	2	3	4
$P(x)$	1/16	4/16	6/16	4/16	1/16

Multiplying the probabilities by 160, the number of families, gives the following expected distribution.

Sex distribution	$4B$	$3B, 1G$	$2B, 2G$	$1B, 3G$	$4G$
Expected frequency	10	40	60	40	10

(b) Thus, the χ^2 value for the data follows:

$$\chi^2 = \sum \frac{(\text{obs} - \text{exp})^2}{\text{exp}} = \frac{(9 - 10)^2}{10} + \frac{(46 - 40)^2}{40} + \frac{(54 - 60)^2}{60} + \frac{(38 - 40)^2}{40} + \frac{(13 - 10)^2}{10}$$

$$= \frac{1}{10} + \frac{36}{40} + \frac{36}{60} + \frac{4}{40} + \frac{9}{10} = 2.9$$

(c) There are df $= 5 - 1 = 4$ degrees of freedom. This is derived from the fact that the sum of the five numbers in the table is 160, so any four of the numbers determine the fifth number.

(d) Table B-1 shows that the critical χ^2 value for df $= 4$ at the $\alpha = 0.10$ significance level is $c = 7.78$. Since $2.9 < 7.78$, we accept the null hypothesis H_0 that the distribution is binomial with $p = 1/2$.

B.4. A resort has 200 cabins which can sleep up to 4 people. Suppose the following table gives the overnight occupancy of the cabins for some night.

Number of people in room	0	1	2	3	4
Number of rooms	7	34	55	80	24

Let H_0 be the null hypothesis that the occupancy distribution is binomial.

(a) Find the expected distribution.   (c) Find the degrees of freedom df.

(b) Find the χ^2 value.   (d) Test H_0 at the $\alpha = 0.10$ significance level.

Here $n = 4$ but p is not given. Thus, we use the sample proportion $\hat{p}$ of the number of occupied beds as an estimate of p. The number s of people in all the rooms follows:

$$s = 0(7) + 1(34) + 2(55) + 3(80) + 4(24) = 480$$

The total number b of beds is $4(200) = 800$. Thus, we set

$$p = \hat{p} = \frac{s}{b} = \frac{480}{800} = 0.6$$

(a) The binomial distribution with $n = 4$, $p = 0.6$ and $q = 1 - 0.6 = 0.4$ follows:

$$P(0) = (0.4)^4 = 0.025\,6, \qquad P(3) = 4(0.6)^3\,(0.4) = 0.345\,6$$
$$P(1) = 4(0.6)(0.4)^3 = 0.153\,6, \qquad P(4) = (0.6)^4 = 0.129\,6$$
$$P(2) = 6(0.6)^2\,(0.4)^2 = 0.345\,6,$$

Multiplying the probabilities by 200, the number of rooms, yields the following expected distribution:

Number of people in room	0	1	2	3	4
Expected number of rooms	5	31	69	69	26

(b) Thus, the χ^2 value for the data follows:

$$\chi^2 = \sum \frac{(obs - exp)^2}{exp} = \frac{(7-5)^2}{5} + \frac{(34-31)^2}{31} + \frac{(55-69)^2}{69} + \frac{(80-69)^2}{69} + \frac{(24-26)^2}{26}$$

$$= \frac{4}{5} + \frac{9}{31} + \frac{196}{69} + \frac{121}{69} + \frac{4}{26} = 5.84$$

(c) There are df $= 5 - 1 = 4$ degrees of freedom. This is derived from the fact that the sum of the five numbers in the table is 200, so any four of the numbers determine the fifth number.

(d) Table B-1 shows that the critical χ^2 value for df $= 4$ at the $\alpha = 0.10$ significance level is $c = 7.78$. Since $5.84 < 7.78$, we accept the null hypothesis H_0 that the distribution is binomial.

GOODNESS OF FIT FOR NORMAL DISTRIBUTION

B.5. Suppose the weights W, in pounds, of male students of 500 students at a college have the following distribution:

Weight	<120	120–140	140–160	160–180	180–200	>200
Number of students	37	91	128	150	75	19

Let H_0 be the null hypothesis that the distribution is normal with mean $\mu = 160$ and standard deviation $\sigma = 25$.

(a) Find the expected distribution.   (c) Find the degrees of freedom df.

(b) Find the χ^2 value.   (d) Test H_0 at the following significance levels:

                                     (i) $\alpha = 0.10$,   (ii) $\alpha = 0.05$.

Using the formula $z = (W - \mu)/\sigma$, we derive the following z values corresponding to the above W values:

W value	120	140	160	180	200
z value	−1.6	−0.8	0	0.8	1.6

Figure B-8 shows the normal curve with the W values, the corresponding z values, and the probability distribution for these z values obtained from Table 6-1 (page 184) of the standard normal distribution.

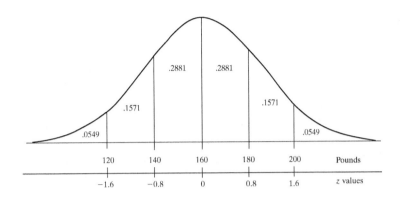

Fig. B-8

(a) Multiplying the probabilities in Fig. B-8 by 500 gives the following expected numbers of students for the given weight intervals:

Weight	<120	120–140	140–160	160–180	180–200	>200
Number of students	28	78	144	144	78	28

(b) The χ^2 value for the data follows:

$$\chi^2 = \sum \frac{(\text{obs} - \text{exp})^2}{\text{exp}} = \frac{(37 - 28)^2}{28} + \frac{(91 - 78)^2}{78} + \frac{(128 - 144)^2}{144}$$

$$+ \frac{(150 - 144)^2}{144} + \frac{(75 - 78)^2}{78} + \frac{(19 - 28)^2}{28}$$

$$= \frac{81}{28} + \frac{169}{78} + \frac{196}{144} + \frac{36}{144} + \frac{9}{78} + \frac{81}{28} = 9.67$$

(c) There are df = 6 − 1 = 5 degrees of freedom. This is derived from the fact that the sum of the six numbers in the table is 500, so any five of the numbers determines the sixth number.

(d) The row df = 4 in Table B-1 shows that the critical χ^2 value is $c = 9.24$ for $\alpha = 0.10$ and $c = 11.07$ for $\alpha = 0.05$. Since $\chi^2 = 9.67$, we: (i) reject H_0 for $\alpha = 0.10$, (ii) accept H_0 for $\alpha = 0.05$.

B.6. Suppose the average hourly daily workload x of 600 American employees yielded the following data:

Hourly workload x	<5	5–6	6–7	7–8	8–9	9–10	>10
Number of employees	8	45	150	210	130	40	17

Suppose $\bar{x} = 7.5$ is the sample mean and $s = 1.2$ is the sample standard deviation. Let H_0 be the null hypothesis that the distribution is normal with the estimation that the mean $\mu = \bar{x} = 7.5$ and the standard deviation $\sigma = s = 1.2$.

(a) Find the expected distribution.   (c) Find the degrees of freedom df.

(b) Find the χ^2 value.   (d) Test H_0 at the following significance levels:

                                                        (i) $\alpha = 0.10$,   (ii) $\alpha = 0.05$.

Using the formula $z = (x - \mu)/\sigma$, we derive the following z values corresponding to the above x values:

x value	5	6	7	8	9	10
z value	−2.08	−1.25	−0.42	0.42	1.25	2.08

Figure B-9 shows the normal curve with the x values, the corresponding z values, and the probability distribution for these z values obtained from Table 6-1 of the standard normal distribution.

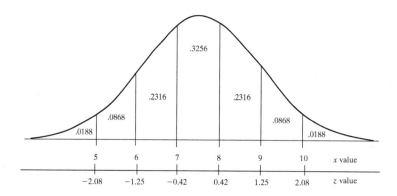

Fig. B-9

(a) Multiplying the probabilities in Fig. B-9 by 600 gives the following expected numbers of employees for the given time intervals:

Hourly workload	<5	5–6	6–7	7–8	8–9	9–10	>10
Number of employees	11	52	139	196	139	52	11

(b) The χ^2 value for the data follows:

$$\chi^2 = \sum \frac{(\text{obs} - \text{exp})^2}{\text{exp}} = \frac{(8 - 11)^2}{11} + \frac{(45 - 52)^2}{52} + \frac{(150 - 139)^2}{139}$$

$$+ \frac{(210 - 196)^2}{196} + \frac{(130 - 139)^2}{139} + \frac{(40 - 52)^2}{52} + \frac{(17 - 11)^2}{11}$$

$$= \frac{9}{11} + \frac{49}{52} + \frac{121}{139} + \frac{196}{196} + \frac{81}{139} + \frac{49}{52} + \frac{36}{11} = 8.43$$

(c) The seven numbers in our table are related by three equations, one determined by the size $n = 600$, one by the sample mean $\bar{x} = 7.5$, and one by the sample standard deviation $s = 1.2$. Thus, there are only df $= 7 - 3 = 4$ degrees of freedom.

(d) The row df $= 4$ in Table B-1 shows that the critical χ^2 value is $c = 7.78$ for $\alpha = 0.10$ and $c = 9.49$ for $\alpha = 0.05$. Since $\chi^2 = 8.43$, we: (i) reject H_0 for $\alpha = 0.10$, (ii) accept H_0 for $\alpha = 0.05$.

INDEPENDENCE

B.8. Voters in a certain town can only register as Democratic, Republican, or Independent. A poll of 800 registered voters yields the following gender distribution:

	Democratic	Republican	Independent	Total
Male	140	192	68	400
Female	160	158	82	400
Total	300	350	150	800

Let H_0 be the hypothesis that the party affiliation is independent of gender.

(a) Find the χ^2 value.   (b) Find the degrees of freedom df.

(c) Test H_0 at the following significance levels: (i) $\alpha = 0.10$, (ii) $\alpha = 0.05$.

The expected six entries in the table, assuming independence, are obtained by the formula:

$$\text{Expected entry} = \frac{(\text{row total})(\text{column total})}{\text{grand total}}$$

For example, the expected number of males to register as Democrats, Republicans, and Independents, respectively, follow:

$$\frac{(400)(300)}{800} = 150, \qquad \frac{(400)(350)}{800} = 175, \qquad \frac{(400)(150)}{800} = 75$$

Adding the six expected values to the above table yields the following:

	Democratic	Republican	Independent	Total
Male	140/150	192/175	68/75	400
Female	160/150	158/175	82/75	400
Total	300	350	150	800

(a) Using the above table for the $(\text{obs} - \text{exp})^2$ values, we obtain the χ^2 value as follows:

$$\chi^2 = \sum \frac{(\text{obs} - \text{exp})^2}{\text{exp}} = \frac{100}{150} + \frac{289}{175} + \frac{49}{75} + \frac{100}{150} + \frac{289}{175} + \frac{49}{75} = 5.94$$

(b) The number of degrees of freedom is obtained by

$$df = (r - 1)(c - 1) = (2 - 1)(3 - 1) = 2$$

This is derived from the fact that the marginal values are given; hence:

(i) Any value in a column will determine the other value.

(ii) Any two values in a row will determine the third value.

Thus, for example, given the first two values in the first row, we can determine the remaining four values.

(c) The row df = 2 in Table B-1 shows that the critical χ^2 value is $c = 4.61$ for $\alpha = 0.10$ and $c = 5.99$ for $\alpha = 0.05$. Since $\chi^2 = 5.94$, we: (i) reject H_0 for $\alpha = 0.10$, (ii) accept H_0 for $\alpha = 0.05$.

B.10. A grocery chain of stores carries four brands A, B, C, D of a certain type of cereal. The chain recorded the brand of the cereal sold and the age of the buyer, where the buyers were divided into three age categories: younger than 20, 20–40, older than 40. The frequency distribution during 1 week follows:

	A	B	C	D	Total
<20	90	64	78	48	280
20–40	88	78	70	64	300
>40	62	58	52	48	220
Total	240	200	200	160	800

Let H_0 be the hypothesis that the cereal choice is independent of age. (a) Find the χ^2 value. (b) Find the degrees of freedom df. (c) Test H_0 at the $\alpha = 0.10$ significance level.

The expected 12 entries in the table, assuming independence, are obtained by the formula:

$$\text{Expected entry} = \frac{(\text{row total})(\text{column total})}{\text{grand total}}$$

Adding the 12 expected values to the above table yields the following:

	A	B	C	D	Total
<20	90/84	64/70	78/70	48/56	280
20–40	88/90	78/75	70/75	64/60	300
>40	62/66	58/55	52/55	48/44	220
Total	240	200	200	160	800

(a) The χ^2 value for the data follows:

$$\chi^2 = \sum \frac{(\text{obs} - \text{exp})^2}{\text{exp}} = \frac{36}{84} + \frac{36}{70} + \frac{64}{70} + \frac{64}{56} + \frac{4}{90} + \frac{9}{75} + \frac{25}{75} + \frac{16}{60}$$

$$+ \frac{16}{66} + \frac{9}{55} + \frac{4}{55} + \frac{16}{44} = 4.61$$

(b) The number of degrees of freedom is obtained by

$$\text{df} = (r - 1)(c - 1) = (3 - 1)(4 - 1) = 6$$

This is derived from the fact that the marginal values are given; hence any two values in a column will determine the third value, and any three values in a row will determine the fourth value. Thus, for example, the first three values in the first two rows, determines the remaining six values.

(c) Table B-1 shows that the critical χ^2 value for df = 6 and $\alpha = 0.10$ is $c = 10.64$. Since $4.61 < 10.64$, we accept the null hypothesis H_0 that the choice of cereal is independent of age.

HOMOGENEITY

B.11. Suppose an opinion poll on a referendum in 4 city districts yields the following data:

	Yes	No	Undecided	Total
District 3	18	20	12	50
District 8	26	16	8	50
District 11	20	24	6	50
District 16	28	12	10	50
Total	92	72	36	200

Let H_0 by the hypothesis that the voter opinion on the referendum is homogeneous in the 4 districts. (a) Find the χ^2 value. (b) Find the degrees of freedom df. (c) Test H_0 at the $\alpha = 0.10$ significance level.

First we find the 12 expected entries in the table. Assuming the combined population of 200 voters will give a better estimate of voter opinion than either individual district, we find the expected entries using the following formula:

$$\text{Expected entry} = \frac{(\text{row total})(\text{column total})}{\text{grand total}}$$

We add the 12 expected values to the above table as follows:

	Yes	No	Undecided	Total
District 3	18/23	20/18	12/9	50
District 8	26/23	16/18	8/9	50
District 11	20/23	24/18	6/9	50
District 16	28/23	12/18	10/9	50
Total	92	72	36	200

(a) The χ^2 value for the data follows:

$$\chi^2 = \sum \frac{(\text{obs} - \text{exp})^2}{\text{exp}} = \frac{25}{23} + \frac{4}{18} + \frac{9}{9} + \frac{9}{23} + \frac{4}{18} + \frac{1}{9} + \frac{9}{23} + \frac{16}{18}$$
$$+ \frac{9}{9} + \frac{25}{23} + \frac{16}{18} + \frac{1}{9} = 7.40$$

(b) The number of degrees of freedom is obtained by

$$\text{df} = (r - 1)(c - 1) = (4 - 1)(3 - 1) = 6$$

This is derived from the fact that the marginal values are given; hence any three values in a column will determine the fourth value, and any two values in a row will determine the third value.

Thus, for example, the first two values in the first three rows determines the remaining six values.

(c)   Table B-1 shows that the critical χ^2 value for df = 6 and $\alpha = 0.10$ is $c = 10.64$.   Since $7.40 < 10.64$, we accept the null hypothesis H_0 that the voter opinion on the referendum is homogeneous in the 4 districts.

B.12.   Suppose the following table gives the distribution of geometry grades in 2 high schools.

	A	B	C	D	F	Total
High school 1	27	39	60	42	32	200
High school 2	28	51	120	68	33	300
Total	55	90	180	110	65	500

Let H_0 be the hypothesis that the grade distributions are homogeneous in the 2 high schools.   (a) Find the χ^2 value.   (b) Find the degrees of freedom df.   (c) Test H_0 at the following significance levels: (i) $\alpha = 0.10$, (ii) $\alpha = 0.05$.

First we find the 10 expected entries in the table.   Assuming the combined population of 500 students will give a better estimate of the grade distribution than either individual school, we find the expected entries using the following formula:

$$\text{Expected entry} = \frac{(\text{row total})(\text{column total})}{\text{grand total}}$$

We add the 10 expected values to the above table as follows:

	A	B	C	D	F	Total
High school 1	27/22	39/36	60/72	42/44	32/26	200
High school 2	28/33	51/54	120/108	68/66	33/39	300
Total	55	90	180	110	65	500

(a)   The χ^2 value for the data follows:

$$\chi^2 = \sum \frac{(\text{obs} - \text{exp})^2}{\text{exp}} = \frac{25}{22} + \frac{9}{36} + \frac{144}{72} + \frac{4}{44} + \frac{36}{26} + \frac{25}{33} + \frac{9}{54} + \frac{144}{108} + \frac{4}{66} + \frac{36}{39} = 8.10$$

(b)   The number of degrees of freedom is obtained by

$$df = (r - 1)(c - 1) = (2 - 1)(5 - 1) = 4$$

This is derived from the fact that the marginal values are given; hence:

(i)   Any value in a column will determine the other value.

(ii)   Any four values in a row will determine the fifth value.

Thus, for example, the first four values in the first row determines the remaining six values.

(c)   The row df = 4 in Table B-1 shows that the critical χ^2 value is $c = 7.78$ for $\alpha = 0.10$ and $c = 9.49$ for $\alpha = 0.05$.   Since $\chi^2 = 8.10$, we: (i) reject H_0 for $\alpha = 0.10$, (ii) accept H_0 for $\alpha = 0.05$.

Supplementary Problems

GOODNESS OF FIT

B.13. A coin is tossed 80 times yielding 48 heads and 32 tails. Let H_0 be the hypothesis that the coin is fair.

(a) Find the χ^2 value and the degrees of freedom df.

(b) Test H_0 at the following significance levels: (i) $\alpha = 0.10$, (ii) $\alpha = 0.05$.

B.14. Suppose the frequency of each digit in the first 100 digits in a random number yields the following distribution:

Digit	0	1	2	3	4	5	6	7	8	9
Frequency	7	11	12	9	6	13	11	10	9	12

Let H_0 be the hypothesis that each digit occurs with the same probability. (a) Find the χ^2 value and the degrees of freedom df. (b) Test H_0 at the $\alpha = 0.05$ significance level.

B.15. The following table lists the grading policy of a department for a sophomore mathematics course and the number of such grades by a professor for 120 of her students.

Grade	A	B	C	D	F
Policy	10%	40%	35%	10%	5%
Course	18	55	34	7	6

Let H_0 be the hypothesis that the professor conforms to department policy. (a) Find the expected distribution. (b) Find the χ^2 value and the degrees of freedom df. (c) Test H_0 at the following significance levels: (i) $\alpha = 0.10$, (ii) $\alpha = 0.05$.

B.16. It is estimated that 60 percent of cola drinkers prefer Coke over Pepsi. In a random poll of 600 cola drinkers, 330 preferred Coke over Pepsi. (a) Find the expected distribution and the χ^2 value. (b) Test the hypothesis H_0 that the estimate is correct at the $\alpha = 0.10$ significance level.

B.17. It is estimated that the political preferences in a certain town are as follows:

35% Democrat, 40% Republican, 15% Independent, 10% other

A random sample of 200 people resulted in the following preferences:

64 Democrat, 76 Republican, 38 Independent, 22 other

Let H_0 be the hypothesis that the estimate is correct

(a) Find the χ^2 value and the degrees of freedom df.

(b) Test H_0 at the $\alpha = 0.10$ significance level.

B.18. The following table gives the age percentages of people living in the United States for some given year (using four age categories: under 20, 20–39, 40–64, 65 and over), and the age distribution of a sample of 500 people living in Florida:

Age	<20	20–39	40–64	≥65
United States	28%	24%	32%	16%
Florida	115	130	155	100

Let H_0 be the hypothesis that the age distribution in Florida is the same as the national distribution.   (a) Find the expected distribution.   (b) Find the χ^2 value and the degrees of freedom df.   (c) Test H_0 at the $\alpha = 0.10$ significance level.

BINOMIAL DISTRIBUTION

B.19. Applicants for a civil service position take a national test with three separate parts. The following table gives the number of parts passed by each of 500 applicants:

Number of parts passed	0	1	2	3
Number of applicants	180	200	100	20

Let H_0 be the hypothesis that the distribution is binomial with $p = 0.3$.   (a) Find the expected distribution. (b) Find the χ^2 value and the degrees of freedom df.   (c) Test H_0 at the $\alpha = 0.10$ significance level.

B.20. A study is made of the number of children in a 4-child family who have attended college.   Interviews with 600 families produced the following data:

Number of children	0	1	2	3	4
Number of families	75	170	150	90	15

Let H_0 be the hypothesis that the distribution is binomial.   (a) Find the population proportion $\hat{p}$ of children attending college.   (b) Find the expected distribution using $p = \hat{p}$.   (c) Find the χ^2 value and the degrees of freedom df.   (d) Test H_0 at the $\alpha = 0.10$ significance level.

NORMAL DISTRIBUTION

B.21. Suppose the following table gives the average daily minutes of time T spent watching television by a sample of 400 10-year-old children.

Time	<600	600–1000	1000–1400	1400–1800	>1800
Number of children	20	82	160	100	38

Let H_0 be the hypothesis that the distribution is normal with mean $\mu = 1200$ and standard deviation $\sigma = 400$.   (a) Find the z values corresponding to $T = 600, 1000, 1400, 1800$.   (b) Find the expected distribution.   (c) Find the χ^2 value and the degrees of freedom df.   (d) Test H_0 at the following significance levels: (i) $\alpha = 0.10$, (ii) $\alpha = 0.05$.

B.22. Suppose the following gives the number x of eggs produced annually by 200 chickens at a farm.

Eggs	<280	280–300	300–320	320–340	340–360	>360
Number of chickens	10	33	70	55	20	12

Furthermore, suppose $\bar{x} = 315$ is the sample mean and $s = 25$ is the sample standard deviation. Let H_0 be the hypothesis that the distribution is normal with the estimation that the mean $\mu = \bar{x} = 315$ and the standard deviation $\sigma = s = 25$.   (a) Find the z values corresponding to $x = 280, 300, 320, 340, 360$. (b) Find the (approximate) expected distribution.   (c) Find the χ^2 value and the degrees of freedom df. (d) Test H_0 at the $\alpha = 0.10$ significance level.

INDEPENDENCE

B.23. Voters in a certain town can only register as Democratic, Republican, or Independent. A poll of 500 registered voters yields the following gender distribution:

	Democratic	Republican	Independent	Total
Male	95	125	40	260
Female	105	100	35	240
Total	200	225	75	500

Let H_0 be the hypothesis that the party affiliation is independent of gender. (a) Find the expected distribution. (b) Find the χ^2 value and the degrees of freedom df. (c) Test H_0 at the $\alpha = 0.10$ significance level.

B.24. Suppose a large university wants to determine the student opinion (favor or oppose) on the requirement of a certain dress code for attending classes. A poll of 100 students per class is taken yielding the following table:

	Freshman	Sophomore	Junior	Senior	Total
Favor	35	42	45	58	180
Oppose	65	58	55	42	220
Total	100	100	100	100	400

Let H_0 be the hypothesis that the opinion is independent of the class of the student. (a) Find the χ^2 value and the degrees of freedom df. (b) Test H_0 at the $\alpha = 0.10$ significance level.

HOMOGENEITY

B.25. A study is made of political party affiliation of voters in three regions of the country, northeast, south, and west. Interviews with 500 voters yielded the following distribution:

	Democrat	Republican	Other	Total
Northeast	105	79	16	200
South	60	82	8	150
West	65	79	6	150
Total	230	240	30	500

Let H_0 be the hypothesis that the distribution is homogeneous.
(a) Find the expected distribution. (b) Find the χ^2 value and the degrees of freedom df.
(c) Test H_0 at the following significance levels: (i) $\alpha = 0.10$, (ii) $\alpha = 0.05$, (iii) $\alpha = 0.025$.

B.26. A study is made of the grades of full-time and part-time students for a freshman mathematics course at a university yielding the following data:

	A	B	C	D	F	Total
Full time	32	58	70	50	30	240
Part time	18	32	50	30	30	160
Total	50	90	120	80	60	400

Let H_0 be the hypothesis that the distribution is homogeneous. (a) Find the expected distribution. (b) Find the χ^2 value and the degrees of freedom df. (c) Test H_0 at the $\alpha = 0.10$ significance level.

Answers to Supplementary Problems

B.13. (a) $\chi^2 = 64/45 + 64/35 = 3.25$, df = 1; (b) (i) no, (ii) yes.

B.14. (a) $\chi^2 = 4.6$, df = 9; (b) yes.

B.15. (a) [12, 48, 42, 12, 6]; (b) $\chi^2 = 36/12 + 49/48 + 64/42 + 25/12 + 0/6 = 7.62$, df = 4; (c) (i) no, (ii) yes.

B.16. (a) [360, 240], $\chi^2 = 900/360 + 900/240 = 6.25$; (b) no.

B.17. (a) $\chi^2 = 36/70 + 16/80 + 64/30 + 4/20 = 3.05$, df = 3; (b) yes.

B.18. (a) [140, 120, 160, 80]; (b) $\chi^2 = 10.45$, df = 3; (c) no.

B.19. (a) [171.5, 220.5, 94.5, 13.5]; (b) $\chi^2 = 72.25/171.5 + 420.25/220.5 + 30.25/94.5 + 36.25/13.5 = 5.33$, df = 3; (c) yes.

B.20. (a) $\hat{p} = 0.4$; (b) [64.8, 172.8, 172.8, 76.8, 12.8];

(c) $\chi^2 = \dfrac{104.04}{64.8} + \dfrac{7.84}{172.8} + \dfrac{519.84}{172.8} + \dfrac{174.24}{76.8} + \dfrac{4.84}{15} = 7.26$, df = 4; (d) yes.

B.21. (a) [−1.5, −0.5, 0.5, 1.5]; (b) [27, 96, 153, 96, 27];
(c) $\chi^2 = 49/27 + 196/96 + 49/153 + 9/96 + 121/27 = 8.75$, df = 4; (d) (i) no, (ii) yes.

B.22. (a) [−1.4, −0.6, 0.2, 1.0, 1.8]; (b) [16, 39, 61, 52, 25, 7];
(c) $\chi^2 = 36/16 + 36/39 + 81/61 + 9/52 + 25/25 + 25/7 = 9.25$, df = 3; (d) no.

B.23. (a) [104, 117, 39; 96, 108, 36]; (b) $\chi^2 = 81/104 + 64/117 + 1/39 + 81/96 + 64/108 + 1/36 = 2.72$, df = 2; (c) yes.

B.24. (a) $\chi^2 = 11.23$, df = 3; (b) no.

B.25. (a) [92, 96, 12; 69, 72, 9; 69, 72, 9];

(b) $\chi^2 = \dfrac{169}{92} + \dfrac{289}{96} + \dfrac{16}{12} + \dfrac{81}{69} + \dfrac{100}{72} + \dfrac{1}{9} + \dfrac{16}{69} + \dfrac{49}{72} + \dfrac{9}{9} = 10.76$, df = 4; (c) no, no, yes.

B.26. (a) [30, 54, 72, 48, 36; 20, 36, 48, 32, 24];

(b) $\chi^2 = \dfrac{4}{30} + \dfrac{16}{54} + \dfrac{4}{72} + \dfrac{4}{48} + \dfrac{36}{36} + \dfrac{4}{20} + \dfrac{16}{36} + \dfrac{4}{48} + \dfrac{4}{32} + \dfrac{36}{24} = 3.92$, df = 4; (c) yes.

INDEX

CPSIA information can be obtained
at www.ICGtesting.com
Printed in the USA
FSHW022049110521
81357FS

9 780071 755610